D1211519

EMC AND THE PRINTED CIRCUIT BOARD

IEEE Press Editorial Board
Roger F. Hoyt, *Editor-in-Chief*

J. B. Anderson
P. M. Anderson
M. Eden
M. E. El-Hawary
S. Furui
A. H. Haddad
R. Herrick
S. Kartalopoulos
D. Kirk
P. Laplante
M. Padgett
W. D. Reeve
G. Zobrist

Kenneth Moore, *Director of IEEE Press*
John Griffin, *Acquisition Editor*
Marilyn G. Catis, *Assistant Editor*
Denise Phillip, *Production Editor*

IEEE Electromagnetic Compatibility Society, *Sponsor*
EMC-S Liaison to IEEE Press, Hugh Denny

Cover design: William T. Donnelly, *WT Design*

Technical Reviewers

Daryl Gerke, P.E., *Kimmel Gerke Associated, Ltd.*
William H. Hubbard, *Purdue University*
Elya B. Joffe, *K.T.M. Project Engineering, Ltd.*
William Kimmel, *Kimmel Gerke Associated, Ltd.*
W. Michael King
Norm Violette, *Violette Engineering Corp.*

Books of Related Interest from IEEE Press

PRINTED CIRCUIT BOARD DESIGN TECHNIQUES FOR EMC COMPLIANCE
Mark I. Montrose
1996 Cloth 256 pp ISBN 0-7803-1131-0

CAPACITIVE SENSORS: Design and Applications
Larry Baxter
1997 Cloth 320 pp ISBN 0-7803-1130-2

EMC AND THE PRINTED CIRCUIT BOARD

Design, Theory, and Layout Made Simple

Mark I. Montrose
Montrose Compliance Services, Inc.

IEEE Electromagnetic Compatibility Society, *Sponsor*

IEEE Press Series on Electronics Technology
Robert Herrick, *Series Editor*

The Institute of Electrical and Electronics Engineers, Inc., New York

A JOHN WILEY & SONS, INC., PUBLICATION
New York • Chichester • Weinheim • Brisbane • Singapore • Toronto

A NOTE TO THE READER
This book has been electronically reproduced from digital information stored at John Wiley & Sons, Inc. We are pleased that the use of this new technology will enable us to keep works of enduring scholarly value in print as long as there is a reasonable demand for them. The content of this book is identical to previous printings.

© 1999 THE INSTITUTE OF ELECTRICAL AND ELECTRONICS ENGINEERS, INC. 3 Park Avenue, 17th Floor, New York, NY 10016-5997
All rights reserved.

No part of this publication may be reproduced, stored in a retrieval system, or transmitted in any form or by any means, electronic, mechanical, photocopying, recording, scanning or otherwise, except as permitted under Sections 107 and 108 of the 1976 United States Copyright Act, without either the prior written permission of the Publisher, or authorization through payment of the appropriate per-copy fee to the Copyright Clearance Center, 222 Rosewood Drive, Danvers, MA 01923, (978) 750-8400, fax (978) 750-4744. Requests to the Publisher for permission should be addressed to the Permissions Department, John Wiley & Sons, Inc., 605 Third Avenue, New York, NY 10158-0012. (212) 850-6011, fax (212) 850-6008, E-mail: PERMREQ@WILEY.COM.

For ordering and customer service, call 1-800-CALL-WILEY.
Wiley-IEEE Press **ISBN 0-7803-4703-X**

Printed in the United States of America.
10 9 8 7 6 5 4

Library of Congress Cataloging-in-Publication Data

Montrose, Mark I.
EMC and the printed circuit board : design, theory, and layout made simple / Mark I. Montrose.
p. cm. — (IEEE Press series on electronics technology)
"IEEE Electromagnetic Compatibility Society, sponsor."
Includes bibliographical references and index.
ISBN 0-7803-4703-X (alk. paper)
1. Printed circuits—Design and construction. 2. Electromagnetic compatibility. I. IEEE Electromagnetic Compatibility Society. II. Title. III. Series.
TK7868.P7M65 1998
621.3815′31—dc21 98-35408
CIP

To my family

Margaret,

Maralena,

and Matthew

Contents

PREFACE xiii

ACKNOWLEDGMENTS xvii

CHAPTER 1 EMC FUNDAMENTALS 1

1.1 Fundamental Definitions 2
1.2 EMC Concerns for the Design Engineer 4
1.2.1 Regulations 4
1.2.2 RFI 4
1.2.3 Electrostatic discharge (ESD) 5
1.2.4 Power disturbances 5
1.2.5 Self-compatibility 6
1.3 The Electromagnetic Environment 6
1.4 The Need to Comply (A Brief History of EMI) 9
1.5 Potential EMI/RFI Emission Levels for Unprotected Products 12
1.6 Methods of Noise Coupling 12
1.7 Nature of Interference 16
1.7.1 Frequency and Time (à la Fourier: time domain ⇔ frequency domain) 17
1.7.2 Amplitude 18
1.7.3 Impedance 18
1.7.4 Dimensions 18
1.8 PCBs and Antennas 18
1.9 Causes of EMI—System Level 19
1.10 Summary for Control of Electromagnetic Radiation 20
References 21

CHAPTER 2 EMC INSIDE THE PCB 23

2.1 EMC and the PCB 23
2.1.1 Wires and PCB-traces 25
2.1.2 Resistors 25
2.1.3 Capacitors 26
2.1.4 Inductors 26
2.1.5 Transformers 27
2.2 Theory of Electromagnetics (Made Simple) 28
2.3 Relationship Between Electric and Magnetic Sources (Made Simple) 30
2.4 Maxwell Simplified—Further Still 34
2.5 Concept of Flux Cancellation (Flux Minimization) 37
2.6 Skin Effect and Lead Inductance 39
2.7 Common-Mode and Differential-Mode Currents 41
2.7.1 Differential-mode currents 42
2.7.2 Differential-mode radiation 42
2.7.3 Common-mode currents 44
2.7.4 Common-mode radiation 46
2.7.5 Conversion between differential and common mode 46
2.8 Velocity of Propagation 47
2.9 Critical Frequency (λ/20) 49
2.10 Fundamental Principles and Concepts for Suppression of RF Energy 49
2.10.1 Fundamental principles 49
2.10.2 Fundamental concepts 49
2.11 Summary 51
References 52

CHAPTER 3 COMPONENTS AND EMC 53

3.1 Edge Rate 53
3.2 Input Power Consumption 56
3.3 Clock Skew 58
3.3.1 Duty cycle skew 59
3.3.2 Output-to-output skew 59
3.3.3 Part-to-part skew 60
3.4 Component Packaging 60
3.5 Ground Bounce 65
3.6 Lead-to-Lead Capacitance 69
3.7 Grounded Heatsinks 70
3.8 Power Filtering for Clock Sources 74
3.9 Radiated Design Concerns for Integrated Circuits 76
3.10 Summary for Radiated Emission Control—Component Level 78
References

CHAPTER 4 IMAGE PLANES 81

4.1 Overview 81
4.2 5/5 Rule 83
4.3 How Image Planes Work 84
4.3.1 Inductance 84
4.3.2 Partial inductance 85
4.3.3 Mutual partial inductance 86
4.3.4 Image plane implementation and concept 88
4.4 Ground and Signal Loops (Not Eddy Currents) 91
4.4.1 Loop area control 92
4.5 Aspect Ratio—Distance Between Ground Connections 95
4.6 Image Planes 97
4.7 Image Plane Violations 99
4.8 Layer Jumping—Use of Vias 102
4.9 Split Planes 104
4.10 Partitioning 106
4.10.1 Functional subsystems 106
4.10.2 Quiet areas 106
4.11 Isolation and Partitioning (Moating) 107
4.11.1 Method 1: Isolation 108
4.11.2 Method 2: Bridging 108
4.12 Interconnects and RF Return Currents 112
4.13 Layout Concerns for Single- and Double-Sided Boards 114
4.13.1 Single-sided PCBs 115
4.13.2 Double-sided PCBs 116
4.13.3 Symmetrically placed components 116
4.13.4 Asymmetrically placed components 118
4.14 Gridded Ground System 119
4.15 Localized Ground Planes 120
4.15.1 Digital-to-analog partitioning 122
4.16 Summary 123
References 124

CHAPTER 5 BYPASSING AND DECOUPLING 125

5.1 Review of Resonance 126
5.1.1 Series resonance 127
5.1.2 Parallel resonance 128
5.1.3 Parallel C—Series RL resonance (antiresonant circuit) 128
5.2 Physical Characteristics 129
5.2.1 Impedance 129
5.2.2 Energy storage 131
5.2.3 Resonance 132
5.2.4 Benefits of power and ground planes 134

5.3 Capacitors in Parallel 136
5.4 Power and Ground Plane Capacitance 138
5.4.1 Buried capacitance 141
5.4.2 Calculating power and ground plane capacitance 142
5.5 Lead-Length Inductance 143
5.6 Placement 144
5.6.1 Power planes 144
5.6.2 Decoupling capacitors 144
5.7 Selection of a Decoupling Capacitor 148
5.7.1 Calculating capacitor values (wave-shaping) 149
5.8 Selection of Bulk Capacitors 152
5.9 Designing a Capacitor Internal to a Component's Package 155
5.10 Vias and Their Effects in Solid Power Planes 157
References 158

CHAPTER 6 TRANSMISSION LINES 159

6.1 Overview on Transmission Lines 159
6.2 Transmission Line Basics 162
6.3 Transmission Line Effects 163
6.4 Creating Transmission Lines in a Multilayer PCB 165
6.5 Relative Permittivity (Dielectric Constant) 166
6.5.1 How losses occur within a dielectric 169
6.6 Routing Topologies 171
6.6.1 Microstrip topology 171
6.6.2 Embedded microstrip topology 172
6.6.3 Single stripline topology 174
6.6.4 Dual stripline topology 175
6.6.5 Differential microstrip and stripline 177
6.7 Routing Concerns 178
6.8 Capacitive Loading 180
References 182

CHAPTER 7 SIGNAL INTEGRITY AND CROSSTALK 185

7.1 Need for Signal Integrity 185
7.2 Reflections and Ringing 188
7.2.1 Identification of signal distortion 191
7.2.2 Conditions that create ringing 192
7.3 Calculating Trace Lengths (Electrically Long Traces) 195
7.4 Loading Due to Discontinuities 200
7.5 RF Current Distribution 202

7.6 Crosstalk 203
7.6.1 Units of measurement—Crosstalk 206
7.6.2 Design techniques to prevent crosstalk 207
7.7 The 3-W Rule 210
References 212

CHAPTER 8 TRACE TERMINATION 215

8.1 Transmission Line Effects 216
8.2 Termination Methodologies 217
8.2.1 Source termination 221
8.2.2 Series termination 221
8.2.3 End termination 226
8.2.4 Parallel termination 227
8.2.5 Thevenin network 230
8.2.6 RC network 234
8.2.7 Diode network 236
8.3 Terminator Noise and Crosstalk 237
8.4 Effects of Multiple Terminations 239
8.5 Trace Routing 241
8.6 Bifurcated Lines 243
8.7 Summary—Termination Methods 244
References 245

CHAPTER 9 GROUNDING 247

9.1 Reasons for Grounding—An Overview 247
9.2 Definitions 247
9.3 Fundamental Grounding Concepts 249
9.4 Safety Ground 253
9.5 Signal Voltage Referencing Ground 254
9.6 Grounding Methods 255
9.6.1 Single-point grounding 256
9.6.2 Multipoint grounding 259
9.6.3 Hybrid or selective grounding 261
9.6.4 Grounding analog circuits 261
9.6.5 Grounding digital circuits 262
9.7 Controlling Common-Impedance Coupling Between Traces 262
9.7.1 Lowering the common-impedance path 262
9.7.2 Avoiding a common-impedance path 264
9.8 Controlling Common-Impedance Coupling in Power and Ground 266
9.9 Ground Loops 268
9.10 Resonance in Multipoint Grounding 271

9.11 Field Transfer Coupling of Daughter Cards to Card Cage 273
9.12 Grounding (I/O Connector) 277
References 277

GLOSSARY 279

BIBLIOGRAPHY 287

APPENDIX

A The Decibel 291
B Fourier Analysis 294
C Conversion Tables 298
D International EMC Requirements 302

INDEX 317

ABOUT THE AUTHOR 325

Preface

EMC and the Printed Circuit Board: Design, Theory, and Layout Made Simple is a companion book to *Printed Circuit Board Design Techniques for EMC Compliance.* When used together, these two books cover all aspects of a PCB design as it relates to both time and frequency domain issues. One must be cognizant that if a PCB does not work as intended in the time domain, frequency domain concerns become irrelevant, especially compliance to international EMC requirements. Time and frequency domain aspects must be considered together.

The intended audience for this book is the same as that for *Printed Circuit Board Design Techniques for EMC Compliance:* those involved in logic design and PCB layout; test engineers and technicians; those working in the areas of mechanical, manufacturing, production, and regulatory compliance; EMC consultants; and management responsible for overseeing a hardware engineering design team.

Regardless of the engineer's specialty, a design team must come up with a product that not only can be manufactured in a reasonable time period, but will also minimize cost during design, test, integration, and production. Frequently, more emphasis is placed on functionality to meet a marketing specification than on the need to meet legally mandated EMC and product safety requirements. If a product fails to meet compliance tests, redesign or rework may be required. This redesign significantly increases costs, which include, but are not limited to engineering manpower (along with administrative overhead), new PCB layout and artwork, prototyping material, system integration and testing, purchase of new components for quick delivery (very expensive), new in-circuit test fixtures, and documentation. These costs are in addition to loss of market share, delayed shipment, loss of customer faith in the company (goodwill), drop in stock price, anxiety attacks, and many other issues. Personal experience as a consultant has allowed me the opportunity to witness these events several times with small startup companies.

My main focus as a consultant is to assist and advise in the design of high-technology products at minimal cost. Implementing suppression techniques into the PCB design saves money, enhances performance, increases reliability, and achieves first-time compliance with emissions and immunity requirements, in addition to having the product function as desired.

Working in this industry has allowed me to participate in state-of-the-art designs as we move into the future. Although my focus is on technology of the future, one cannot forget that simple, low-technology products are being produced in ever increasing numbers. Although the thrust of this book is toward high-end products, an understanding of the fundamental concept of EMC suppression techniques will allow *any* PCB being designed to pass EMC tests. The key words here are "fundamental concepts." When one does not understand fundamental concepts, compliance and functional disaster may await.

When management decides to bring in a consultant after production has started, having failed an EMI test, causing a stop-ship situation, it is too late for efficiency. Generally, nothing can be done without major expenses being incurred. I have watched small companies go bankrupt because they invested all their capital in a product for quick shipment and then had to redesign everything from scratch. Those who control the finances of a company by mandating cost over compliance have frequently been spotted working at a different company every year. Accountants who do not understand what it takes to be a hardware or PCB designer engineer can doom a company to failure.

Sometimes, use of a single component (filter) costing $0.50 is too much for management to accept on a $1000 product. Engineers may be able to implement a redesign to prevent use of this inexpensive filter. This redesign may cost the company tens of thousands of dollars (including new compliance tests) for a production build of a few hundred units. Although the accountant may receive bonus pay for keeping the cost of the PCB down, the Return-On-Investment (ROI) will never be achieved. I do not advocate adding cost to a design unless it is mandatory. High-technology products now require use of additional power and ground planes, filter components, and the like, all at a cost for both functionality and compliance.

Detailed definitions of various terms are presented within specific chapters of this book. Before we proceed, an important distinction is in order. EMC stands for Electromagnetic Compatibility. This means that electrical equipment must work within an intended environment. We can have EMI (Electromagnetic Interference) problems due to incompatibilities between equipment. EMC is *achieved;* EMI *occurs.* According to common usage, EMC refers to the total discipline concerned with achieving electromagnetically compatible equipment and systems. EMI refers to the event or episode indicating an incompatibility, for example, a *lack* of EMC. EMI refers to all events experienced across the frequency spectrum. Radio Frequency Interference (RFI) originally referred to those incompatibilities arising between radio sets. During the 1970s and 1980s, RFI was generally not used because it failed to indicate the problems that can arise from Electromagnetic Pulse (EMP), lightning, Electrostatic Discharge (ESD), and so on. Over the past few years, however, RFI has been creeping back into our vocabulary. Caution should be used with the acronym RFI, however, for its meaning is unclear in the field of EMC.

The main differences between my two books on EMC and PCBs are as follows.

Printed Circuit Board Design Techniques for EMC Compliance provides information for those who have to get a product designed and shipped within a reasonable time frame and within budget. It illustrates that a PCB may exhibit an EMI problem, it briefly explains why the problem occurs, and it shows how to solve the design flaw during layout. It does not go into detail on how and why EMI occurs, theoretically.

University textbooks are available (listed in the References sections) that cover all aspects of theoretical physics related to EMC. Numerous other publications present EMC concepts in a brief manner, giving just enough detail to make one aware of theory with minimal mathematical analysis. Many managers and some engineers do not care about why something happens. *Printed Circuit Board Design Techniques for EMC Compliance* has compiled a track record of successful results.

EMC and the Printed Circuit Board: Design, Theory, and Layout Made Simple is a companion book targeted at those designers who want to know how and why EMI occurs within a PCB. These designers may not be directly responsible for the actual PCB layout, but they may have to oversee the end product. Engineers generally want to understand technical concepts. This book is written for ease of understanding a subject that is generally not taught in universities or other educational environments, again using a minimal amount of math.

In the present book, we examine two sides of the coin—time domain (signal functionality and quality) and frequency domain (EMC). A signal that is present within the PCB may be viewed in both domains. No difference exists between the two; rather, only the way one examines a signal. Test instrumentation also differs. Chapter 2 illustrates using simplified physics, with minimal mathematical analysis, how these two domains exist simultaneously. Theory is presented in a format that is easy to comprehend in the limited time one has to read and study a book on EMC and PCB, especially when work needs to be done at the office.

The focus of this book is *strictly* on the PCB. Discussion of containment techniques (box shielding), internal and external cabling, power supply design, and other system-level subassemblies that use PCBs as subcomponents will not be discussed in depth. Again, excellent reference material is listed in the References on these aspects of EMC system-level design engineering.

The incentive for writing this book has come from my numerous seminar and workshop students in the United States, Europe, and Asia. These students ask, "How and why does EMI get developed within a PCB?" Recognizing a need to fill a gap that currently does not exist within the published literature in the public domain worldwide (at time of writing), I want to enlighten the reader to a field of engineering that is considered to be a *Black Magic* art. Those who do not take electromagnetic compatibility seriously provide job security for EMC engineers. EMC engineers know various tricks of the trade on how to apply rework or a quick fix to a PCB to pass a particular test. These under-the-pressure enhancements implemented during compliance testing are identified as *Band-Aid* techniques. These PCBs could have been designed properly from the start. The concept advanced is to change design habits and thinking from Band-Aids to low-cost suppression layout techniques during the design cycle.

Mark I. Montrose
Santa Clara, California

Acknowledgments

I want to acknowledge the following individuals who played a part in the review cycle of this book.

Bill Kimmel and Daryl Gerke, Kimmel & Gerke Associated, Ltd., both of whom performed a thorough and technical review not only of this book, but also *Printed Circuit Board Design Techniques for EMC Compliance.* As always, both gentlemen provided excellent input to ensure accuracy.

Elya B. Joffe, K.T.M. Project Engineering, Ltd., Israel, who was a technique reviewer scrutinizing everything for accuracy, in addition to providing numerous comments in the interest of enhancing the reader's ability to understand complex concepts.

Bob Herrick, Department of Electrical Engineering, Purdue University, who as editor of the IEEE Press Series on Electronic Technology, served as my direct interface with IEEE Press.

David Angst, TCAD, provided a sanity check on the time-domain portion of this book.

William Hubbard, Department of Electrical Engineering, Purdue University, for a comprehensive review on all subject matter.

Norm Violette, Violette Engineering Corporation, who performed a technical review and also provided insights into unique aspects of EMC theory with a different point of view.

Hugh Denny, Georgia Technical Research Institute, who as the IEEE EMC Society's liaison to IEEE Press, and a technical reviewer, took me to task on various issues presented to guarantee the accuracy of my discussions.

Hans Melberg, EMC Consultant, who gave me my first opportunity to document my knowledge of EMC and PCBs into written form at Wyse Technology. This document inspired me to write my first book, *Printed Circuit Board Design Techniques for EMC Compliance.*

A special acknowledgment is given to W. Michael King, my mentor in the field of EMC. Without his friendship, expertise, and the ability to teach me complex aspects of EMC, along with providing a comprehensive technical review of the material, this book could not have been written.

My most special acknowledgment is to my family. Margaret, my wife, and my two children, Maralena and Matthew. After surviving the time and effort it took me to write my first book, my family stood by me again as I spent months to write another. Their understanding and support are beyond belief.

Mark I. Montrose
Santa Clara, California

1

EMC Fundamentals

This book seeks primarily to help engineers minimize harmful interference between components, circuits, and systems. These interferences include not only radiated and conducted radio frequency (RF) emissions, but also the influences of electrostatic discharge (ESD), electrical overstress (EOS), and radiated and conducted susceptibility (immunity). Meeting these requirements will satisfy legally mandated international and domestic regulatory requirements and governmental regulations. A companion book, *Printed Circuit Board Design Techniques for EMC Compliance,* presents design rules and layout concepts that assist in achieving an EMC-compliant product using suppression design techniques.

One of the engineer's goals is to meet design requirements in order to satisfy both international and domestic regulations and voluntary industrial standards related to EMC compliance.

The information presented in this book is intended for

- Non-EMC engineers who design and layout printed circuit boards (PCBs).
- EMC engineers and consultants who must solve design problems at the PCB level.
- Design engineers who want to understand fundamental concepts related to how electromagnetic interference (EMI) exists within a PCB.
- Those who want a comprehensive understanding of how PCB design and layout techniques work within a PCB.

This book is applicable for use as a reference document throughout any design project.

With these considerations in mind, the reader should understand that *EMC and the Printed Circuit Board* is written for the engineer who never studied applied electromagnetics in school, requires a refresher course, or has minimal hands-on experience with high-speed, high-technology product designs. As we well know, technology is advancing at a rapid rate. Design techniques that worked several years ago are no longer effective in today's products with high-speed digital design requirements. Because EMC may be insufficiently covered in engineering schools, training courses and seminars are now being held all over the country and internationally to provide this information.

Only a minimal amount of mathematical analysis is presented here because the intent of this book is to present *a basic understanding and analysis of how a PCB creates RF energy, and the manner in which RF energy is propagated.* The information presented is therefore in a format that is easy both to understand and to implement.

Since World War II, controlling emissions from a product has been a necessity for acceptable performance of an electronic device in both the civilian and military environment. It is more cost-effective to design a product with suppression at the source than to "build a better box." Containment measures are not always economically justified and may degrade as the life cycle of the product is extended beyond the original design specification. For example, the end user often removes covers from enclosures for ease of access to repair or upgrade. In many cases, sheet metal covers are never replaced, particularly those internal subassembly covers that act as partition shields. The same is true for blank metal panels or faceplates on the front of a system that contains a chassis or backplane assembly. As a result, containment measures become compromised. Proper layout of a PCB with suppression techniques also promotes EMC compliance with use of cables and interconnects, whereas box shielding (containment) does not. In addition to EMC compliance, signal functionality concerns exist. It does us no good if a product passes EMC tests and then fails to operate as designed.

This book provides details on why a variety of design techniques work for most PCB layout applications. It is impossible to anticipate every possible application or design concern. The concepts presented are *fundamental* in nature and are applicable to all electronic products. While every design is different, the basics of product design rarely change unless new components and materials become available.

Herein we discuss high-technology, high-speed designs that require new and expanded techniques for EMC suppression at the PCB level. Many traditional PCB techniques are not effective for proper signal functionality and compliance. Components have become faster and more complex. Use of custom gate array logic and application-specific integrated circuits (ASICs) presents new and challenging opportunities. The design and layout of a PCB to suppress EMI at the source can be realized while maintaining systemwide functionality.

Why worry about EMC compliance? After all, isn't speed the most important design parameter? Legal requirements dictate the maximum permissible interference potential of digital products. These requirements are based on experiences in the marketplace related to emission and immunity complaints. Often, suppression techniques on a PCB will aid in improving signal quality and signal-to-noise performance.

1.1 FUNDAMENTAL DEFINITIONS

The following basic terms are used throughout this book.

Electromagnetic Compatibility (EMC). The capability of electrical and electronic systems, equipment, and devices to operate in their intended electromagnetic environment within a defined margin of safety, and at design levels or performance, without suffering or causing unacceptable degradation as a result of electromagnetic interference. (ANSI C64.14-1992)

Electromagnetic Interference (EMI). The lack of EMC, since the essence of interference is the lack of compatibility. EMI is the process by which disruptive electromagnetic energy is transmitted from one electronic device to another via radiated or conducted paths (or both). In common usage, the term refers particularly to RF signals, but EMI can occur in the frequency range from "DC to daylight."

Radio Frequency (RF). A frequency range containing coherent electromagnetic radiation of energy useful for communication purposes—roughly the range from 10 kHz to 100 GHz. This energy may be transmitted as a byproduct of an electronic device's operation. RF is transmitted through two basic modes:

Radiated Emissions. The component of RF energy that is transmitted through a medium as an electromagnetic field. Although RF energy is usually transmitted through free space, other modes of field transmission may occur.

Conducted Emissions. The component of RF energy that is transmitted through a medium as a propagating wave, generally through a wire or interconnect cables. LCI (Line Conducted Interference) refers to RF energy in a power cord or AC mains input cable. Conducted signals do not propagate as fields but may propagate as conducted waves.

Susceptibility. A relative measure of a device or a system's propensity to be disrupted or damaged by EMI exposure to an incident field of signal. It is the lack of immunity.

Immunity. A relative measure of a device or system's ability to withstand EMI exposure while maintaining a predefined performance level.

Electrostatic Discharge (ESD). A transfer of electric charge between bodies of different electrostatic potential in proximity or through direct contact. This definition is observed as a high-voltage pulse that may cause damage or loss of functionality to susceptible devices. Although lightning qualifies as a high-voltage pulse, the term *ESD* is generally applied to events of lesser amperage, and more specifically to events that are triggered by human beings. However, for the purposes of discussion, lightning is included in the ESD category because the protection techniques are very similar, though different in magnitude.

Radiated Immunity A product's relative ability to withstand electromagnetic energy that arrives via free-space propagation.

Conducted Immunity. A product's relative ability to withstand electromagnetic energy that penetrates it through external cables, power cords, and I/O interconnects.

Containment. A process whereby RF energy is prevented from exiting an enclosure, generally by shielding a product within a metal enclosure (Faraday cage or Gaussian structure) or by using a plastic housing with RF conductive paint. Reciprocally, we can also speak of containment as preventing RF energy from entering the enclosure.

Suppression. The process of reducing or eliminating RF energy that exists without relying on a secondary method, such as a metal housing or chassis. Suppression may include shielding and filtering as well.

1.2 EMC CONCERNS FOR THE DESIGN ENGINEER

Within the field of EMC, multiple design concerns exist. Most items identified here are *not* obvious. Past experience determines the amount of effort required to address these issues as they relate to EMC compliance along with signal functionality. Awareness of five key areas is mandatory for understanding why electromagnetic compatibility is required. With an understanding of these five areas, we can reduce difficult problems to simple applications of design techniques and implementations. About 95% of all EMC issues encountered are associated with the following. Each will be discussed separately [2].

1. Regulations
2. RFI
3. Electrostatic discharge
4. Power disturbances
5. Self-compatibility

1.2.1 Regulations

Part of the need for regulations stems from complaints regarding interference to electronic products used in both residential and commercial applications and part from the requirement to protect vital communication services. Without regulations, the "electromagnetic environment" in which we live would be crowded with interference and only a few electronic devices could survive and operate.

Regulations protect the radio spectrum and limit "spurious" radiation from both intended radiators (such as transmitters) and unintended radiators (most electronic equipment). Numerous consumer complaints developed basically over interference to television and radio reception. In addition, aeronautical communication systems started to break down; police and fire units were unable to use their radios for emergency purposes; and commercial and residential electronic products were failing in the field owing to the presence of other electronic equipment located in the general vicinity. With these complaints, the Federal Communications Commission (FCC) developed a set of requirements for electronic equipment that would limit the amount of interference polluting the electromagnetic environment. The FCC followed the lead of Germany's Verband Deutscher Electrotechniker (VDE), which implemented mandatory requirements shortly after World War II. Other countries worldwide have followed the VDE and FCC in developing requirements for digital products.

Regulations control not only emissions but also susceptibility (or immunity). Europeans have taken the lead in mandating immunity tests; in North America, however, these same tests are only voluntary at the time of this writing.

1.2.2 RFI

Radio Frequency Interference (RFI) poses a threat to electronic systems due to the proliferation of radio transmitters that exist. Cellular phones, handheld radios, wireless remote control units, pagers, and the like are now quite widespread. It does not take a great deal of radiated power to cause harmful interference. Typical equipment failures occur in the electric field level range of 1 to 10 volts/meter. For example, a 1 watt radio transmitter at 1 meter distance from an electronic device has a field strength of approximately 5 V/m, depending on the frequency and antenna used for measurement purposes. Preventing RFI from corrupting a device has become legally mandatory for all products used within Europe, North America, and many Asian countries.

1.2.3 Electrostatic Discharge (ESD)

ESD technology has progressed to the point where components have become extremely dense along with small geometries (0.18 micron). The sensitivity of high-speed, multimillion transistor microprocessors is easily damaged by external ESD events. These events can be caused by either direct or radiated means. Direct contact ESD events generally cause permanent damage of the device or create a latent failure mode that will trigger permanent damage sometime in the future. Radiated ESD events (caused, for instance, by furniture moving in a room, reflected ESD energy off a structure, or a person walking across a carpet) can cause an upset in the device that may result in improper operation without leading to permanent damage to the system.

An ESD event is considered to be a broadband high-frequency problem with edge rates that are usually less than 1 nanosecond. This translates to a spectral bandwidth problem that can approach 1 GHz. It is not uncommon to observe ESD in the sub-nanosecond time period. This faster edge rate becomes a problem well into the gigahertz spectral bandwidth.

ESD is treated under the immunity requirements for compliance with the EU's (European Union's) EMC Directive. Most manufacturers worldwide recognize this problem. These manufacturers must design suppression and layout techniques into their products to guarantee that failure will not occur in the field.

1.2.4 Power Disturbances

With more and more electronic equipment being plugged into the power mains network, potential interference occurs. These problems include power-line disturbances, electrical fast transients (EFT), power sag and surges, voltage variations (high/low voltage levels), lightning transients, and power-line harmonics. Older products and power supplies were generally not affected by these disturbances. With newer, high-frequency switching power supplies, these disturbances are starting to become noticeable as the switching components consume AC voltage generally on the crest of the waveform, not the complete waveform.

Analog and digital devices respond differently to power-line disturbances. Digital circuits are affected by spikes on the power system (EFT and lightning), as well as failure due to excessively high or low voltage levels. Analog devices generally operate on voltage levels, which may be degraded by a disturbance changing the reference level of the system's power source.

Power-line harmonics have become a major concern, especially in Europe. Nonlinear loads (switching power supplies) consume AC mains power at the peak of the cycle rather than over the entire sine wave. This varying load generates harmonics and waveform distortions that affect the power distribution network. For example, it is common to see 230 VAC, 150 Hz (third harmonic), or 250 Hz (fifth harmonic) present in a power system that is intended to operate at 50 Hz consuming various levels of input current at these higher frequencies.

1.2.5 Self-Compatibility

A commonly overlooked issue is self-compatibility. A digital partition or circuit can interfere with analog devices, create crosstalk between traces and wires, or a fan motor may cause an upset with digital circuits. While most of these concerns are known to the system designer, these failures are not recognized as an EMI event. Recognition of this concern, along with design implementations that prevent internal system failures from occurring, will result in a less expensive and more robust system.

1.3 THE ELECTROMAGNETIC ENVIRONMENT

A product must operate within a particular environment compatible with other electronic equipment. To understand the need for compatibility in an environment where products must operate, we now examine this environment.

Any periodic signal (clock) generates a wide spectrum of RF energy when viewed in the frequency domain. Figure 1.1 illustrates a spectral plot of a nonsinusoidal oscillator in the frequency range between 30 and 200 MHz. In studying this plot, we observe not only the fundamental frequency of the oscillator (1.8432 MHz), but also all the harmonics created across the 170 MHz window. A low-frequency oscillator was chosen to illustrate this wide harmonic spectrum. The spectral bandwidth of the oscillator is determined by the "edge rate" of the oscillator, not the "clock rate." A detailed discussion of why the "edge rate" of the digital pulse signal is of more concern than operating "frequency" is presented in Chapter 3.

Using this same oscillator waveform, we examine a very narrow frequency range. Figure 1.2 shows that both even and odd harmonics of the primary oscillator in the frequency range of 88–108 MHz are present.

The FM radio band (88–108 MHz) is allocated to a specific range of pre-assigned frequencies. Many digital products produce unintentional radiated RF energy within this frequency spectrum, especially lower order harmonics. In Fig. 1.3, we observe two traces. The upper trace displays FM radio signals. For this example, the spectrum analyzer was configured to make the FM radio signal appear similar to the signature characteristics of our clock signal. To help differentiate between the clock harmonics and the FM radio signals, a 10-dB displacement is observed in Fig. 1.3 with the FM signals shown 10 dB higher above the oscillator. The lower trace is a narrow-band view of the oscillator in the same frequency range. Notice that the signals measured are harmonics from the 1.8432-MHz oscillator. With this situation, potential interference between the oscillator and FM signal may exist. This scenario can be applied to any communications system, such as between a nonintentional radiator (digital device) and aeronautical communications, or an emergency services broadcast.

To illustrate the effects of a design change, the lower trace in Fig. 1.4 represents a compliant product. Changing just one component, moving a single trace, or using an alternate manufacturer of a logic family for the same function (74Fxx in place of a 74LSxx)

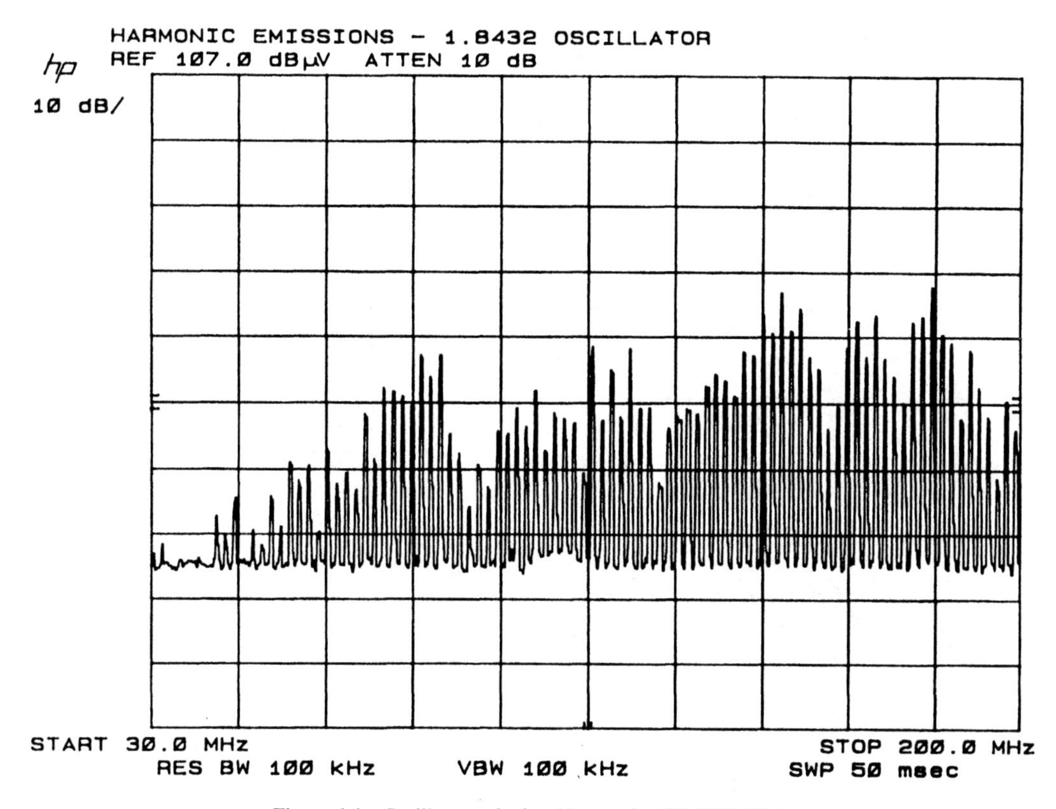

Figure 1.1 Oscillator and related harmonics (30–200 MHz).

now makes a compliant product noncompliant. This plot should enlighten those skeptics who believe that an alternate device that is identical in form, fit, and function can be easily substituted. Although the component may be functionally 100% compatible, its effects on changing the overall EMC characteristics may be radically different. The edge rate of the source driver may differ between vendors, although functionality remains the same. Not all components are the same, EMI considered. Chapter 3 provides details on how and why different components with the same function can cause functionality and EMI concerns.

Although difficult to observe in Fig. 1.4, we are able to distinguish the effects of a simple change to the circuit, especially in the middle frequency range of the plot.

Designing products that will pass legally required EMI tests is not as difficult as one might expect. Engineers often strive to design elegant products, but elegance sometimes must give way to product safety, manufacturing, cost, and, of course, EMC. Such abstract problems can be challenging, particularly if the engineer is unfamiliar with compliance or manufacturing requirements. We must remove the mystery from the "Hidden Schematic" syndrome.

When an EMI problem occurs, the engineer should approach the situation logically. A simple EMI model has three elements:

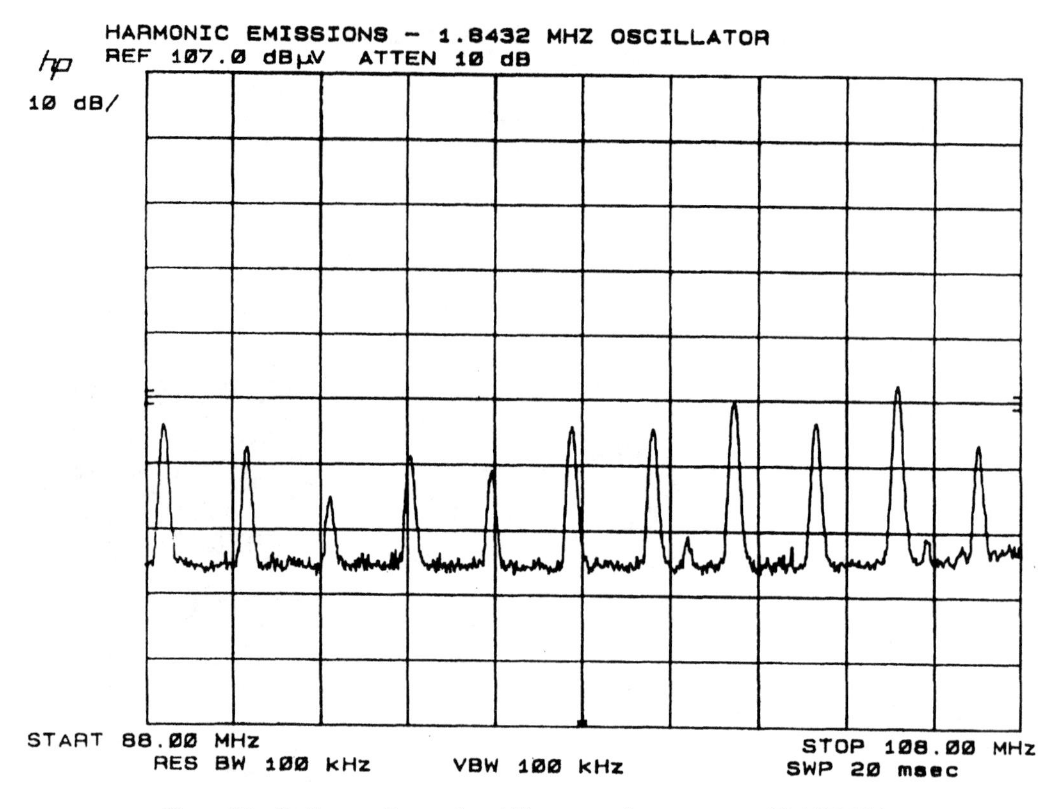

Figure 1.2 Oscillator and harmonics within a narrow frequency range (88–108 MHz).

1. There must be a source of energy.
2. There must be a receptor that is upset by this energy when the intensity of the electromagnetic interference is above a tolerable limit.
3. There must be a coupling path between the source and receptor for the unwanted energy transfer.

For interference to exist, all three elements have to be present. If one of the three elements is removed, there can be no interference. It therefore becomes the engineer's task to determine which is the easiest element to remove. Generally, designing a PCB that eliminates most sources of RF interference is the most cost-effective approach (called suppression). The source of interference is the active element producing the original waveform. What is required is to design the PCB to keep the RF energy created to only those sections of the board which require this energy. The second and third elements tend to be addressed with containment techniques. Figure 1.5 illustrates the relationship between these three elements and presents a list of products associated with each element.

With respect to PCBs, we observe the following.

- Noise sources are clock generation circuits, component radiation within a plastic package, incorrect trace routing, electrically long trace lengths, poor impedance control, internal cable interconnects, and the like.

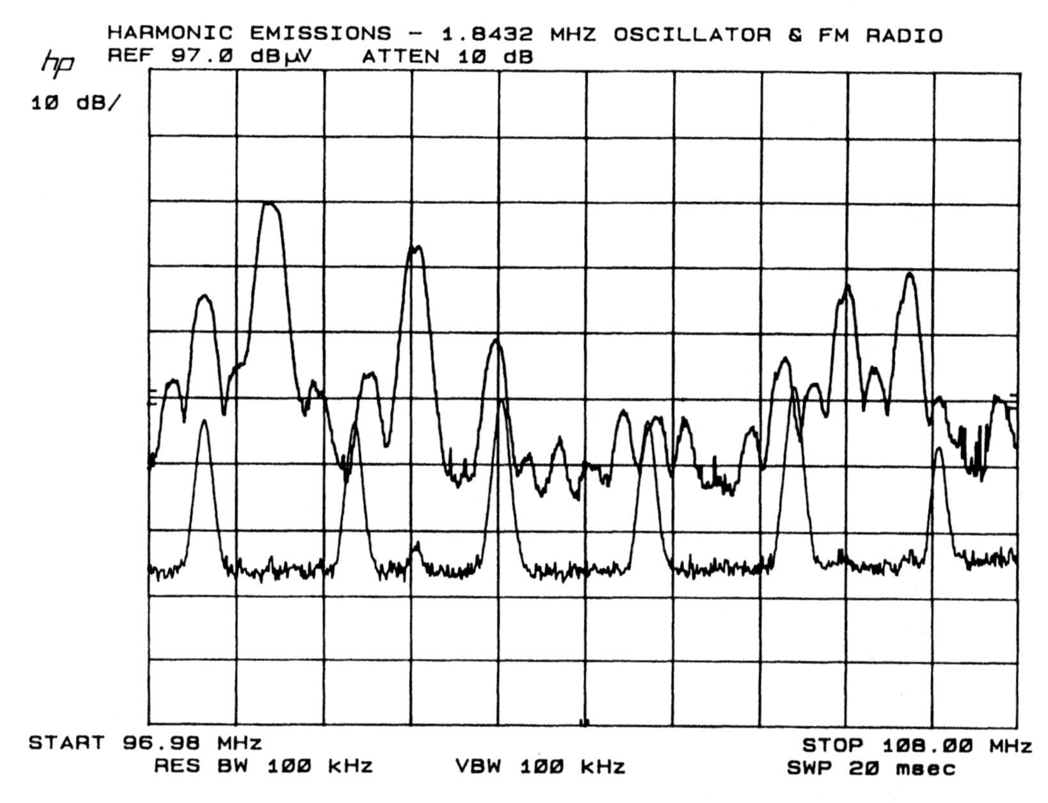

Figure 1.3 Oscillator's harmonics and FM radio signals superimposed. *Note:* FM radio stations are plotted 10 dB above oscillator for clarity.

- The propagation path is the medium that carries the RF energy, such as free space or interconnect cabling (common impedance coupling).
- Receptors can be components on the PCB that easily accept harmful radiated interference from I/O cables and transfer this harmful energy to circuits and devices susceptible to disruption.

1.4 THE NEED TO COMPLY (A BRIEF HISTORY OF EMI)

In North America, interference to communication systems became a concern in the 1930s, whereupon the United States Congress enacted the Communications Act of 1934. The Federal Communications Commission was created to oversee enforcement and administration of this act. Harmful effects were being observed with the technology of this time period, enough to cause the U.S. government to take action.

EMI was also recognized as a problem during World War II with vacuum tubes. The terminology used was Radio Frequency Interference (RFI). During this period, spectrum signatures of communication transmitters and receivers were developed along with radar systems. Because of the size and expense of these devices, the military owned the

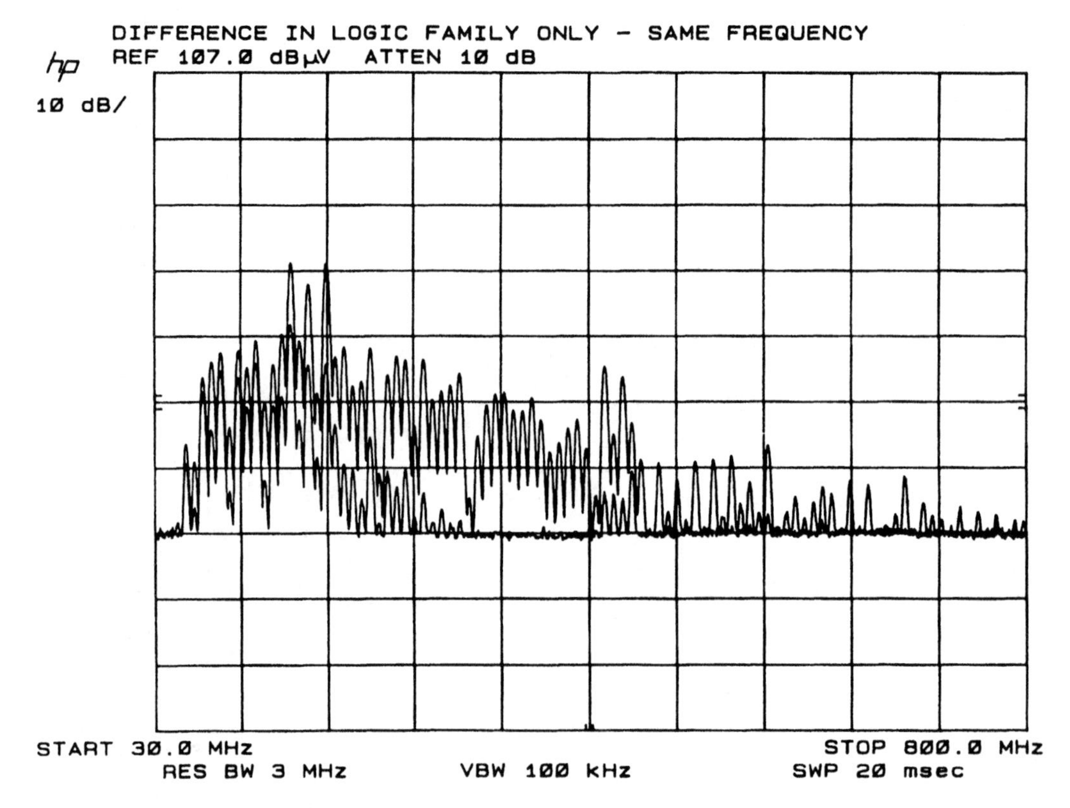

Figure 1.4 Effects of a single change to a circuit related to RF emissions.

majority of high-technology electronic systems. Research and information on EMC was kept from the general public under the guise of national security.

Following the Korean War, most EMC work was not classified unless it dealt with the specifics of a particular tactical or strategic system such as the Minuteman rocket, B-52 bombers, and similar military and espionage equipment. Conferences on EMI began to be held in the mid-1950s where unclassified information was presented. The first conference on EMI (RFI) was held in 1956 sponsored by the IEEE (Institute of Electrical and Electronic Engineers) and the IRE (Institute of Radio Engineers). During this time frame, the Army Signal Corps of Engineers and the United States Air Force created strong ongoing programs dealing with EMI, RFI, and related areas of EMC.

In the 1960s, NASA (National Aeronautical and Space Administration) began stepped-up EMI control programs for their launch vehicles and space system projects. Governmental agencies and private corporations became involved with combating EMI in equipment such as security systems, church organs, HI-FI amplifiers, and the like. All of these devices were analog-based systems.

As digital logic devices were developed, EMI became a greater concern. In approximately 1970, research was started to characterize EMI in consumer electronics, which included TV sets, common AM/FM radios, medical devices, audio and video recorders, and similar products. Very few of these products were digital but were becoming so. Analog

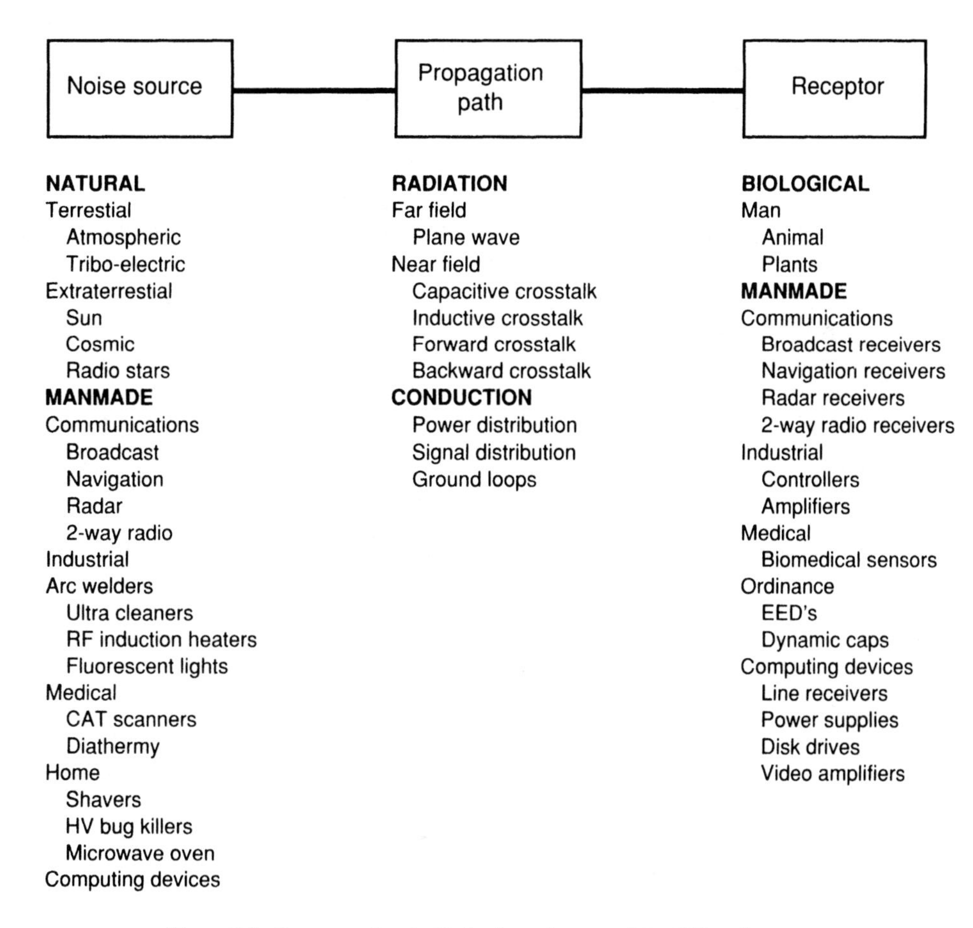

Figure 1.5 Items associated with the three elements of the EMI environment.

systems are more susceptible to problems than digital equipment because the threshold of susceptibility in a switch or control circuit is higher than that of a high gain amplifier.

In the late 1970s, problems associated with EMC compatibility became an issue for products that were beginning to be used within the commercial marketplace. These products include home entertainment systems (TVs, VCRs, camcorders), personal computers, communication equipment, household appliances with digital logic, intelligent transportation systems, sophisticated commercial avionics, control systems, audio and video displays, and numerous other applications. During this time period, the general public became aware of EMC and the threats associated with it.

After the general public became involved with EMI associated with digital equipment used within residential areas, the Federal Communications Commission (FCC) in the mid- to late 1970s began to promulgate an emissions standard for personal computers and similar equipment. Since personal computers comprised such a huge market, commercial entities became involved in the field of EMC. This was due in part to military and space funding being tapered off during this time period. Now almost all electronic equipment is digital, whether or not it needs to be.

The focus of electronic equipment has now shifted from analog to digital. Another factor that pushed digital devices into regulation status was that in the early days of digital, the prevailing wisdom was that digital devices were "not susceptible" to EMI. Because of this perception, the commercial community was totally shocked that digital devices were actually susceptible to disruption.

If it had not been for the personal computer, commercial manufacturers would not be paying much attention to the threat of EMI. The Food and Drug Administration (FDA), however, recognized the threat posed by EMI. The issue of compliance became a concern when the European Union (EU), through its EMC Directive 89/336/EEC, imposed emissions and immunity requirements. Another forcing function of EMC compliance is the increasing role played by electronics in power conversion, communications, and control systems where electromechanical systems once were primarily used. Since a lot of things are now done electronically that once were not, the opportunities and consequences of EMC are much greater. Because of this issue, a lot of people have had to deal with EMC for only 20 years, whereas the military, NASA, and RF engineers have been dealing with this issue from day one.

EMC is now a major factor in the design of all electrical products; emissions, and immunity. Digital logic has fallen below the 1-nanosecond edge rate, while clock frequencies approach 1 GHz.

1.5 POTENTIAL EMI/RFI EMISSION LEVELS FOR UNPROTECTED PRODUCTS

The complexity of a product is dependent on processing speed and other factors. The more complex a product is, the more likely it will both radiate and be susceptible to RF energy. This is shown in Table 1.1. This table defines a matrix in which product size and complexity are plotted against processing speed.

In the upper left corner of the table, products with processing speeds of less than 10 MHz are commonly found. As we proceed toward the bottom right corner, systems are much faster and more complex. These high-technology systems include RISC (Reduced Instruction Set Computing) CPUs, Pentium[1] processors, and similar products. Most of these products have edge rates in the sub-nanosecond range and operate above 100 MHz.

1.6 METHODS OF NOISE COUPLING

A product must be designed for two levels of performance: one is to minimize RF energy exiting an enclosure (emissions), and the other is to minimize the amount of RF energy entering the enclosure (susceptibility or immunity). Both emissions and immunity travel by either radiated or conductive paths. This relationship is shown in Fig. 1.6.

[1]Pentium is a registered trademark of Intel Corporation.

TABLE 1.1 Matrix of Potential Emission Levels for Various Products

Processing speed \ Product size/complexity	Low Single board	Medium Mother/daughter board	Large Multiple modules
Slow < 10 MHz	Low	Medium	Large
Medium 10 MHz < f < 100 MHz	Medium	Medium to high	High to very high
High > 100 MHz	High	High to very high	Very high

To further examine coupling paths, it must be realized that the propagation path contains multiple transfer mechanisms. These are detailed in Fig. 1.7 and include

1. Direct radiation from source to receptor (path 1).
2. Direct RF energy radiated from the source transferred to AC mains cables or signal/control cables of the receptor (path 2).
3. RF energy radiated by AC mains, signal, or control cables from source to receptor (path 3).
4. RF energy conducted by common electrical power supply lines or by common signal/control cables (path 4).

In addition to the four coupling paths, there are four transfer mechanisms that exist for each coupling path. These four mechanisms are [5]

1. Conductive.
2. Electromagnetic.
3. Magnetic field (subset of electromagnetic identified separately in Fig. 1.7).
4. Electric field (subset of electromagnetic identified separately in Fig. 1.7).

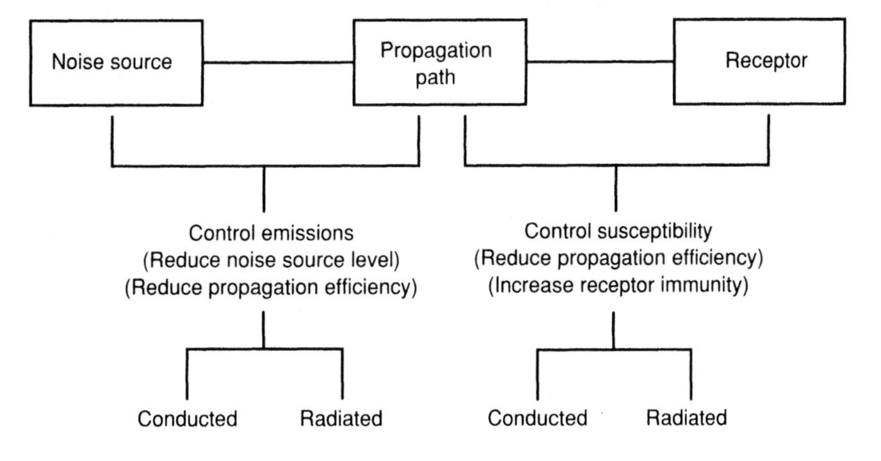

Figure 1.6 Coupling paths.

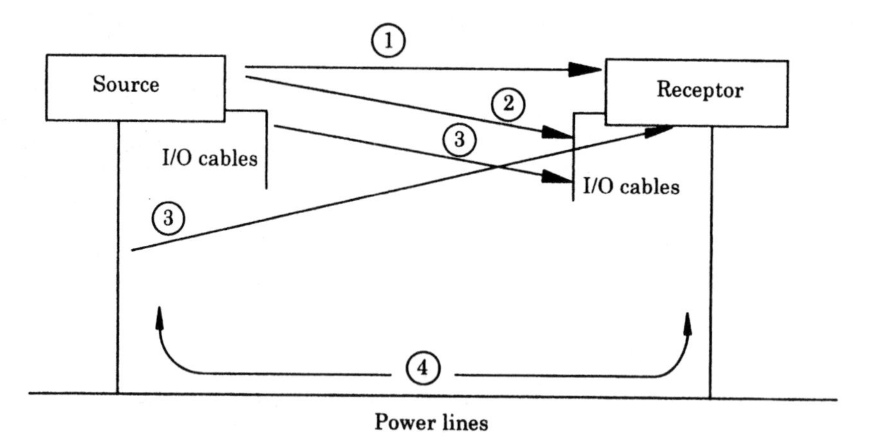

Figure 1.7 Coupling path mechanisms.

If the noise coupling mechanism can be ascertained, a logical solution can be determined to reduce the coupling.

Conductive coupling is identified as common impedance coupling. This coupling occurs when both noise source and susceptible circuits are connected by a mutual impedance. A minimum of two connections is required. This is because the noise current must flow from a source to load and then return to the source. Figure 1.8 illustrates two circuits and a power source. Current from each circuit flows through both the shared impedance of the power subsystem and interconnect wiring, all caused by shared metallic connections. For this figure, the shared connection is the return line.

Magnetic coupling occurs when a portion of magnetic flux created by one current loop passes through a second loop formed by another current path. Magnetic flux coupling is represented by mutual inductance between the two loops. The noise voltage inducted in the second loop is $V_2 = M_{12}dI_1/dt$, where M_{12} = mutual coupling factor and dI_1/dt = the time rate of change of current in the trace. Magnetic flux coupling is shown in Fig. 1.9.

Electric field coupling occurs in low-impedance circuits. The effects are small relative to other coupling that may occur. In a circuit there is mutual capacitance if we have high Z_S in parallel with Z_L (see Fig. 1.10). Capacitive coupling occurs when a portion of the electric flux created by one circuit terminates on the conductors of another circuit. Electric flux coupling between two circuits can be represented by mutual capacitance. The noise current injected into the susceptible circuit is approximately $I = CdV/dt$.

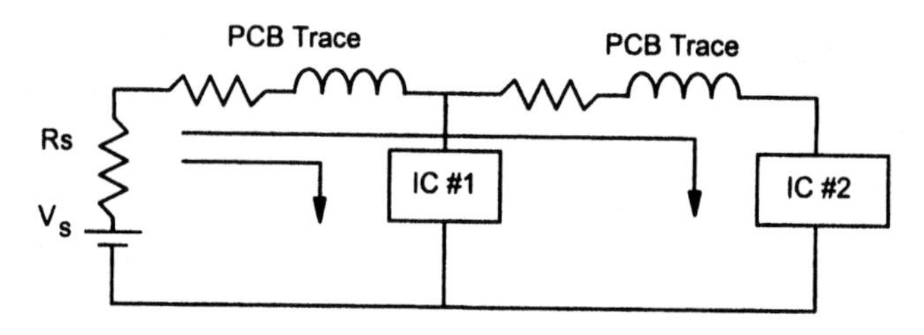

Figure 1.8 Conductive transfer mechanism.

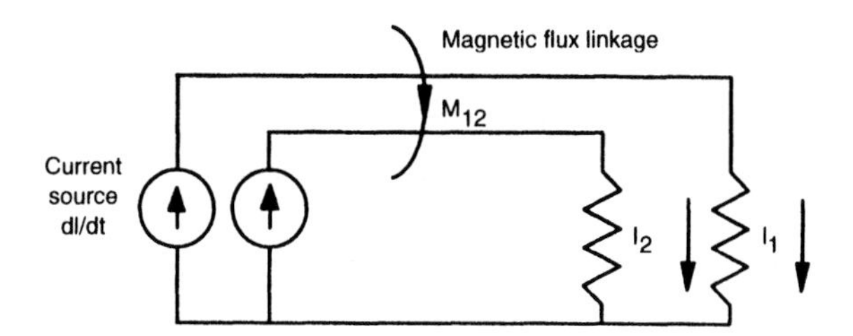

Figure 1.9 Magnetic field coupling.

Electromagnetic field coupling is a combination of both magnetic and electric fields affecting a circuit simultaneously. Depending on the distance between source and susceptor, the electric field (***E***) and magnetic field (***H***) effects may be different, depending on whether we are in the near field or far field. This is the most common transfer mechanism observed.

When dealing with emissions, the general rule of thumb is:

> The higher the frequency, the greater the efficiency of there being a radiated coupling path; the lower the frequency, the greater the efficiency that a conducted coupling path will be the cause of EMI. The probability of coupling is "1." The extent of coupling depends on frequency.

The most overlooked noise coupling method is through a conductor, a wire, or a PCB trace. This conductor may pick up RF noise from a culprit device and transfer this noise to a victim circuit. The easiest way to prevent this transfer from occurring is either to remove the noise from the culprit trace or to prevent the victim trace from receiving this RF energy.

What happens to a signal that is propagating down a PCB trace from source to destination? Figure 1.11 illustrates a model of one propagation path. The signal line connects directly between a source and a destination. With this circuit we have both inductive coupling (L) and capacitive coupling (C) between adjacent circuits.

Looking at Fig. 1.11, notice that the output capacitance siphons off a certain percentage of output drive current. The inductance of the line attenuates the signal, which also couples to adjacent traces. The capacitance between signal traces also shunts RF energy in addition to corrupting the signal through crosstalk. Finally, the capacitance of the load shunts energy away from the input source. The load capacitance thus couples the electromagnetic signal energy to ground.

If the signal trace is long compared to the rise time of the signal, distributed effects are observed.[2] Energy, which has been propagating down the trace with a characteristic

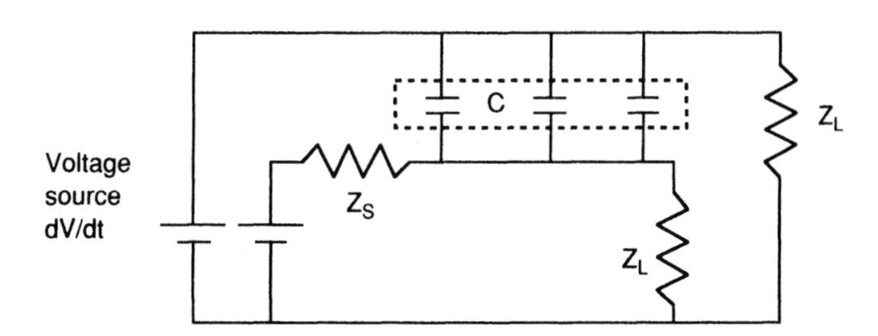

Figure 1.10 Electric field transfer coupling.

[2]An electrically long trace is defined as a routed trace containing a signal with an edge rate that is faster than the time it takes for a signal to travel from source to load, and return from load to source, causing functionality concerns that include ringing and reflections.

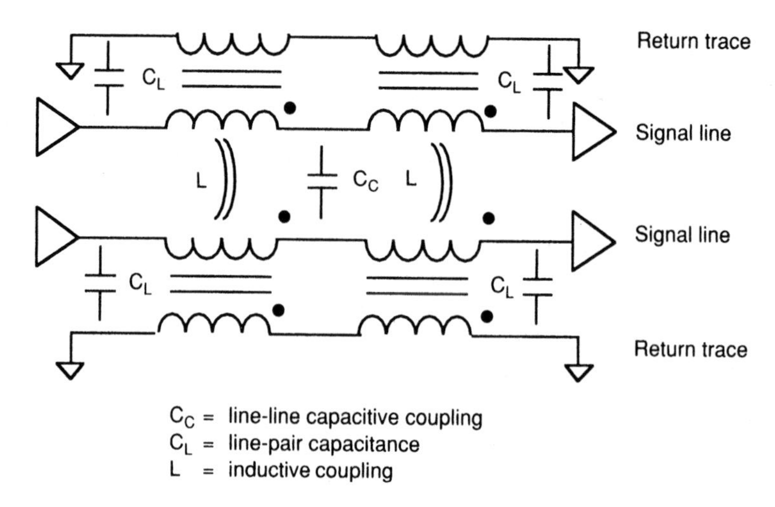

Figure 1.11 Coupling model for traces.

impedance Z_o, will arrive at the load. If the load impedance, Z_o, is the same as the source impedance, all energy will be absorbed in the load. If the load impedance is high, the signal will reflect back to the source since the signal cannot be consumed by the circuit. This reflection can be identified as ringing or over/undershoots. Transfer impedance, both inductive and capacitive, also exists inside components (Chapter 3). Internal ground bounce further degrades the quality of the transmitted signal (Chapter 7).

1.7 NATURE OF INTERFERENCE

Interference can be grouped into two categories, internal and external. The internal problem can be due to signal degradation along the transmission path along with parasitic coupling between adjacent circuits, in addition to field coupling between internal subassemblies, such as a power supply to a disk drive. Stated more specifically, the problems are signal losses and reflections along the path and crosstalk between adjacent signal traces.

External problems are divided into emissions and susceptibility concerns. Emissions are primarily from harmonics of clocks or other periodic signals. Remedies concentrate on containing the periodic signals to as small an area as possible and blocking the parasitic coupling paths to the outside world.

Susceptibility to external influences, such as ESD or radio frequency interference, are related primarily to energy which couples onto I/O lines and then becomes transferred to the inside of the unit. The principal recipients are high-speed input lines and sensitive adjacent traces, particularly those terminated with edge-triggered devices.

There are five major considerations in EMC analysis [2].

1. *Frequency.* Where in the frequency spectrum is the problem observed?
2. *Amplitude.* How strong is the source energy level, and how great is its potential to cause harmful interference?

3. *Time.* Is the problem continuous (periodic … ist only during certain cycles of operation (e …
4. *Impedance.* What is the imped… r units and the impedance of the transfer mec…
5. *Dimensions.* What are the physi… e that can cause emissions to occur? RF curr… ields that will exit an enclosure through chas… ons of a wavelength or significant fractions … ngths on a PCB have a direct relationship as tra… RF currents.

Whenever an EMI problem is approached, it is helpful to review the above list based on product application. Understanding these five items will remove much of the mystery of how EMI exists within a PCB. Applying these five major considerations teaches that design techniques make sense in certain contexts but not in others. For example, single-point grounding is excellent when applied to low-frequency applications but is completely inappropriate for radio frequencies, which is where most of the EMI problems exist. Many engineers blindly apply single-point grounding for all product designs without realizing that additional and more complex problems are created using this grounding methodology.

How does one make use of the above list? It is common to think of a current source being created from a voltage applied across an impedance (Thevenin equivalent). It is, however, more advantageous to consider voltage as a result of current traveling through an impedance (Norton equivalence). Using the Norton network, many EMI questions are answered as it is easier to visualize EMI using the Norton configuration. *E*-field coupling involves the induction of common-mode voltage sources, whereas *H*-field coupling can end up with either common- or differential-mode currents (depending on victim wiring).

Current is preferable to voltage for a simple reason: current always travels around a closed-loop circuit following one or more paths. It is to our advantage to direct or steer this current in the manner that is desired for proper system operation. To control the path that current flows, we must provide a low-impedance RF return path back to the original source of interference. We must also divert interference current away from the load. For those applications that require a high-impedance path from source to the load, we must consider all possible paths through which the return currents may travel.

1.7.1 Frequency and Time (à la Fourier: time domain ⇔ frequency domain)

It is common for design engineers to think in terms of the time frame. EMI is generally viewed in a frequency frame. RF energy is a periodic wave front that propagates through various mediums. Different wavelengths of the sine wave are recorded as EMI for those products that are not designed to be intentional radiators. It is difficult to understand an EMI problem in the time domain alone. (Conversion between the time and frequency domain is detailed in Appendix B using Fourier analysis.)

Baron Jean Baptiste Joseph Fourier (1768–1830), a French mathematician and physicist, formulated a method for analyzing periodic functions. Fourier proved that any periodic waveform can be decomposed into an infinite series of sine waves, each at an integral multiple or harmonic of a fundamental frequency. The composition of the harmonics is determined during a mathematical operation known as a Fourier transform. Fourier transforms can easily be calculated for simple waveforms and displayed with modern instrumentation.

1.7.2 Amplitude

The impact of amplitude is obvious. The higher the amplitude, the more interference one may encounter. It becomes important to limit the peak amplitude of the RF energy to only that necessary for circuit, device, and system performance.

1.7.3 Impedance

If both source and receptor are not the same impedance, one should expect greater interference problems than a source and receptor with identical impedance. This is because, for example, high-impedance sources can have minimal impact on low-impedance receptors and vice versa. Similar rules apply to radiated coupling. High impedances are associated with electric fields. Low impedances are associated with magnetic fields.

1.7.4 Dimensions

Physical dimensions play a significant factor related to the wavelength of an RF wave. When dealing with physical dimensions of a PCB trace, or the slot (aperture) opening of an enclosure, this aspect of EMC comes into view. Circuit analysis can no longer be assumed with lumped circuit parameters if numerical modeling is used during the design cycle.

The need to minimize the physical length of a trace or aperture opening relates to the electrical parameters of high-speed digital devices. When the speed of propagation becomes a significant portion of the propagational delay from source to load, we start to observe effects where field coupling becomes noticeable. When the trace length becomes physically long relative to a wavelength of a particular frequency, or in time domain terms, when the rise time becomes less than the propagational delay between source and load, the trace assumes the characteristics of a transmission line. All transmission lines must be terminated in their characteristic impedance for optimal transfer of the signal. While this practice is related primarily to preserving signal integrity, it also helps to control EMI.

Regarding EMI, we must concern ourselves with preventing creation of an antenna by having a PCB trace (or wire) approach a dimension that is the same as the offending source. When the trace approaches a particular wavelength of the offending signal (or portion of a wavelength), an efficient antenna will exist.

1.8 PCBs AND ANTENNAS

A PCB can act as an antenna to radiate RF energy through free space or couple through a cable interconnect. When we talk about the PCB acting as an antenna, what exactly do we mean? An antenna is an efficient and integral part of radio frequency communication. We need antennas to operate as intentional radiators. Most PCBs act as an unintentional radiator and are regulated by international EMC requirements unless design requirements include it as being a transmitter. Transmitters are regulated by regulatory requirements. If the PCB is an efficient unintentional radiator and suppression techniques cannot be implemented, containment measures must be provided.

Antennas exhibit various efficiencies as a function of frequency, whether intentional or accidental. When an antenna is driven by a voltage source, its impedance varies dramatically. When an antenna is in resonance, its impedance will be high and mostly reactive. The resistive portion (R) of the impedance equation ($Z = R+j\omega L$) is called "radiation resistance." This radiation resistance is a measure of the antenna's propensity to radiate RF energy at a specific frequency.

Most antennas are efficient radiators at a specific frequency spectrum. These frequencies are typically below 200 MHz, because I/O cables are approximately 2 to 3 meters in length and are sometimes long relative to a wavelength. At higher frequencies, significant radiation is generally observed directly from the unit due to the contribution of apertures in the enclosure.

When it is possible to isolate where the antenna exists, as in common-mode cable radiation, a reduction in the drive voltage is the easiest suppression technique to implement. RF voltages exist because of

- Impedance of circuit traces (which in turn is derived from lead inductance).
- Ground bounce (a point of uniform potential).
- Bypassing or shielding with respect to ground to reduce the drive voltage available to the unintentional antenna.

A schematic representation of an antenna is shown in Fig. 1.12. The antenna presents a specific impedance to the source driver that varies with frequency. At resonance, the reactive components, L and C, cancel out. Radiation resistance is maximum. RF energy is thus radiated.

To minimize the effects of an unwanted antenna existing in a PCB, EMC design and suppression techniques are required. These include establishing a good ground system in the layout in addition to the use of a Faraday cage (to contain RF emission). RF filters also reduce unwanted RF signals with minimal effects on the desired data, as long as the filter is properly chosen for its intended function.

1.9 CAUSES OF EMI—SYSTEM LEVEL

This book deals with the concepts that cause or are related to the creation of EMI within a PCB. A product is not complete without mentioning other aspects of a design that affects compliance. These other aspects are associated with system-level design. Detailed discus-

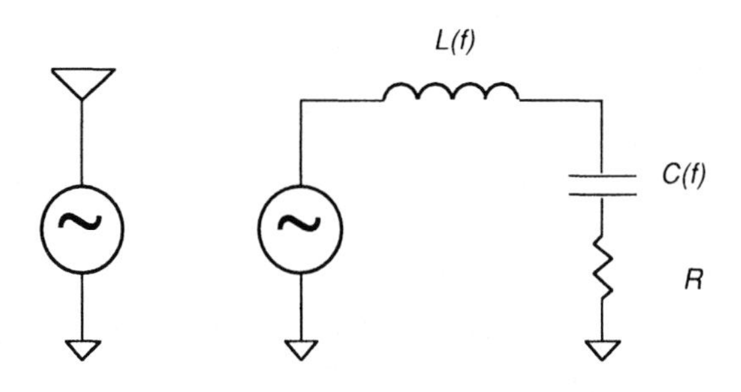

Figure 1.12 Representation of an antenna.

sion of the following is beyond the scope of this book; the focus is on PCBs. Excellent reference material is provided in the References sections for those interested in other aspects of EMC system-level design.

The most common areas of concern for system-level EMC compliance involve

- Improper use of containment measures (metal versus plastic housing)
- Poor design, implementation, and grounding of cables and connectors
- Incorrect PCB layout (includes)
  1. clocks and periodic signal trace routing
  2. stackup arrangement of the PCB and signal routing layer allocation
  3. selection of components with high spectral RF energy distribution
  4. common-mode and differential-mode filtering
  5. ground loops
  6. insufficient bypassing and decoupling

To implement system-level suppression, the following techniques are generally required:

- Shielding
- Gasketing
- Grounding
- Filtering
- Decoupling
- Proper trace routing
- Isolation and separation
- Circuit impedance control
- I/O interconnect design
- PCB suppression techniques designed internal to a component package

Even with all of these items, multiple techniques of both suppression and containment to achieve a compliant product can be required. Depending on the complexity of the system, speed of operation, and EMC requirements where shielding is needed, proper PCB layout will minimize shielding requirements.

1.10 SUMMARY FOR CONTROL OF ELECTROMAGNETIC RADIATION

In order to reduce or eliminate the potential for electromagnetic radiation, several basic concepts must be understood. These are listed below and detailed in later chapters.

1. Reduce the intensity of the RF source (voltage and current drive levels).
2. Provide differential- and common-mode filtering for high-speed signals or use balanced differential pairs with impedance-matched signals.

3. Reduce the energy being coupled to the antenna structure; use self-shielded trace routing and reduce differential-mode to common-mode conversion.
4. Reduce the effectiveness of the antenna's propensity to radiate RF energy.

High-frequency currents on an antenna are necessary to cause electromagnetic radiation. These RF currents can be reduced by differential-mode filtering and slowing down the edge rate of digital logic devices. We can reduce the conversion of differential-mode (DM) to common-mode (CM) currents by improving the impedance balance of the circuit. Since we generally cannot control the length of an external I/O cable, reducing the length of one-half of the antenna will make this radiator less efficient.

For RF energy to exist, there must be a voltage reference difference between two circuits. As a result, maintaining all metal structures (ground planes, ground traces, chassis, etc.) at a uniform or equivalent potential eliminates this voltage reference difference. This voltage reference difference is often due to inductance within the circuit and structure. Regardless of how well we design a product, a finite amount of inductance will always be present. If this inductance is added to the antenna structure, along with mutual capacitance, the antenna becomes an efficient radiator. A few nano-Henries (nH) of inductance or a few pico-Farads (pF) of capacitance are significant at higher frequencies, related to RF emissions.

REFERENCES

[1] Montrose, M. I. 1996. *Printed Circuit Board Design Techniques for EMC Compliance.* Piscataway, NJ: IEEE Press.

[2] Gerke, D., and W. Kimmel. 1994. "The Designers Guide to Electromagnetic Compatibility." EDN (January 20).

[3] Mardiguian, M. 1993. *Controlling Radiated Emissions by Design.* New York: Van Nostrand Reinhold.

[4] Van Doren, T. 1995. *Circuit Board Layout to Reduce Electromagnetic Emission and Susceptibility.* Seminar Notes.

[5] Hartal, O. 1994. *Electromagnetic Compatibility by Design.* W. Conshohocken, PA: R&B Enterprises.

[6] Ott, H. 1988. *Noise Reduction Techniques in Electronic Systems.* 2nd ed. New York: John Wiley & Sons.

2

EMC Inside the PCB

In today's international marketplace, products must conform to a host of regulations and requirements mandated by government agencies, private standards organizations, or voluntary councils. Mandatory compliance exists for North America, the European Union (EU), and numerous countries worldwide. These requirements relate to Electromagnetic Compatibility (EMC) and product safety. EMC refers to the ability of a product to coexist in its intended electromagnetic environment without causing or suffering functional degradation or damage. EMC comprises two main areas, emissions and immunity. This chapter investigates both aspects of EMC and how EMC can exist within a printed circuit board (PCB).

2.1 EMC AND THE PCB

Traditionally, EMC has been considered *Black Magic*; in reality, EMC can be explained by mathematical concepts. Some of the relevant equations and formulas are complex and beyond the scope of this book. Even if mathematical analysis is applied, the equations become too complex for practical applications. Fortunately, simple models can be formulated to describe how, but do not directly explain why, EMC compliance can be achieved.

Many variables exist that cause EMI. This is because EMI is often the result of exceptions to the normal rules of passive component behavior. A resistor at high frequency acts like a series combination of inductance with resistance in parallel with a capacitor. A capacitor at high frequency acts like an inductor and resistor in series-parallel combination with the capacitor plates. An inductor at high frequencies performs like an inductor with a capacitor in parallel across the two terminals. These expected behaviors of passive components at both high and low frequencies are illustrated in Fig. 2.1.

For example, when designing with passive components, we must ask ourselves this question, "When is a capacitor not a capacitor?" The answer is simple. The capacitor does

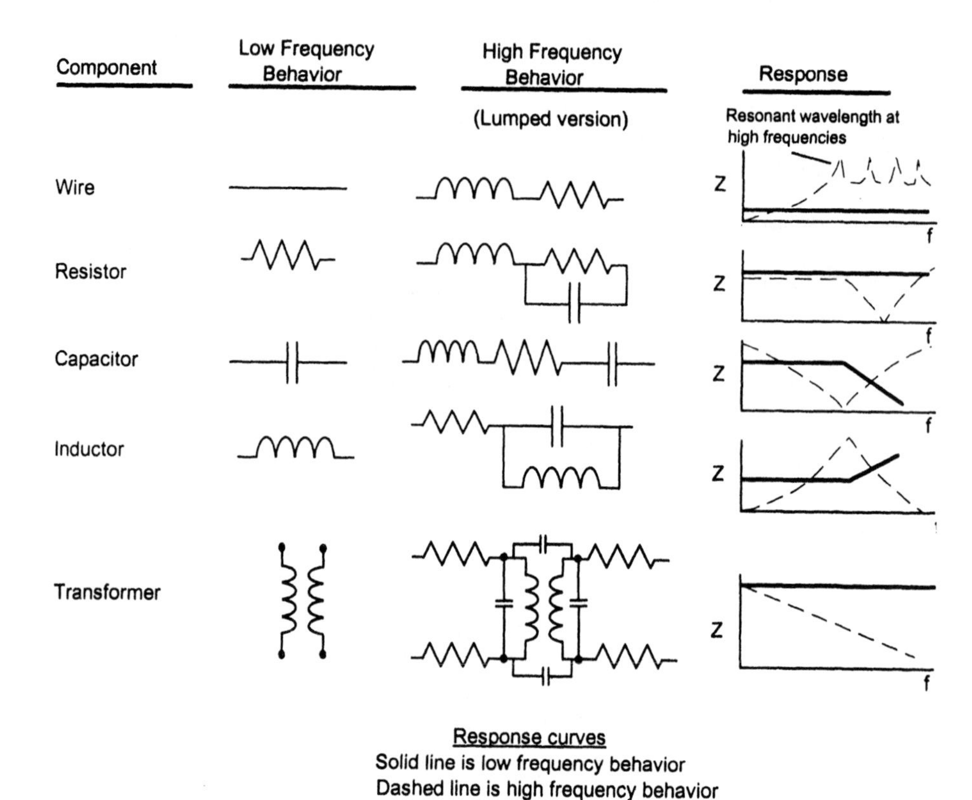

Figure 2.1 Component characteristic at RF frequencies. (*Source*: *Designers Guide to Electromagnetic Compatibility*, EDN. © 1994, Cahners Publishing Co. Reprinted with permission.)

not function as a capacitor because it has changed its functional characteristics to appear as an inductor due to lead-length inductance at high frequencies. Conversely, "When is an inductor not an inductor"? An inductor appears to function as a capacitor due to parasitic wire coupling at high frequencies. To be a successful designer, one must recognize the limitations of passive components. Use of proper design techniques to accommodate for these hidden features becomes mandatory, in addition to designing a product to meet a marketing functional specification.

These behavioral characteristics are referred to as the "hidden schematic." Digital engineers generally assume that components have a single-frequency response. As a result, passive component selection is based on functional performance in the time domain without regard to the characteristics exhibited in the frequency domain. Many times, EMI exceptions occur if the designer bends or breaks the rules, as seen in Fig. 2.1.[1]

To restate the complex problems that exist, consider the field of EMC as "*Everything that is not on a schematic or assembly drawing.*" This statement explains why the field of EMC is considered to be an art of Black Magic.

[1]Daryl Gerke and Bill Kimmel, "The Designers Guide to Electromagnetic Compatibility." Reprinted from *EDN Magazine* (January 20, 1994). © Cahners Publishing Company, 1994. A Division of Reed Publishing USA.

Once the hidden behavior of components is understood, it becomes a simple process to design products that pass EMC requirements. Hidden behavior also takes into consideration the switching speed of active components along with their unique characteristics, which also have hidden resistive, capacitive, and inductive components. We now examine each passive device separately.

2.1.1 Wires and PCB Traces

One does not generally consider the internal wiring, harnesses, and traces of a product as efficient radiators of RF energy. Every component has lead-length inductance, from the bond wires of the silicon die to the leads of resistors, capacitors, and inductors. Each wire or trace contains hidden parasitic capacitance and inductance. These parasitic components affect wire impedance and are frequency sensitive. Depending on the LC value (self-resonant frequency) and the length of the PCB trace, a self-resonance may occur between a component and trace, thus creating an efficient radiating antenna.

At low frequencies, wire is primarily resistive. At higher frequencies, the wire takes on the characteristics of being an inductor. This impedance changes the relationship that the wire (or PCB trace) has with grounding strategies, leading us into use of ground planes and ground grids. The major difference between a wire and a PCB trace is that wire is round while a trace is rectangular. The impedance of wire contains both resistance, R, and inductive reactance, ($X_L = 2\pi fL$), and is defined by $Z = R + jX_L \approx j2\pi fL$ at high frequencies. Capacitive reactance, $Xc = 1/2\pi fC$ is not a part of this equation for the high-frequency impedance response of the wire. For DC and low-frequency applications, the wire (or trace) is essentially resistive. At higher frequencies, the wire (or trace) becomes the important part of this impedance equation. Above 100 kHz, inductive reactance ($j2\pi fL$) exceeds resistance. As a result, the wire (or trace) is no longer a low-resistive connection but rather an inductor. As a general rule of thumb, any wire (or trace) operating above the audio frequency range is inductive, not resistive, and may be considered to be an efficient antenna to radiated RF energy.

Most antennas are designed to be an efficient radiator at one-fourth or one-half wavelength (λ) of a particular frequency of interest. Within the field of EMC, design recommendations are to design a product that does not allow a wire (or trace) to become an unintentional radiator below $\lambda/20$ of a particular frequency of interest. Inductive and capacitive elements can result in efficiencies through circuit resonance that mechanical dimensions do not describe.

For example, assume a 10-cm trace has $R = 57$ mΩ. Assuming 8 nH/cm, 80 nH total (details on derivation are presented in Chapter 6), we achieve an inductive reactance of 50 mΩ at 100 kHz. For those traces with frequencies above 100 kHz, the trace becomes inductive. The resistance becomes negligible and is no longer part of the equation. This 10-cm trace is calculated to be an efficient radiator above 150 MHz ($\lambda/20$ of 100 kHz).

2.1.2 Resistors

Resistors are one of the most commonly used components on a PCB. Resistors also have a limitation related to EMI. Depending on the type of material used for the resistor (carbon composition, carbon film, mica, wire-wound, etc.), a limitation exists related to frequency domain requirements. A wire-wound resistor is not suitable for high-frequency applications due to excessive inductance in the wire. Film resistors contain some inductance and are sometimes acceptable for high-frequency applications due to low lead-length inductance.

A commonly overlooked aspect of resistors deals with package size and parasitic capacitance. Capacitance exists between the two terminals of the resistor. This parasitic capacitance can play havoc with extremely high-frequency designs, especially those in the GHz range. For most applications, parasitic capacitance between resistor leads is not a major concern compared to the lead-length inductance that is present.

One major concern for resistors lies in the overvoltage stress condition to which the device may be subjected. If an ESD event is presented to the resistor, interesting results occur. If the resistor is a surface-mount device, chances are this component will arc-over (or self-destruct) upon observance of the event. For resistors with leads, ESD will see a high resistive (and inductive) path and be kept from entering the circuit protected by the resistor's hidden inductive and capacitive characteristics.

2.1.3 Capacitors

Chapter 5 presents a detailed discussion of capacitors. This section, however, provides a brief overview on the hidden attributes of capacitors.

Capacitors are generally used for power bus decoupling, bypassing, and bulk applications. An actual capacitor remains capacitive up to its self-resonant frequency. Above this self-resonant frequency, the capacitor exhibits inductive effects. This is described by the formula $X_C = 1 / (2\pi fC)$ where X_C is capacitive reactance (unit of ohms), f is frequency in hertz, and C is capacitance in farads. To illustrate this formula, a 10 μf electrolytic capacitor has a capacitive reactance of 1.6 Ω at 10 kHz, which decreases to 160 μΩ at 100 MHz. At 100 MHz, a short-circuit condition would exist which is wonderful for EMI. However, electrical parameters of electrolytic capacitors with high values of equivalent series inductance (ESL) and equivalent series resistance (ESR) limit the effectiveness of this particular type of capacitor to operation below 1 MHz.

Another aspect of capacitor usage lies in lead-length inductance and body structure. This subject is discussed in detail in Chapter 5 and will not be examined at this time. To summarize, parasitic inductance in the capacitor's wire bond leads causes the capacitor to function as an inductor above self-resonance and ceases to function as a capacitor for its intended function.

2.1.4 Inductors

Inductors are used for EMI control within a PCB. For an inductor, inductive reactance increases linearly with increasing frequency. This is described by the formula $X_L = 2\pi fL$, where X_L is inductive reactance (Ohms), f is frequency (hertz), and L inductance (henries).

For example, an "ideal" 10 mH inductor has a reactance of 628 ohms at 10 kHz. This inductive reactance increases to 6.2 MΩ at 100 MHz. The inductor now appears to be an open circuit at 100 MHz. If we want to pass a signal at 100 MHz, great difficulty will be present related to signal quality (time domain concern). Like a capacitor, the electrical parameters of this inductor (parasitic capacitance between windings) limits this particular device to less than 1 MHz.

The question now at hand is what to do at high frequencies when an inductor cannot be used. Ferrite beads can become the savior. Ferrite materials are alloys of iron/magnesium or iron/nickel. These materials have high permeability that provides for high-frequency and high-impedance with a minimum of capacitance that is observed between windings in an inductor. Ferrites are generally used in high-frequency applications because at low frequencies they are basically inductive and thus impose few losses on the line. At high frequencies, they are basically reactive and frequency dependent. This is graphically shown in Fig. 2.2. In reality, ferrite beads are high-frequency attenuators of RF energy.

Ferrites are, in fact, better represented by a parallel combination of a resistor and inductor. At low frequencies, the resistor is "shorted out" by the inductor, whereas at high frequencies, the inductive impedance is so high that it forces the current through the resistor.

The fact is that ferrites are "dissipative devices" where they dissipate high-frequency energy as heat. This can only be explained by the resistive, not the inductive, effect.

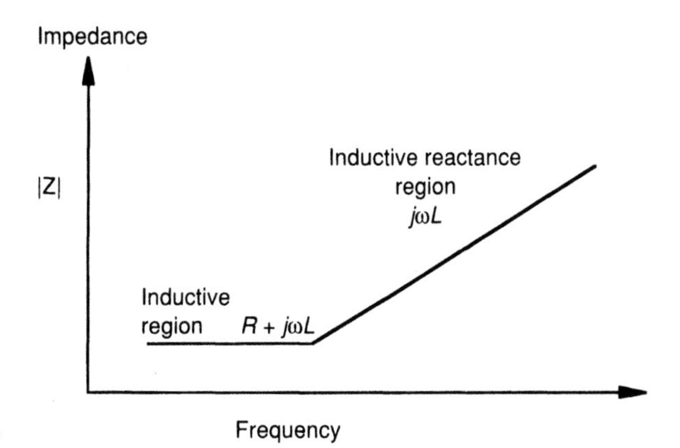

Figure 2.2 Characteristics of ferrite material.

2.1.5 Transformers

Transformers are generally found in power supply applications in addition to being used for isolation for data signals, I/O connections, and power interfaces. Depending on the type and application of the transformer, a shield may be provided between the primary and secondary windings. This shield, connected to a ground reference source, is designed to prevent against capacitive coupling between the two sets of windings.

Transformers are also widely used to provide common-mode (CM) isolation. These devices depend on a differential-mode transfer (DM) across their input to magnetically link the primary windings to the secondary windings in their attempt to transfer energy. As a result, CM voltage across the primary winding is rejected. One flaw that is inherent in the manufacturing of transformers is signal source capacitance between the primary and secondary windings. As the frequency of the circuit increases, so does capacitive coupling; circuit isolation is now compromised. If enough parasitic capacitance exists, high-frequency RF energy (fast transients, ESD, lighting, etc.) may pass through the transformer and cause an upset in the circuits on the other side of the isolation gap that received this transient event.

Having examined the hidden behavior characteristics of components, we now explore why these hidden features create EMI within a PCB.

2.2 THEORY OF ELECTROMAGNETICS (MADE SIMPLE)

Since we know that hidden behavioral characteristics of components exist, we now investigate how RF energy is created within a PCB. To understand the hidden characteristics and aspects of these components, we need to understand Maxwell's equations. Maxwell's four equations describe the relationship of electric and magnetic fields and are derived from Ampere's law, Faraday's law, and two equations from Gauss's law. These equations describe the field strength and current density within a closed-loop environment and require extensive knowledge of higher order Calculus. Since Maxwell's equations are extremely complex, we will present only a brief overview of this material. For a rigorous presentation of Maxwell's equations, refer to the reference material listed in the References. A list of Maxwell's equations is shown in Eq. (2.1) for completeness. A detailed knowledge of Maxwell is not a prerequisite for PCB design and layout.

To discuss Maxwell's equations in *simple* terms, a few fundamental principles are examined. The letters ***J, E, B,*** and ***H*** refer to vector quantities. Basically,

- Maxwell's equations describe the interaction of electric charges, currents, magnetic fields, and electric fields.
- The Lorentz force relation describes the physical forces imposed by both electric and magnetic fields on charged particles.
- All materials have a constitutive relationship to other materials. These include
  1. conductivity—relates current flow to electric field (Ohm's law in materials): $\boldsymbol{J} = \sigma \boldsymbol{E}$.
  2. permeability—relates magnetic flux to magnetic field: $\boldsymbol{B} = \mu \boldsymbol{H}$.
  3. dielectric constant—relates charge storage to an electric field: $\boldsymbol{D} = \varepsilon \boldsymbol{E}$.

where $\boldsymbol{J}$ = conduction-current density, A/m^2
σ = conductivity of the material
$\boldsymbol{E}$ = electric field intensity, V/m
$\boldsymbol{D}$ = electric flux density, $coulombs/m^2$
ε = permittivity of vacuum, 8.85 pF/m
$\boldsymbol{B}$ = magnetic flux density, $Weber/m^2$ or Tesla
$\boldsymbol{H}$ = magnetic field, A/m
μ = permeability of the medium, H/m

Maxwell's first equation is known as the divergence theorem based on Gauss's law. This applies to the accumulation of an electric charge that creates an electrostatic field, ***E***. This is best observed between two boundaries, conductive and nonconductive. The boundary-condition behavior referenced in Gauss's law causes the conductive enclosure (also called a Faraday cage) to act as an electrostatic shield. At the boundary, electric charges are kept on the inside of the boundary. Electric charges that exist on the outside of the boundary are excluded from internally generated fields.

Maxwell's second equation illustrates that there are no magnetic charges (no monopoles), only electric charges. These electric charges are either positively charged or negatively charged. Magnetic monopoles do not exist. Magnetic fields are produced through the action of electric currents and fields. Electric currents and fields emanate as a point source. Magnetic fields form closed loops around the current that generates these fields.

First Law: Electric Flux (from Gauss)

$$\nabla \bullet D = \rho \qquad \varphi_e = \oint_s D \bullet ds = \int_v \rho \, dv = 0$$

Second Law: Magnetic Flux (from Gauss)

$$\nabla \bullet B = 0 \qquad \varphi_m = \oint_s B \bullet ds = 0$$

Third Law: Electric Potential (from Faraday) (2.1)

$$\nabla \times E = -\frac{\partial B}{\partial t} \qquad \oint E \bullet dl = -\int_s \frac{\partial B}{\partial t} \bullet ds$$

Fourth Law: Electric Current (from Ampere)

$$\nabla \times H = J + \frac{\partial D}{\partial t} \qquad \oint H \bullet dl = \int_s \left(J + \frac{\partial D}{\partial t} \right) \bullet ds = I_{\text{total}}$$

Maxwell's third equation, also called Faraday's Law of Induction, describes a magnetic field traveling in a closed-loop circuit, generating current. The third equation has a companion equation (fourth equation). The third equation describes the creation of electric fields from *changing* magnetic fields. Magnetic fields are commonly found in transformers or windings, such as electric motors, generators, and the like. The interaction of the third and fourth equations is the primary focus for electromagnetic compatibility. Together, they describe how coupled electric and magnetic fields propagate (radiate) at the speed of light. This equation also describes the concept of "skin effect," which predicts the effectiveness of magnetic shielding. In addition, inductance is described which allows antennas to exist.

Maxwell's fourth equation is also identified as Ampere's law. This equation states that magnetic fields arise from two sources. The first source is current flow in the form of a transported charge. The second source describes how the changes in electric fields traveling in a closed-loop circuit create magnetic fields. These electric and magnetic sources describe the actions of inductors and electromagnetics. Of the two sources, the first is the description of how electric currents create magnetic fields.

To summarize, Maxwell's equations describe the root causes of how EMI is created within a PCB—time-varying currents. Static-charge distributions produce static electric fields, not magnetic fields. Constant currents produce both static magnetic and electric fields. Time-varying currents produce both electric and magnetic fields.

Static fields store energy. This is the basic function of a capacitor: accumulation of charge and retention. Constant current sources are a fundamental concept for the use of an inductor.

2.3 RELATIONSHIP BETWEEN ELECTRIC AND MAGNETIC SOURCES (MADE SIMPLE)

Having examined the process whereby changing currents create magnetic fields and static-charge distributions create electric fields, we will next determine the relationship between currents and radiated fields. We must look at the geometry of the current source and how it affects the radiated signal. In addition, we must also be aware that signal strength falls off with the distance from the source.

Time-varying currents exist in two configurations:

- Magnetic sources (which are closed loops)
- Electric sources (which are dipole antennas)

To investigate these two configurations in more detail, we first examine magnetic sources.

Consider a circuit containing a clock source (oscillator) and a load (Fig. 2.3). We observe current flowing in this circuit around a closed loop (trace and RF current return path). We can assess the radiated field generated by modeling this signal trace using simulation software with discrete parts. The field produced by this loop is a function of four variables.

1. *Current amplitude in the loop.* The field is proportional to the current that exists in the signal trace.
2. *Orientation of the source loop antenna relative to the measuring device.* For a signal to be measured or observed, the polarization of the source loop current should match that of the measuring device if the measuring antenna is also a loop. If the measuring antenna is a dipole, it must be in the same polarization rather than cross polarized. For example, if a loop antenna is horizontally polarized, it must be in an identical polarization; however, if the measuring antenna is a dipole, it must be vertically polarized!
3. *Size of the loop.* If the loop is electrically small (much less than the wavelength of the generated signal or frequency of interest), the field strength will be pro-

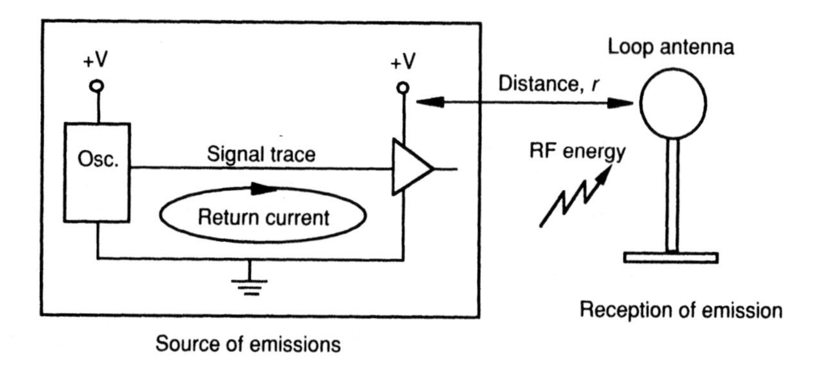

Figure 2.3 RF transmission of a magnetic field.

portional to the area of the loop. The larger the loop, the lower the frequency that is observed at the terminals of the antenna. For a particular physical dimension, the antenna will be resonant for that particular frequency.

4. *Distance.* The rate at which the field strength drops off from the source depends on the distance between the source and antenna. In addition, this distance also determines whether the field created is magnetic or electric. When the distance is electrically *"close"* to the loop source, the magnetic field falls off as the square of the distance. When the distance is electrically *"far,"* we observe an electromagnetic plane wave. This plane wave falls off inversely with increasing distance. The point where the magnetic and electric field vectors cross occurs at approximately one-sixth of a wavelength (which is also identified as $\lambda/2\pi$). The wavelength at this distance is the speed of light divided by the frequency. This formula can be simplified to $\lambda = 300/f$ where λ is in meters and f is in MHz. This one-sixth wavelength applies to a point source, which is what we usually assume in the EMI world. This distance can be farther for larger antennas.

For the electric source, in contrast to the closed-loop magnetic source, the electric source is modeled by a time-varying electric dipole. This means that two separate, time-varying point charges of opposite polarity exist in close proximity. The ends of the dipole contain this change in electric charge. This change in electric charge is accomplished by current flowing throughout the dipole's length. Using the circuit described above, we can represent the electric source by an oscillator's output driving an unterminated antenna. When examined in the context of low-frequency circuit theory, we discover that this circuit is not valid. We did not take into account the finite propagation velocity of the signal in the circuit (based on the dielectric constant of the nonmagnetic material), in addition to the RF currents that are created herein. This is because propagation velocity is *finite*, not *infinite*! The assumptions made are that the wire, at all points, contains the same voltage potential and that the circuit is at equilibrium at all points instantaneously. The fields created by this electric source are a function of four variables.

1. *Current amplitudes in the loop.* The fields created are proportional to the amount of current flowing in the dipole.
2. *Orientation of the dipole relative to the measuring device.* This is equivalent to the magnetic source variable described above.
3. *Size of the dipole.* The fields created are proportional to the length of the current element. This is true if the length of the trace is a small fraction of a wavelength. The larger the dipole, the lower the frequency that is observed at the terminals of the antenna. For a particular physical dimension, the antenna will be resonant for a particular frequency.
4. *Distance.* Electric and magnetic fields are related to each other. Both field strengths fall off inversely with distance. In the far field, the behavior is similar to that of the loop source. When we move in close to the point source, both magnetic and electric fields have a greater dependence on the distance from the source.

The relationship between near-field (magnetic and electric components) and far-field is illustrated in Fig. 2.4. All waves are a combination of both electric and magnetic field components. We generally call this combination of electric and magnetic field com-

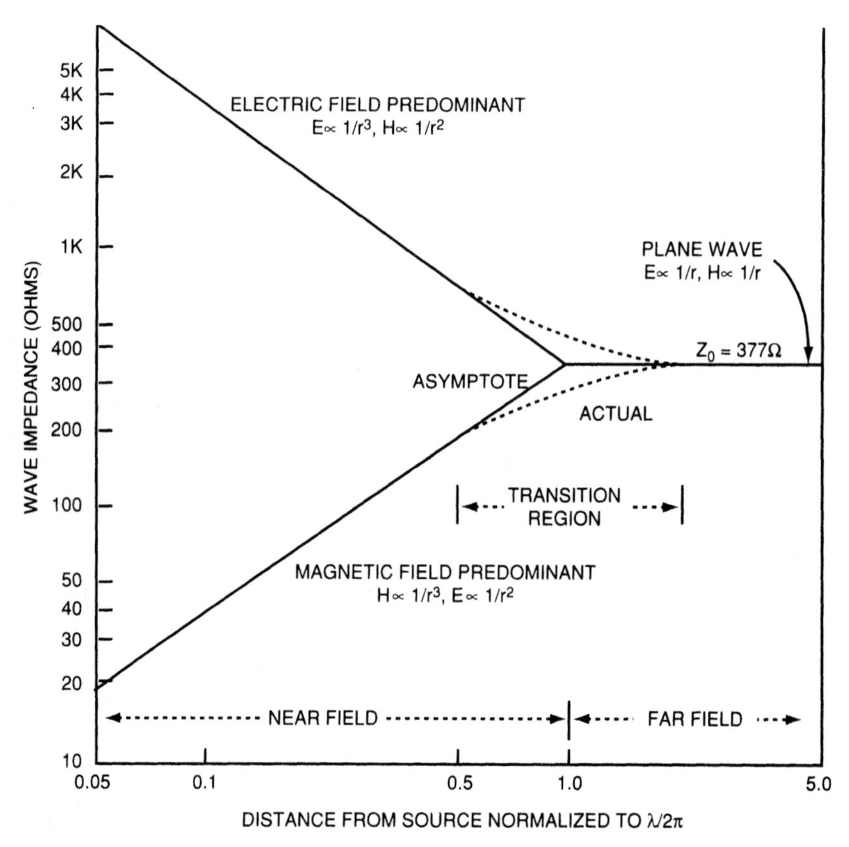

Figure 2.4 Wave impedance versus distance from ***E*** and ***H*** dipole sources. (*Source*: *Noise Reduction Techniques in Electronics Systems*. 2nd edition. H. Ott, © 1988. Reprinted by permission of John Wiley & Sons, Inc.)

ponents a Poynting vector. There is no such thing as an electric wave or magnetic wave. The reason we see a plane wave is that to a small antenna, several wavelengths from the source, the wavefront looks nearly plane. This appearance is due to the physical profile that would be observed at the antenna (like ripples in a pond some distance from the source charge). Fields propagate radially from the field point source at the velocity of light, $c = 1 / \sqrt{\mu_o \varepsilon_o} = 3 \times 10^8$ m/s, where $\mu_o = 4\pi * 10^{-7}$ H/m and ε_o=8.85 $* 10^{-12}$ F/m. The electric field component is measured in volts/meter, while the magnetic field component is in amps/meter. The ratio of both electric field (***E***) to magnetic field (***H***) is identified as the impedance of free space. The point to emphasize here is that in the plane wave, the wave impedance, Z_o, the characteristic impedance of free space, is independent of the distance from the source, and does not hinge on the characteristics of the source. For a plane wave in free space.

$$Z_o = E/H = \sqrt{\mu_o/\varepsilon_o} = \sqrt{\frac{4\pi 10^{-7}\ \text{H/m}}{\frac{1}{36\pi}(10^{-9})\ \text{F/m}}} \tag{2.2}$$

$$= 120\pi \text{ or } 377 \text{ ohms (exactly } 376.99 \text{ ohms)}$$

Energy carried in the wave front is measured in watts/meter^2.

For most applications of Maxwell, noise coupling methods are represented as equivalent component models. For example, a time-varying *electric* field between two conductors can be represented as a capacitor. A time-varying *magnetic* field between these same two conductors is represented by mutual inductance. Figures 2.5a and 2.5b illustrates these two noise coupling mechanisms. A discussion of mutual inductance is presented in Chapter 4.

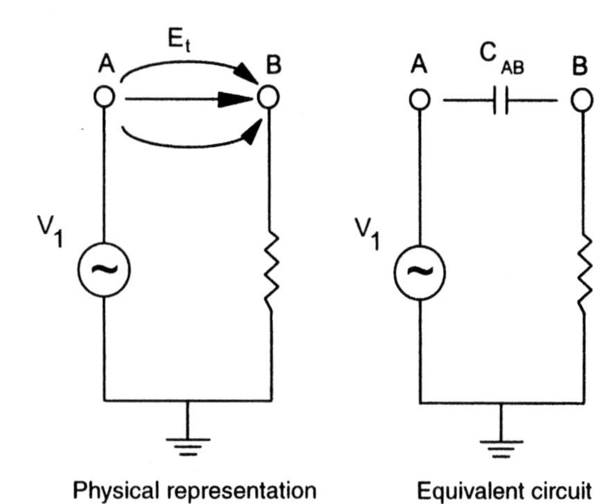

Figure 2.5a Noise coupling method—electric field. (*Source: Noise Reduction Techniques in Electronics Systems.* H. Ott, © 1988, Reprinted by permission of John Wiley & Sons, Inc.)

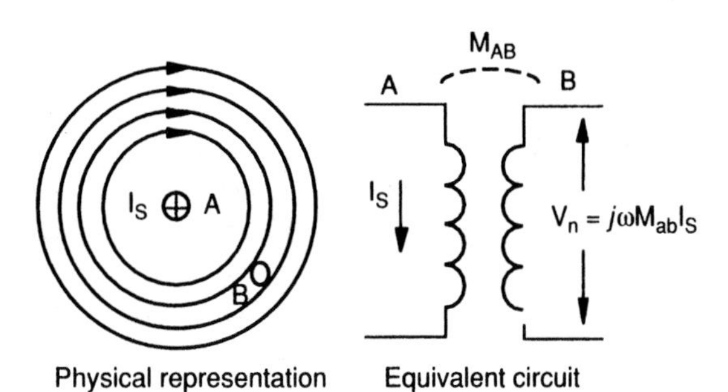

Figure 2.5b Noise coupling method—magnetic field. (*Source: Noise Reduction Techniques in Electronics Systems.* 2nd ed. H. Ott, © 1988, Reprinted by permission of John Wiley & Sons, Inc.)

For this noise coupling model to be valid, the physical dimensions of the circuits must be small compared to the wavelengths of the signals involved. When the model is not truly valid, we can still use lumped component representation to explain EMC for the following reasons.

1. Maxwell's equations cannot be applied directly for most real-world situations due to complicated boundary conditions. If we have no level of confidence in the validity of the approximation of the lumped modeling, then the model is invalid.
2. Numerical modeling does not show how the noise generated is dependent on system parameters. Even if a modeling answer is possible, system-dependent parameters are not clearly known, identified, or shown, along with the explanation in item 1 above.

Why is this theory and discussion about Maxwell's equations important for PCB design and layout? The answer is simple. We need to know how fields are created so that we can reduce these RF-generated fields within a PCB. This reduction applies to reducing the amount of RF current in the circuit. The RF current in the circuit directly relates to signal distribution networks along with bypassing and decoupling. RF currents are ultimately generated as harmonics of clock and other digital signals. Signal distribution networks must be as small as possible to minimize the loop area for the RF return currents. Bypassing and decoupling relate to the current draw that must occur through a power distribution network, which has by definition, a large loop area for RF return currents.

In addition to the loop areas that must be reduced, electric fields are created by unterminated transmission lines and excessive drive voltage. Electric fields can be reduced through use of proper termination, grounding, filtering, and shielding (containment).

2.4 MAXWELL SIMPLIFIED—FURTHER STILL

Now that the fundamental concept of Maxwell's equations has been reviewed, how do we relate all this physics and advanced calculus to EMC within the PCB. To acquire a full comprehension, we must "overly simplify Maxwell" as it applies to a PCB layout. In order to apply Maxwell, we relate his equations to Ohm's law.

Ohm's Law (time domain)	*Ohm's Law (frequency domain)*
$V = I * R$	$V_{rf} = I_{rf} * Z$

where V is voltage, I is current, R is resistance, Z is impedance ($R + jX$), and the subscript rf refers to radio frequency energy. To associate *Maxwell Made Simple to Ohm's Law*, if RF current exists in a PCB trace which has a "fixed impedance value," an RF voltage will be created that is proportional to the RF current. Notice that in the electromagnetics model, R is replaced by Z, a complex number that contains both resistance (real component) and reactance (a complex component).

For the impedance equation, various forms exist depending on whether we are examining plane wave impedance, circuit impedance, and the like. For wire, or a *PCB trace*, use Eq. (2.3).

$$Z = R + jX_L + \frac{1}{jXc} = R + j\omega L + \frac{1}{j\omega C} \tag{2.3}$$

where $X_L = 2\pi fL$ (the component in the equation that relates only to wires or PCB traces)

$$Xc = 1/(2\pi fC)$$
$$\omega = 2\pi f$$

When a *component* has a known resistive and inductive element, such as a ferrite bead-on-lead, a resistor, a capacitor, or other device with parasitic components, Eq. (2.4) is applicable, as the magnitude of impedance versus frequency must be considered.

$$|Z| = \sqrt{R^2 + jX^2} \tag{2.4}$$

For frequencies greater than a few kHz, the value of inductive reactance typically exceeds R; in some cases this might not happen. Current takes the path of least impedance. Below a few kHz, the path of least impedance is resistance; above a few kHz, the path of least reactance is dominant. Because most circuits operate at frequencies above a few kHz, the belief that current takes the path of least resistance provides an incorrect concept of how current flow occurs within a transmission line structure.

Since current always takes the path of least impedance for wires carrying currents above 10 kHz, the impedance is equivalent to the path of least reactance. If the load impedance connects to wiring, a cable, or a trace, and is much greater than the shunt capacitance of the transmission line path, inductance becomes the dominant element. If the wiring conductors have approximately the same cross-sectional shape, the path of least inductance is the one with the smallest loop area.

Each and every trace has a finite impedance value. Trace inductance is only one of the reasons why RF energy is produced within a PCB. Even the lead bond wires that connect a silicon die to its mounting pads may be sufficiently long to cause RF potentials to exist. Traces routed on a board can be highly inductive, especially traces that are electrically long. Electrically long traces are those that are physically long in routed length such that the round-trip propagation delayed signal on the trace does not return to the source driver before the next edge-triggered event occurs when viewed in the time domain. In the frequency domain, an electrically long transmission line (trace) exceeds approximately $\lambda/10$ of the frequency that is present within the trace. Basically, if an RF voltage traverses through an impedance, we end up with RF current. It is this RF current that radiates into free space and causes noncompliance to emission requirements. These examples help us to understand Maxwell's equations and PCBs in extremely simple terms.

It is understood that a moving electrical charge in a trace generates an electric current that creates a magnetic field. *Magnetic fields*, created by this moving electrical charge, are also identified as magnetic lines of flux. Magnetic lines of flux can easily be visualized using the Right-Hand Rule, graphically shown in Fig. 2.6. To observe this rule,

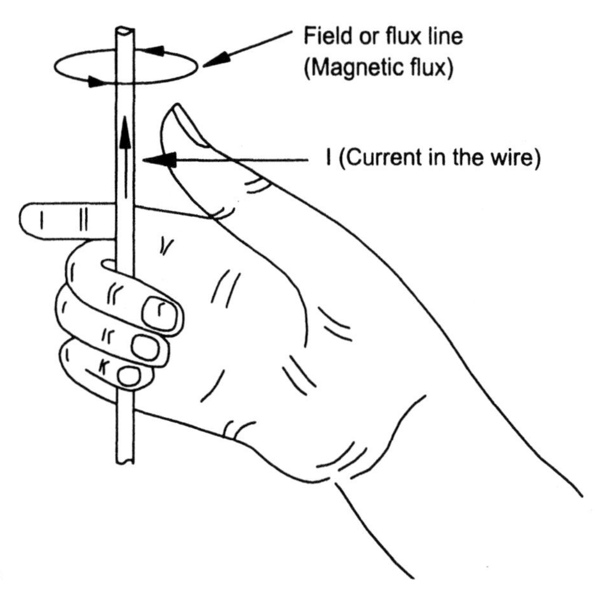

Figure 2.6 Right-hand rule.

make your right hand into a loose fist with your thumb pointing straight up. Current flow is in the direction of the thumb (upwards), simulating current flowing in a wire or PCB trace. Your curved fingers encircling the wire point in the direction of the magnetic field or lines of magnetic flux. Time-varying magnetic fields create a transverse orthogonal electric field. RF emissions are a combination of both magnetic and electric fields. These fields will exit the PCB structure by either radiated or conducted means.

Notice that the magnetic field travels around a closed-loop boundary. In a PCB, RF currents are generated by a source driver and transferred to a load through a trace. RF currents must return to their source (Ampere's law) through a return system. As a result, an RF current loop is developed. This loop does not have to be circular and is often a convoluted shape. Since this process creates a closed loop within the return system, a magnetic field is developed. This magnetic field creates a radiated electric field. In the near field, the magnetic field component will dominate, whereas in the far field the ratio of the electric to magnetic field (wave impedance) is approximately $120\pi\ \Omega$ or $377\ \Omega$, independent of the source. Obviously, in the far field, magnetic fields can be measured using a loop antenna and a sufficiently sensitive receiver. The reception level will simply be $E/120\pi$ (A/m, if E is in V/m). The same applies to electric fields, which may be observed in the near field with appropriate test instrumentation.

Another simplified explanation of how RF exists within a PCB is depicted in Figs. 2.7 and 2.8. Here we examine a typical circuit in both the time and frequency domain. According to Kirchhoff's and Ampere's laws, a closed-loop circuit must exist if the circuit is to work. Kirchhoff's voltage law states that the algebraic sum of the voltage around any closed path in a circuit must be zero. Ampere's law describes the magnetic induction at a point due to given currents in terms of the current elements and their positions relative to that point.

Without a closed-loop circuit, a signal would never travel through a transmission line from a source to a load. When the switch is closed, the circuit is complete, and AC or DC current flows. In the frequency domain, we observe the current as RF energy. There are *not* two types of currents, time domain or frequency domain. There is only one current, which may be represented in *either* the time domain or frequency domain! The RF return path from load to source must also exist, or the circuit would not work. Hence, a PCB structure must conform to Maxwell's equations, Kirchhoff's voltage law, and Ampere's law.

Maxwell, Kirchhoff, and Ampere all state that if a circuit is to function or operate as intended, a closed-loop network must exist. Figure 2.7 illustrates a typical circuit. When a trace goes from source to load, a return current path must also be present, as required by both Kirchhoff and Ampere.

Consider a typical circuit with a switch in series with a source driver (Fig. 2.8). When the switch is closed, the circuit operates as desired; when the switch is opened, nothing happens. For the time domain, the desired signal component travels from source to load. This signal component must have a return path to complete the circuit, generally

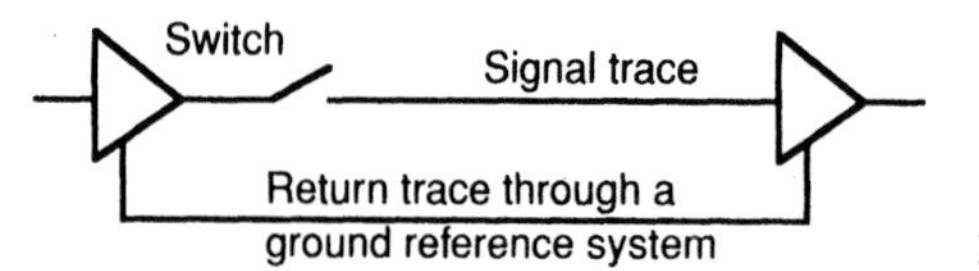

Figure 2.7 Closed-loop circuit.

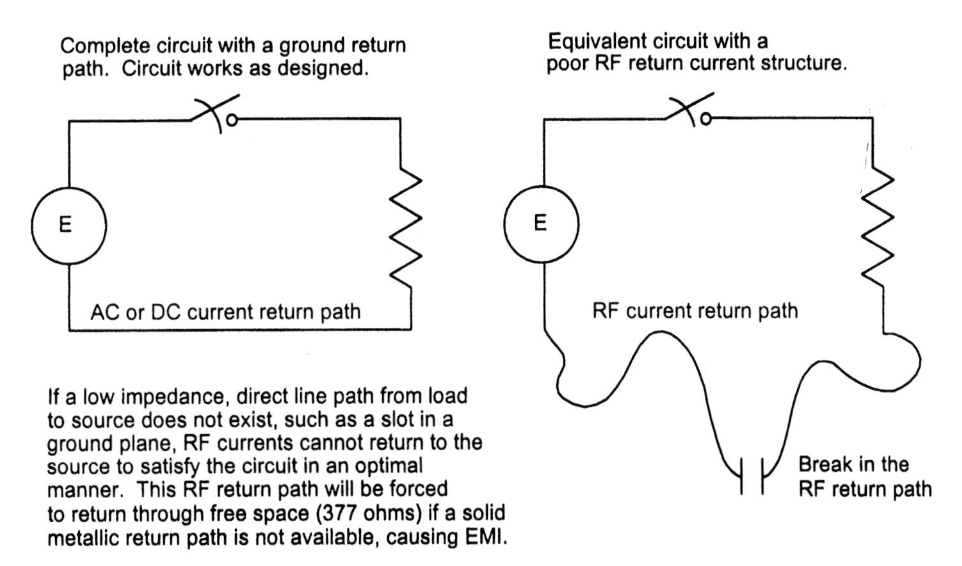

Figure 2.8 Representation of a closed-loop circuit.

through a 0V (ground) return structure (Kirchhoff's law). RF current travels from source to load and must return by the lowest impedance path possible, usually a ground trace or ground plane (also referred to as an image plane). The RF current that exists is best described by Ampere.

2.5 CONCEPT OF FLUX CANCELLATION (FLUX MINIMIZATION)

To review *one* fundamental concept regarding how EMI is created within a PCB, we examined the basic mechanism of how magnetic lines of flux are created within a transmission line. Magnetic lines of flux are created by a current flowing through an impedance, either fixed or variable. Impedance in a network will always exist within a trace, component bond lead wires, vias, and the like. If magnetic lines of flux exist in a PCB, defined by Maxwell, various transmission paths for RF energy must also exist. These transmission paths may be either radiated through free space or conducted through cable interconnects.

To eliminate RF currents within a PCB, the concept of *flux cancellation* or *flux minimization* needs to be discussed. Although the term *cancellation* is used throughout this chapter, we may substitute the term *minimization*. Because magnetic lines of flux travel counterclockwise within a transmission line, if we bring the RF return path parallel and adjacent to its corresponding source trace, the magnetic flux lines observed in the return path (counterclockwise field), related to the source path (clockwise field), will be in the opposite direction. When we combine a clockwise field with a counterclockwise field, a cancellation effect is observed. If unwanted magnetic lines of flux between a source and return path are canceled or minimized, then a radiated or conducted RF current cannot exist except within the minuscule boundary of the trace. The concept of implementing flux cancellation is simple. However, one must be aware of many pitfalls and oversights that may occur when implementing flux cancellation or minimization techniques. With one small mistake, many additional problems will develop creating more work for the EMC engineer to diagnose and debug. The easiest way to implement flux cancellation is to use *image planes*.[2] Regardless of how well we design and lay out a PCB, magnetic and electric fields will always be present. If we cancel out magnetic lines of flux, then EMI cannot exist. It's that simple!

How do we cancel or minimize magnetic lines of flux during PCB layout? This is easier said than done. Various design and layout techniques are available to the design engineer [1]. A brief summary of some of these techniques is presented below. Not all techniques are involved with flux cancellation/minimization. Although the following items have not yet been discussed, each is described in detail within various chapters of this book. These techniques include and are not limited to those listed here.

- Having proper stackup assignment and impedance control for multilayer boards.
- Routing a clock trace adjacent to a return path ground plane (multilayer PCB), ground grid, or use of a ground or guard trace (single- and double-sided boards).
- Capturing magnetic lines of flux created internal to a component's plastic package into the 0V reference system to reduce component radiation.
- Carefully choosing logic families to minimize RF spectral distribution from component and trace radiation (use of slower edge rate devices).
- Reducing RF currents on traces by reducing the RF drive voltage from clock generation circuits, for example, Transistor-Transistor Logic (TTL) versus Complimentary Metal Oxide Semiconductor (CMOS).
- Reducing ground noise voltage in the power and ground plane structure.
- Providing sufficient decoupling for components that consume power when all device pins switch simultaneously under maximum capacitive load.
- Properly terminating clock and signal traces to prevent ringing, overshoot, and undershoot.
- Using data line filters and common-mode chokes on selected nets.
- Making proper use of bypass (not decoupling) capacitors when external I/O cables are provided.
- Providing a grounded heatsink for components that radiate large amounts of internal generated common-mode RF energy.

As seen in this list, magnetic lines of flux are only part of the reason on how EMI is created within a PCB. Other major areas of concern are as follows.

- Existence of common-mode (CM) and differential-mode (DM) currents between circuits and I/O cables.
- Ground loops creating a magnetic field structure.
- Component radiation.
- Impedance mismatches.

[2] R. F. German, H. Ott, and C. R. Paul. 1990. "Effect of an image plane on PCB radiation." *Proceedings of the IEEE International Symposium on Electromagnetic Compatibility*, New York: IEEE, pp. 284–291.

Remember that the majority of EMI emissions are caused by common-mode levels. These common-mode levels are developed as a result of minimized fields in the board or circuit design. These areas of concern are discussed later in this chapter.

2.6 SKIN EFFECT AND LEAD INDUCTANCE

A consequence of Maxwell's third and fourth equations is skin effect related to a voltage charge imposed on a homogeneous medium where current flows, such as a wire lead bond from a component or a PCB trace. If voltage is maintained at a constant DC level, current flow will be uniform throughout the transmission path. A finite period of time is required for uniformity to occur. The current first flows on the outside edge of the conductor and then diffuses inward.

When the source voltage is *not* DC, but high-frequency AC, current flow tends to be concentrated in the outer portion of the conductor. The magnitude of this occurrence is identified as skin effect. Skin depth is defined as the distance to the point inside the conductor at which the electromagnetic field, and hence current, is reduced to 37% of the surface value.

We can define skin depth (δ) by Eq. (2.5)

$$\delta = \sqrt{\frac{2}{\omega\mu_0\sigma}} = \sqrt{\frac{2}{2\pi f\mu_0\sigma}} = \frac{1}{\sqrt{\pi f\mu_0\sigma}} \tag{2.5}$$

where ω = angular (radian) frequency ($2\pi f$)
μ_0 = material permeability ($4\pi \cdot 10^{-7}$ H/m)
σ = material conductivity ($5.82 \cdot 10^7$ mho/m for copper)
f = frequency (Hertz)

Table 2.1 presents an abbreviated table of skin depth values at various frequencies for a 1-mil thick copper substrate (1 mil = 0.001 inch = 2.54×10^{-5} m).

As any of the three parameters of Eq. (2.5) increases, skin depth decreases. The skin depth of conductors at high frequencies is very thin, typically observed at 0.0066 mils or $6.6 \cdot 10^{-6}$ inch (0.0017 mm) at 100 MHz. Current tends to be dominant in a strip near the

TABLE 2.1 Skin Depth for Copper Substrate.

f	δ (copper)
60 Hz	0.335 in. (335 mil, 8.5 mm)
100 Hz	0.260 in. (260 mil, 6.6 mm)
1 kHz	0.082 in. (82 mil, 2.1 mm)
10 kHz	0.026 in. (26 mil, 0.66 mm)
100 kHz	0.008 in. (8 mil, 0.20 mm)
1 MHz	0.0026 in. (2.6 mil, 0.06 mm)
10 MHz	0.0008 in. (0.8 mil, 0.02 mm)
100 MHz	0.00026 in. (0.26 mil, 0.006 mm)
1 GHz	0.00008 in. (0.08 mil, 0.002 mm)

surface of the conductor at a depth of δ. When high-frequency RF currents are present, current flow is concentrated into a narrow strip near the conductor surface, identified as the skin.

The wire's internal inductance equals its DC resistance independent of the wire radius up to the frequency where the wire radius is on the order of a skin depth. Below this particular frequency, the wire's resistance *increases* as $\sqrt{f}$ or 10 dB/decade. Internal inductance is the portion of the magnetic field internal to the wire per-unit-length where the transverse magnetic field contributes to the per-unit-length inductance of the line. The portion of the magnetic flux external to the transmission line contributes to a portion of the total per-unit-length inductance of the line and is referred to as external inductance. Above this particular frequency, the wire's internal inductance *decreases* $\sqrt{f}$ as or −10 dB/decade.

For a solid round copper wire, the effective DC resistance is described by Eq. (2.6). Table 2.2 provides details on some of the parameters used in Eq. (2.6). Signals may be further attenuated by the resistance of the copper used in the conductor and by skin effect losses resulting from the finish on the copper surface. The resistance of the copper may reduce steady-state voltage levels below functional requirements for noise immunity. This condition is especially true of high-frequency differential mode devices (such as Emitter Coupled Logic [ECL]) where a voltage divider is formed by termination resistors and line resistance.

$$R_{dc} = \frac{L}{\sigma \pi r_w^2}\ \Omega \tag{2.6}$$

where L is the length of the wire, r_w is the radius (Table 2.2), and σ is conductivity. The units must be appropriate for the equation to work. As the frequency is increased, the current over the wire cross section will tend to crowd closer to the outer periphery of the conductor. Eventually, the current will be concentrated on the wire's surface equal to the thickness of the skin depth as described by Eq. (2.7) when the skin depth is less than the wire radius.

$$\delta = \frac{1}{\sqrt{\pi f \mu_0 \sigma}} \tag{2.7}$$

where at various frequencies

δ = skin depth
μ_0 = permeability of copper ($4\pi * 10^{-7}$ H/meter)
σ = the conductivity of copper (5.8×10^7 mho/meter),
w = $2\pi f$

A first approximation for inductance of a conductor at high frequency is

$$L = 0.0051 * l\left(2.38 \ln \frac{4l}{d} - 0.75\right) \tag{2.8}$$

where l is the conductor length and d is the diameter in the same units (inches or centimeters). Because of the logarithmic relationship of the ratio l/d, the reactive component of impedance for large-diameter wires dominates the resistive component above only a few

TABLE 2.2 Physical Characteristics of Wire

Wire Gage (AWG)	Solid Wire Diameter (mils)	Stranded Wire Diameter (mils)	R_{dc}—solid wire (Ω/1000 ft) @ 25 °C
28	12.6	16.0 (19×40)	62.9
		15.0 (7×36)	
26	15.9	20.0 (19×38)	39.6
		21.0 (10×36)	
		19.0 (7×34)	
24	20.1	24.0 (19×36)	24.8
		23.0 (10×34)	
		24.0 (7×32)	
22	25.3	30.0 (26×36)	15.6
		31.0 (19×34)	
		30.0 (7×30)	
20	32.0	36.0 (26×34)	9.8
		37.0 (19×32)	
		35.0 (10×30)	
18	40.3	49.0 (19×30)	6.2
		47.0 (16×30)	
		48.0 (7×26)	
16	50.8	59.0 (26×30)	3.9
		60.0 (7×24)	

hundred hertz. Thus, it is impractical to obtain a truly low-impedance connection between two points, such as grounding a circuit using only wire. Such a connection would permit coupling of voltages between circuits due to current flow through an appreciable amount of common impedance.

2.7 COMMON-MODE AND DIFFERENTIAL-MODE CURRENTS

In any circuit there exist both common-mode (CM) and differential-mode (DM) currents. Both common-mode and differential-mode currents determine the amount of RF energy that is propagated. There is a major difference between the two. Given a pair of wires or traces and a return reference source, one or the other mode will exist, usually both. Generally speaking, differential-mode signals carry data or the signal of interest (information). Common mode is a side effect of differential-mode and is most troublesome for EMC compliance. (A representation of common-mode and differential-mode currents is shown later in this chapter in Fig. 2.10.)

Common-mode currents, which are considerably less in magnitude than differential-mode currents, can produce excessive levels of radiated electric fields. The radiated emissions of the differential-mode currents subtract, but do not exactly cancel, since the two transmission paths are not 100% coincident. On the other hand, the emissions of common-mode currents add. In fact, it can be calculated that for a 1-meter length of cable whose wires are separated by 50 mils (a typical ribbon cable), a differential-mode current of

20 mA, or 8 μA common-mode current at 30 MHz, will produce a radiated electric field at 3 meters of 100 μV/m, which just meets the FCC Class B limit [4, 5]. This is a ratio of 2500, or 68dB, between the two modes. This small amount of common-mode current is capable of producing significant radiated emission levels. A number of factors such as distance to conducting planes and other structural symmetries can create common-mode currents. Much *less* common-mode current will produce the same amount of RF propagated energy than a *larger* amount of differential-mode current because common-mode currents do not cancel out within the RF return path.

When using simulation software to predict emissions from I/O interconnects that are driven from a PCB, differential-mode analysis is usually performed. It is impossible to predict radiated emissions based solely on differential-mode (transmission-line) currents. These calculated currents can severely underpredict the radiated emissions of PCB traces, since numerous factors and parasitic parameters are involved in the creation of common-mode currents from differential-mode voltage sources. These parameters usually cannot be anticipated and are present within a PCB structure dynamically in the formation of power surges in the planes during edge-switching times.

2.7.1 Differential-Mode Currents

Differential-mode current is the component of RF energy that is present on both signal and return paths that are opposite to each other. If a 180° phase shift is established precisely, RF differential-mode current will be canceled. Common-mode effects may, however, be created as a result of ground bounce and power plane fluctuation caused by components drawing current from a power distribution network.

Differential-mode signals

1. Convey desired information.
2. Cause minimal interference as the fields generated oppose each other and cancel out if properly set up.

With differential mode, a circuit device sends out a current that is received by a load. An equal value of return current must be present. These two equal currents, traveling in opposite directions, represent standard differential-mode operations. We do not want to eliminate differential-mode performance. Because a circuit board can only be made to emulate a perfect self-shielding environment (e.g., a coax), complete E-field capture and H-field cancellation are not achieved. The remaining fields which are not coupled to each other are the source of differential-mode EMI. In the battle to control EMI and crosstalk in the common mode, the key is to control excess energy fields through proper source control and careful handling of the energy-coupling mechanisms.

2.7.2 Differential-Mode Radiation

Differential-mode radiation is caused by the flow of RF current loops within a system's structure. For a small-loop receiving antenna when operating in a field above a ground plane (free space is not a typical environment), this RF energy is described approximately as [3]

$$E = 263*10^{-16}\,(f^2 * A * I_s)\left(\frac{1}{r}\right) \text{ volts per meter} \tag{2.9}$$

where A = loop area in m^2, f is the frequency (Hz), I_s is the source current in A, and r is the distance from the radiating element to the receiving antenna (meters).

The extra ground reflection can increase measured emissions by as much as 6 dB.

In most PCBs, primary emission sources are created from currents flowing between components and in the power and 0V planes. Radiated emissions can be modeled as a small-loop antenna carrying interference RF currents (see Fig. 2.9). When the signal travels from a source to load, a return current must be present in the power return system. A small loop is one whose dimensions are smaller than a quarter wavelength ($\lambda/4$) at a particular frequency of interest which is illuminated by RF current flowing within its structure. For most PCBs, loops exist with small dimensions for frequencies up to several hundred MHz.

The maximum loop area that will not exceed a specific specification level is described by Eq. (2.10).

$$A = \frac{380r\,E}{f^2 I_s} \tag{2.10}$$

Or, conversely, the maximum field strength created from a closed loop boundary area is

$$E = \frac{A f^2 I_s}{380r} \tag{2.11}$$

where E = radiation limit (μV/meter)
r = distance between the loop and measuring antenna (meters)
f = frequency (MHz)
I_s = current (mA)
A = loop area (cm^2)

In free space, radiated energy falls off inversely proportional (distance) between source and antenna. The loop area formed by a specific current component on the PCB must be known, and is the total area of the specific circuit loop between the trace and current return path. Equations (2.10) and (2.11) are for a single frequency. The equation must be solved for each and every loop (different loop-size area) and for each frequency of interest.

Using Eq. (2.10), we can determine if a particular routing topology needs to have special attention as it relates to radiated emissions. This special attention may involve re-

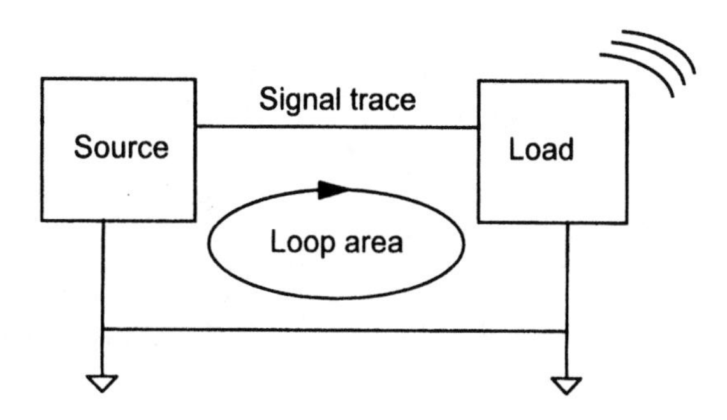

Figure 2.9 Loop area between components.

routing the trace stripline, changing routing topology, locating source and load components closer to each other, or providing external shielding of the assembly (containment).

EXAMPLE

Assume that a convoluted shape exists between two components located on a PCB as a dipole antenna without an RF current return path: $A = 4\ \text{cm}^2$, $I_s = 5$ mA, $f = 100$ MHz. The field strength is 52.6 μV/m at 10-meter distance. Radiated emission limits for EN 55022,[3] Class B is 30 μV/m (quasi-peak). This loop area, which is a typical trace route on many high-technology PCB designs, is 22.6 μV/m above the limit!

2.7.3 Common-Mode Currents

Common-mode current is the component of RF energy that is present on both signal and return paths, usually in a common phase. The measured RF field due to common-mode current will be the sum of the currents that exist in both the signal trace and return trace. This summation could be substantial and is the major cause of RF emissions, especially from I/O cables. Common-mode current is created by poor differential-mode cancellation. This is due to the imbalance between two transmitted signal paths. If the differential signals are not exactly opposite and in phase, their currents will not cancel out. The portion of RF current that is not canceled out is "common-mode" current.

Common-mode signals

1. are the major source of radiation
2. contain no useful information

Common mode begins as the result of currents mixing in a shared metallic structure, such as power and ground planes. Typically, this happens because of the currents flowing through unintentional paths in the planes. Common-mode currents will occur when return currents lose their pairing with their original signal path (e.g., splits or breaks in planes) or when several signal conductors share common areas of the return plane. Since planes have a finite impedance, these common-mode currents set up RF transient voltages on the planes. These RF transients set up currents in other conductive surfaces and signal lines that act as antennas to radiate EMI. The most common cause is the establishment of common-mode currents in conductors and shields of cables running to and from the PCB or enclosure. The key to prevent common-mode EMI is to understand and control the paths of power supply and return currents in the board, by controlling the position of the power and ground planes and the currents within the planes, and to provide proper RF grounding to the case of the system or product.

In Fig. 2.10, current source, I1, represents the flow of current from source, E, to load, Z. Current flow, I2, is current that is observed in the return system, usually identified as an image plane, ground plane, or 0V reference. The measured radiated electric field of the common-mode currents is caused by the summed contribution of both I1 and I2 current produced fields.

[3]EN 55022, "Limits and methods of measurement of radio disturbance characteristics of information technology equipment," an international test specification.

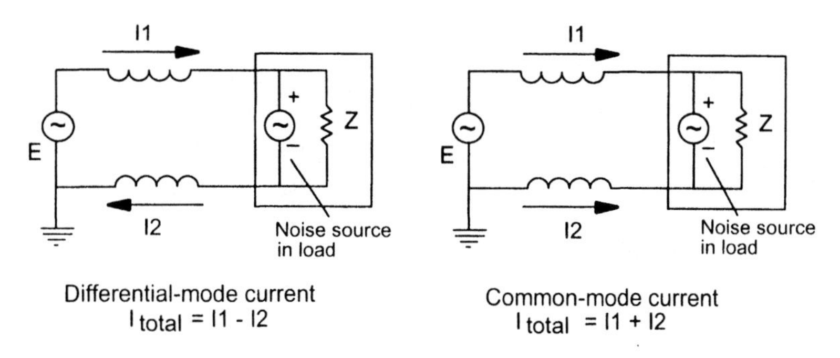

Figure 2.10 Common- and differential-mode current configurations.

With differential-mode currents, the electric field component is the difference between I1 and I2. If I1 = I2 exactly, there will be no radiation from differential-mode currents that emanate from the circuit (assuming the distance from the point of observation is much larger than the separation between the two current-carrying conductors), hence, no EMI. This occurs if the distance separation between I1 and I2 is electrically small. Design and layout techniques for cancellation of radiation emanating from differential-mode currents are easily implemented in a PCB with an image plane or RF return path such as a guard trace (see Chapter 4, Section 13). On the other hand, RF fields created by common-mode currents are harder to suppress. Common-mode currents are the main source of EMI. Fields due to differential mode currents are rarely observed as a significant radiated electromagnetic field.

An RF current return path is best achieved with a ground plane (or ground trace for single- and double-sided boards). The RF current in the return path will couple with the RF current in the source path (magnetic flux lines traveling in opposite direction to each other). The flux that is coupled due to opposite fields will cancel each other out and approach zero (flux cancellation or minimization). However, if the current return path is not provided through a path of least impedance, residual common-mode RF currents will be developed. There will always be some common-mode currents in a PCB, for a finite distance spacing must exist between the signal trace and return path (flux cancellation almost approaches 100%). The portion of the differential-mode return current that does not get canceled out becomes residual RF common-mode current. This situation will occur under many conditions, especially when a ground reference difference exists between circuits. This includes ground bounce, trace impedance mismatches, and lack of decoupling.

It is possible to relate differential-mode voltage to common-mode currents based on the relationship of the magnetic/closed-loop and electric field source. The relationship between magnetic/closed-loop and electric field source was discussed earlier in this chapter.

To make this differential/common-mode comparison to both magnetic/closed-loop and electric field sources, consider a pair of parallel wires carrying a differential-mode signal. Within this wire, RF currents flow in opposite directions (coupling occurs). As a result, the RF fields created are contained. In reality, this coupling cannot be 100%, as a finite distance will exist between the two wires. This finite distance is insignificant related to the overall concept being discussed. This parallel wire set will act as a balanced transmission line that delivers a clean differential (signal-ended) signal to a load.

Using this same wire pair, look at what happens when common-mode voltage is placed on this wire. No useful information is transmitted to the load since the wires carry

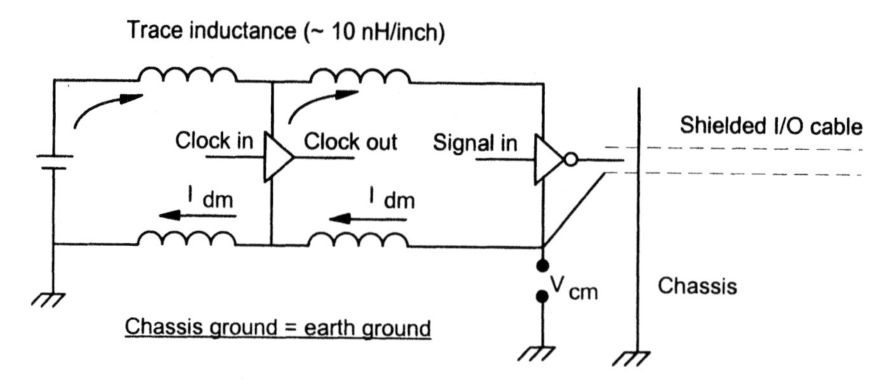

Figure 2.11 System equivalent circuit of differential- and common-mode currents.

the same voltage. This wire pair now functions as a driven antenna with respect to ground. This driven antenna radiates unwanted (or unneeded) common-mode voltage with extreme efficiency. Common-mode currents are generally observed in I/O cables. This is why I/O cables radiate. The mechanism of *how* differential-mode currents create common-mode voltages is detailed in Chapter 3. An illustration of how a PCB and an interconnect cable allow CM and DM current to exist is shown in Fig. 2.11.

2.7.4 Common-Mode Radiation

Common-mode (CM) radiation is caused by unintentional voltage drops in a circuit which cause some grounded parts of the circuit to rise above the referenced real ground potential. Cables connected to the affected ground system act as an antenna and will radiate field components of the CM potential. The far-field electric term is described by Eq. (2.12).

$$E \approx (f\, I_{cm}\, L)\, /\, R \text{ (V/m)} \tag{2.12}$$

where
L = antenna length (m)
I_{cm} = common-mode current (A)
f = frequency (MHz)
R = distance (m)

With a constant current and antenna length, the electric field at a prescribed distance is proportional to the frequency. Unlike differential-mode radiation, which is easy to reduce using proper design techniques, common-mode radiation is a more difficult problem to solve. The only variable available to the designer, if it can be determined, is the common path impedance for the common-mode current. In order to eliminate or reduce common-mode radiation, common-mode fields must approach zero. This is achieved using a sensible grounding scheme.

2.7.5 Conversion Between Differential and Common-Mode

Common-mode currents may be unrelated to the intended signal source (e.g., they may be from other devices). There may also be a component of common-mode current that *is* related to the signal current.

Conversion between differential and common mode occurs when two signal traces (or conductors), both with different impedances, exist. These impedances are dominated at RF by stray capacitance and inductance related to the physical routing of a trace (or interconnect cable). For the majority of layouts, the PCB designer has control over minimizing capacitance and inductance within a network, thus keeping differential- and common-mode currents from being created.

To illustrate this effect, Fig. 2.12 shows differential-mode current, I_{dm}. This is the desired signal of interest across R_L. Common-mode current, *Icm*, will not flow through R_L directly. This common-mode current will flow through impedance Z_a, and Z_b and will return through the return structure. Impedances Z_a and Z_b are not physical components. This is the stray parasitic capacitance or parasitic transfer impedance that exists within the network. This parasitic capacitance exists as a result of a trace located against an RF return path. This parasitic capacitance includes the distance separation between the power and ground plane, decoupling capacitors, input capacitance of devices, an interconnect cable, or other numerous factors that are present within a product design. If $Z_a = Z_b$, no voltage is developed across R_L by I_{cm}. If any inequality results in the network $(Z_a \neq Z_b)$, a voltage difference will be present proportional to the difference in impedance.

$$V_{cm} = I_{cm} * Z_a - I_{cm} * Z_b = I_{cm}(Z_a - Z_b) \tag{2.13}$$

An example of how differential-mode to common-mode conversion occurs with stray capacitance is shown in Fig. 2.12. Because of the need for balanced voltage and ground references, circuits with high-frequency signals that tend to corrupt other signal traces or radiate RF energy (video, high-speed data, etc.), or traces susceptible to external influences must be balanced in such a way that stray and parasitic capacitances of each conductor are identical.

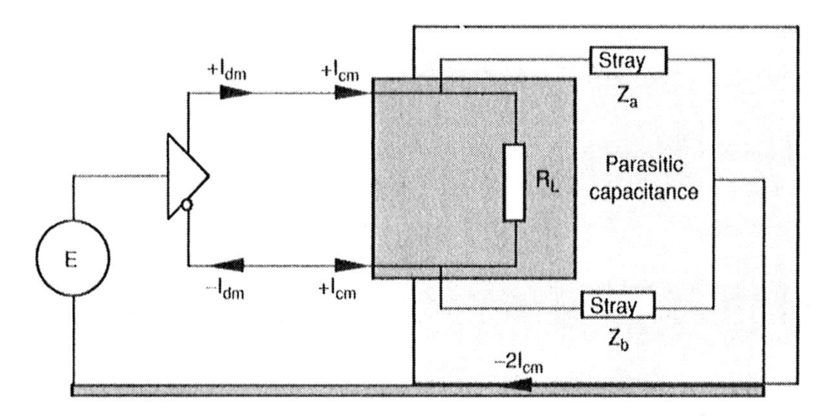

Figure 2.12 Differential to common-mode conversion.

2.8 VELOCITY OF PROPAGATION

This section is provided as background discussion for use throughout this book. Velocity of propagation, *Vp*, is the speed at which data is transmitted through a conductive medium. In air, the velocity of propagation is the speed of light. In a dielectric material, the velocity is slower (at approximately 0.6 the speed of light, depending on the ε_r of the material) and is given by Eq. (2.14).

$$Vp = \frac{C}{\sqrt{\varepsilon_r}} \tag{2.14}$$

where C = $3 * 10^8$ meters per second, or about 11.81 inches/ns (30 cm/ns)
ε_r = relative dielectric constant (compares air to PCB material)

The dielectric constant of various materials used to manufacture a PCB is provided in Table 2.3. Notice that FR-4, the most common material used in the fabrication of a PCB, has a dielectric constant, ε_r of 4.1 at 100 MHz. It was generally assumed that ε_r was in the range of 4.5 to 4.7. This higher value, used by designers for many years, was based on measurements taken with a 1-MHz signal at the time the original measurement was made, and not on how the material works under actual operating conditions.

In reality, a 1-MHz test signal is not appropriate for today's high-technology products. For this reason ε_r is higher in reference material used by designers. A more accurate value of ε_r may be determined by measuring the actual propagation delay of a signal within a trace using a Time Domain Reflectometer (TDR). The values in Table 2.3 are based on a typical, high-speed edge rate signal recorded on a TDR. The effective ε_r for air ≈ 1, much lower than PCB material commonly used in the majority of products designed.

For microstrip topology, the relative dielectric constant may be higher than the number provided by the manufacturer of the material. This is because part of the energy flow is in air and part in the dielectric medium. Microstrip topology and the explanation for why this dielectric constant difference exists are detailed in Chapter 6.

TABLE 2.3 Dielectric Constants and Wave Velocities of PCB Materials

Material	Relative dielectric constant ε_r	Velocity (in/ns)	Velocity (ps/in)
Air	1.0	11.81	84.7
PTFE/glass (Teflon)	2.2	7.96	125.6
Rogers RO 2800	2.9	6.94	144.2
CE/Goreply (Cyanide ester)	3.3	6.50	153.8
GETEK	3.6	6.23	160.6
CE/Glass	3.7	6.14	162.8
Silicon dioxide	3.9	5.98	167.2
BT/glass	4.0	5.90	169.3
Polyimide/glass	4.1	5.83	171.4
FR-4 Glass	4.1	5.83	171.4
Glass cloth	6.0	4.82	207.4
Alumina	9.0	3.93	254.0

Note: Values measured at TDR frequencies using velocity techniques. Values are not measured at 1 MHz which provides higher ε_r.

TABLE 2.4 Frequency/Wavelength Conversions

Frequency	λ	λ/2π	λ/20 Wavelength
10 MHz	30.0 m	4.8 m	1.5 m (5 ft)
27 MHz	11.1 m	1.8 m	0.56 m (1.8 ft)
35 MHz	8.57 m	1.4 m	0.43 m (1.4 ft)
50 MHz	6.00 m	95 cm	0.3 m (12 in.)
80 MHz	3.75 m	60 cm	0.19 m (7.5 in.)
100 MHz	3.00 m	48 cm	0.15 m (5.9 in.)
160 MHz	1.88 m	30 cm	9.4 cm (3.7 in.)
200 MHz	1.50 m	24 cm	7.5 cm (3 in.)
400 MHz	75 cm	12 cm	3.6 cm (1.4 in.)
600 MHz	50 cm	7.9 cm	2.5 cm (1.0 in.)
1000 MHz	30 cm	4.8 cm	1.5 cm (0.6 in.)

2.9 CRITICAL FREQUENCY (λ/20)

Critical frequency refers to a portion of the RF current waveform that subjects a product to RF corruption. Any wavelength less than λ/20 of its respective frequency may be of concern if compliance to EMC standards is required. To determine the frequency, f, of a signal and its related wavelength, λ, use the following conversion equations.

$$f\,(\text{MHz}) = \frac{300}{\lambda(\text{m})} = \frac{984}{\lambda\,(\text{ft})}$$

$$\lambda\,(m) = \frac{300}{f\,(\text{MHz})} \tag{2.15}$$

$$\lambda\,(\text{ft}) = \frac{984}{f\,(\text{MHz})}$$

Throughout this book, reference is made to critical frequencies or high-threat clock and periodic signal traces that have a length greater than λ/20. Miscellaneous frequencies and their respective wavelength distance are summarized in Table 2.4 based on Eq. (2.15).

2.10 FUNDAMENTAL PRINCIPLES AND CONCEPTS FOR SUPPRESSION OF RF ENERGY

2.10.1 Fundamental Principles

The fundamental principles related to radiated emissions deal with common-mode noise created within a PCB at RF frequencies. This fundamental principle deals with energy transferred from a source to load. Common-mode currents are generated everywhere in a circuit, not necessarily in the power distribution system. Common-mode currents by definition are common to both power, return, and other conductors. To close the loop for common-mode currents, a chassis is commonly provided. Since the movement of a charge occurs through an impedance (trace, cable, wire, etc.), a voltage will be developed across this impedance. This voltage will cause radiated emissions to occur if trace stubs, I/O cables, enclosure apertures, and slots are present.

The following principles are discussed in future chapters.

1. For high-speed logic, higher frequency components will be present due to higher fundamental frequencies and shorter rise times (Chapter 3).
2. To minimize the *distribution* of RF currents, proper layout of PCB traces, component placement, and provisions to allow RF currents to return to their source must be provided in an efficient manner to keep RF energy from being propagated throughout the structure (Chapter 4).
3. To minimize development of *common-mode* RF currents, proper decoupling of switching devices along with minimizing ground bounce and ground noise voltage within a plane structure must exist (Chapter 5).
4. To minimize *propagation* of RF currents, proper termination of transmission line structures must occur. At low frequencies, RF currents are not a major problem. At higher frequencies, RF currents will exist and radiate more readily within the structure (Chapter 8).
5. Provide for an optimal 0V reference system. An appropriate grounding methodology needs to be implemented (Chapter 9).

2.10.2 Fundamental Concepts

One of the fundamental concepts for suppressing RF energy within a PCB deals with *flux cancellation* or *minimization*. As discussed earlier, current that travels in a trace (or interconnect structure) causes magnetic lines of flux to exist. These lines of magnetic flux create an electric field. Both field structures allow RF energy to radiate. If we cancel or minimize magnetic lines of flux, RF energy will not be present other than within the boundary between the trace and image plane. Flux cancellation or minimization virtually guarantees compliance with regulatory requirements.

The following two concepts must be understood to minimize radiated emissions.

1. Minimize common-mode currents created as a result of a voltage traveling across an impedance.
2. Minimize the distribution of common-mode currents throughout the network.

Flux cancellation or minimization within a PCB is necessary because of the following sequence of events.

1. Current transients are caused by the production of high-frequency signals (based on a combination of periodic signals (e.g., clocks) and nonperiodic signals (e.g., high-speed data busses) demanded from the power and ground plane structure.
2. RF voltage, in turn, is the product of current transients and the return path provided (Ohm's law).

3. Common-mode RF currents are created from the RF voltage drop between two devices which builds up on inadequate RF return paths between source and load (insufficient differential-mode cancellation of RF currents).
4. Radiated emissions will propagate as a result of these common-mode RF currents.

To summarize what is to be presented in Chapters 3 and 4.

Multilayer boards provide superior signal quality and EMC performance since signal impedance control through stripline or microstrip is observed. The distribution impedance of the power and ground planes must be dramatically reduced. These planes contain RF spectral current surges caused by logic crossover, momentary shorts, and capacitive loading on signals with wide buses. Central to the issue of microstrip (or stripline) is understanding flux cancellation or flux minimization that minimizes (controls) inductance in any transmission line. Various logic devices may be quite asymmetrical in their pull-up/pull-down current ratios.

Asymmetrical current draw in a PCB causes an imbalanced situation to exist. This imbalance relates to flux cancellation or minimization. Flux cancellation will occur through return currents present within the ground or power plane, or both, depending on stackup and component technology. Generally, ground (negative) returns for TTL is preferred. For ECL, positive return is preferred. This is why ECL generally runs on 25.2V; with the more positive line at ground potential. CMOS is more or less symmetrical so that on the average, little difference exists between the ground and voltage planes. One must look at the entire equivalent circuit before making a judgment.

Where three or more solid planes are provided in a multilayer stackup assembly (e.g., one power and two ground planes), optimal flux cancellation may be achieved when the RF flux return path is adjacent to the solid return planes at a common potential throughout the entire trace route. The reason for this statement is one of the *basic fundamental concepts* of implementing flux cancellation within a PCB.

To briefly restate this important concept related to flux cancellation or minimization, it is noted that not all components behave the same way on a PCB related to their pull-up/pull-down current ratios. For example, some devices have 15 mA pull-up/65 mA pull-down. Other devices have 65 mA pull-up/pull-down values (or 50%). When many components are provided within a PCB, asymmetrical power consumption will occur when all devices switch simultaneously. This asymmetrical condition creates an imbalance in the power and ground plane structure. The fundamental concept of board-level suppression lies in flux cancellation (minimization) of RF currents within the board related to traces, components, and circuits referenced to a 0V reference. Power planes, due to this flux phase shift, may not perform as well for flux cancellation as ground planes due to the asymmetry noted above. As a result, optimal performance may be achieved when traces are routed adjacent to 0V reference planes rather than adjacent to power planes.

2.11 SUMMARY

The key points regarding how EMC is created within the PCB are as follows.

1. Current transients exist from the production of high-frequency periodic signals.
2. RF voltage drops between components are the product of currents traveling through a common return impedance path.

3. Common-mode currents are created by unbalanced differential-mode currents, which are created by an inadequate ground return/ground reference.
4. Radiated emissions observed are generally caused by common-mode currents.

REFERENCES

[1] Montrose, M. I. 1996. *Printed Circuit Board Design Techniques for EMC Compliance*. Piscataway, NJ: IEEE Press.

[2] Gerke, D., and W. Kimmel. 1994. "The Designers Guide to Electromagnetic Compatibility." *EDN* (January 20).

[3] Ott, H. 1988. *Noise Reduction Techniques in Electronic Systems*. 2nd ed. New York: John Wiley & Sons.

[4] Paul, C. R. 1992. *Introduction to Electromagnetic Compatibility* New York: John Wiley & Sons.

[5] Paul, C. R. 1989. "A Comparison of the Contributions of Common-Mode and Differential-Mode Currents in Radiated Emissions." *IEEE Transactions on EMC* 31,2:189–193.

3

Components and EMC

It is a well-known fact that RF energy spectra is created as a result of switching current within a PCB. These currents are created as a byproduct of digital components. Each logic state transition produces a transient surge within the power distribution system. Most of the time, these logic transitions do not produce enough ground-noise voltage to be of any functional concern. It is when the edge rate (rise and fall time) of a component becomes extremely fast that RF energy is produced.

Transient spikes placed on a power distribution system creates ground-noise voltage. This ground-noise voltage is first observed as differential-mode (DM) noise. Differential-mode noise is then converted to common-mode (CM) currents. Common-mode currents are the main cause of radiated RF energy. By minimizing production of DM noise in the power distribution network, less CM current results.

This chapter investigates active components (digital logic) along with their relationship to the creation of DM noise. Passive components were discussed in Chapter 2. In addition to the parameters and behaviors that are present within digital logic related to emissions, susceptibility and self-compatibility concerns exist.

3.1 EDGE RATE

When choosing digital components for a particular application, design engineers are generally interested only in functionality and operating speed, basing their selection on the propagation delay of the internal logic gates as published by the manufacturer, not necessarily the actual edge rate of input and output signals.

As the speed of components accelerates (faster internal propagation time), increases in DM currents, crosstalk, and ringing potentially can occur. There is an inverse relationship between operating speed and EMI. Many components have internal logic gates that operate at a faster edge rate than the propagation delay required for functionality. As a result, slower logic families (internal gates) are preferred for EMI since propagation delay is the primary function of the circuit. Figure 3.1 illustrates the relationship between the internal switching speed of a basic inverter gate compared to propagation delay.

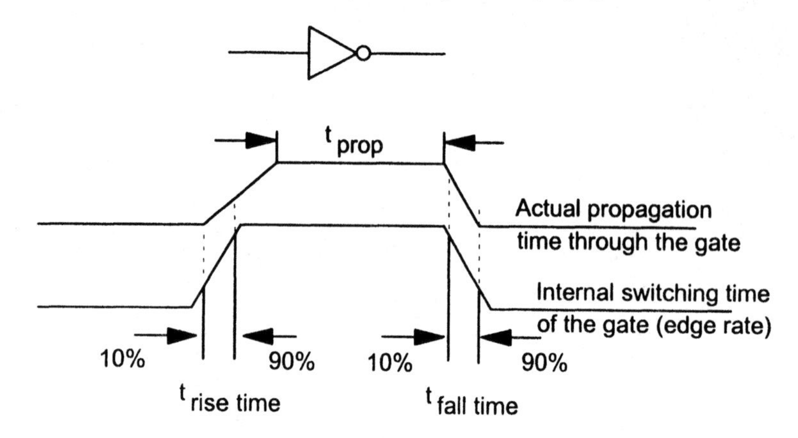

t prop is the propagation delay of the device.

Note: Manufacturer may claim a rise/fall time at 2 ns max. In reality, this value may be well into sub-nanosecond values.

Figure 3.1 Output switching time versus propagation delay.

Speed is important only when the edge rate of a signal (rise or fall time) is fast enough that the desired signal changes logic state in the same or less time than it takes to allow the signal to travel the length of the trace or wire. The actual clock frequency is a secondary concern for EMC compliance, whereas the actual edge rate is the primary concern.

Various logic families are available with different design features. These features vary between CMOS, TTL, and ECL and include input power, package outline, speed-power combinations, voltage swing levels, and edge rates. Certain logic devices are now available with clock skew circuitry to slow down the internal edges of the internal logic gates while maintaining accurate propagation delay.

One extremely important device parameter (for EMC) usually not specified by device manufacturers is *power peak inrush surge current into the power pins.* These peak power currents are the result of logic crossover currents, device capacitive overheads, and capacitance caused by surge currents from trace capacitance and loading device junctions. These surge currents may exhibit levels that are many multiples of the actual signal currents that are injected into a transmission line (trace).

Selection of the slowest logic family possible while maintaining adequate timing margins minimizes EMI effects and enhances signal quality. We should note here that use of standard and low-power Schottky TTL logic, (e.g., 74LS series) is becoming less common in today's marketplace. Moreover, specialized design techniques during layout are usually not required when using slower speed logic families. However, today's high-speed, high-technology products require use of extremely fast-edge logic in the 1.5–5 ns range, for example, 74ACT and 74F series. Use of a 74HCT could be provisionally adequate for replacement of a 74ACT for most applications, with the added benefit of less RF emissions generated. As a general consideration, *do not use faster devices than the functional timing diagram or what the circuit actually demands.*

If timing requires fast logic families, the designer must address individually the issues of decoupling, routing, and handling of clock traces. (See Table 3.1 for details on the EMI characteristics of different logic families.)

Fast switching times (edges) cause proportional increases in problems related to return currents, crosstalk, ringing, and reflections, and only increased attention to meticulous design can alleviate these problems. These problems are independent of device propagation delay. This is because logic families have edge rates that are faster than the propagation delay inherent in the device. No two logic families are the same. Even the same components from different manufacturers may differ in construction and edge rates. Edge rate is defined as the rate of voltage or current change per unit time (volt/ns or amperes/ns).

When selecting a logic family, manufacturers will specify in their data book the *maximum* or *typical* edge rate t_{max} or t_{typ} of the clocks and I/O pins. This specification is usually 2–5 ns maximum. It is observed that the *minimum* edge rate t_{min} may not be published. A device with a 2 ns maximum edge rate specification may in reality be 0.5 to 1.0 ns. The significant contributor to the creation of RF energy is the edge rate, *not* actual operating frequency. A 5 MHz oscillator driving a 74F04 driver (with a 1-ns edge) will generate larger amounts of RF spectral energy over the frequency spectrum than a 100-MHz oscillator driving a 74ALS04 (with a 4-ns edge). ***This one component specification is the most frequently overlooked and forgotten parameter in printed circuit board design. However, this is the most critical aspect of which design engineers must be con-***

TABLE 3.1 Chart of Logic Families

Logic Family	Published Rise/Fall Time (Approx. T_r/T_f	Principal Harmonic Content/ $F = (1/\pi t_r)$	Typical Frequencies Observed as EMI (10th harmonic) $F_{max} = 10*F$
74L xxx	31–35 ns	10 MHz	100 MHz
74C xxx	25–60 ns	13 MHz	130 MHz
74HC xxx	13–15 ns	24 MHz	240 MHz
74 xxx	10–12 ns	32 MHz	320 MHz
(flip-flop)	15–22 ns	21 Mhz	210 MHz
74LS xxx	9.5 ns	34 MHz	340 MHz
(flip-flop)	13–15 ns	24 MHz	240 MHz
74H xxx	4–6 ns	80 MHz	800 MHz
74S xxx	3–4 ns	106 MHz	1.1 GHz
74HCT xxx	5–15 ns	64 MHz	640 MHz
74ALS xxx	2–10 ns	160 MHz	1.6 GHz
74ACT xxx	2–5 ns	160 MHz	1.6 GHz
74F xxx	1.5–1.6 ns	212 MHz	2.1 GHz
ECL 10K	1.5 ns	212 MHz	2.1 GHz
ECL 100K	0.75 ns	424 MHz	4.2 GHz
BTL	1.0 ns*	318 MHz	3.2 GHz
LVDS	0.3 ns*	1.1 GHz	11 GHz
GaAs	0.3 ns*	1.1 GHz	11 GHz
GTL + (Pentium Pro)	0.3 ns*	1.1 GHz	11 GHz

*These are minimum edge rate values.

TABLE 3.2 Selected Characteristics of Logic Families

Logic Family	Voltage Swing (V)	Input Capacitance (pF)	DC Noises Margin (V)	Typical Output Resistance R_o Low/High) Ω
CMOS, 5V	5	5	1.2	300/300
CMOS, 12V	12	5	3	300/300
TTL	3.3	5	0.4	30/150
TTL-LS	3.3	5.5	0.4	30/160
HCMOS	5	4	0.7	160/160
S-TTL	3.3	4	0.3	15/50
FAST & AS-TTL	3.3	4.5	0.3	15/40
ECL, 10k	0.8	3	0.1	7/7
GaAs	1	≈ 1	0.1	—

cerned with to ensure an EMI-compliant product. The frequently heard statement, "Use the slowest logic family possible," is a result of the minimum edge rate parameter ***not being specified or published*** by a component manufacturer in their data books. Edge rates of digital devices are the source of most RF energy created within a PCB.

The reason to use the slowest logic family stems from the relationship between time domain and frequency domain. Fourier analysis of signal edges in the time domain shows that as the slope (edge rate) of the signal becomes faster, a greater amount of spectral bandwidth of RF energy is created. A detailed discussion of Fourier transforms and analysis is presented in Appendix B.

Table 3.1 provides information on the harmonic spectrum of digital logic families. This chart can be used as a reference in determining an optimal logic family to use for minimizing EMI emissions while allowing for proper functionality of the design. In examining Table 3.1, the spectral distribution (bandwidth) is shown as $1/\pi t_r$, with t_r being the edge rate. For bipolar technologies, the rise and fall times are generally different, with t_f the faster of the two. The equivalent bandwidth presented in Table 3.1 is calculated on the shorter (or faster) edge rate, either t_r or t_f (fall time or high-low transition). Rise and fall time is commonly referred to as the "edge rate" and is dependent on the device loading (output) pin. Table 3.1 considers capacitive loading at approximately 20–40 pF to represent reasonably fast conditions and typical board layout.

Table 3.2 shows selected characteristics of several logic families. Here it is observed that the output resistance, R_o, of the logic gate is the current limiting parameter. This defines how much drive current is possible under a heavy capacitive load, or the equivalence of a shorted trace at a specific resonant frequency. Even a shorted gate output (time of transition) cannot deliver a current greater than V/R_o.

3.2 INPUT POWER CONSUMPTION

Power supply transition currents to a gate's input is a major contributor to noise generation on the board by either the power plane (or trace) or ground plane/ground trace (0V reference). The 0V reference refers to the source that is at ground potential relative to the power source. It is common to refer to 0V reference as being either the power return plane, image plane, or ground plane. Transition currents are the main source of differential-mode currents and, hence, RF energy.

1. Examining Table 3.1, we notice that the shorter the transition time, or the faster edge rate, a larger EMI spectral profile exists. EMI increases in severity with frequency, f (for conducted EMI and crosstalk) and often f^2 (for radiated EMI).
2. The power supply transition current demands during component switching can be quite large. These currents have no relationship with the quiescent current required to establish a "1" or "0" signal state in digital logic. For TTL and some CMOS technologies, inrush of surge current is created owing to partial conduction overlap of the output drive transistors. During the time that the crossover between logic high and logic low occurs, the power bus is virtually shorted to ground through two partially saturated transistors, as well as a current limiting resistor. This resistor is designed to keep the inrush of surge current to a level that prevents damage to the drive circuitry.
3. To avoid crossover conduction currents, manufacturers are providing Schottky barrier diodes to prevent the output transistors from going into excessive saturation. Other design techniques used within the fabrication of the component include "output edge rate control." This is accomplished by replacing one large-output transistor with several smaller ones. The peak current surge is still significant, along with potential on-chip problems related to functionality if the component has a large number of output drivers.
4. RF voltages and capacitive crosstalk can exist during the voltage swing between logic low and logic high.
5. The current that is required to change logic state from low to high or high to low is also larger than the quiescent current. This load current is calculated to be

$$I_t = C\frac{dV}{dt} \tag{3.1}$$

where C is the sum of the distributed capacitance of the load, *plus* the trace capacitance to ground. For single-sided boards, C is 0.1 to 0.3 pF/cm. For multilayer, C is 0.3 to 2 pF/cm, and input capacitance is as shown in Table 3.2.

For example, if we have a 3.5-V, 2-ns edge rate signal, with a 7-cm-long trace on a single-layer board, with a fanout of 5 gates, the transient output current is

$$I_t = (7\text{ cm} * 0.3 \times 10^{-12} F/\text{cm} + 5 * 5 \times 10^{-12} F/\text{gate})\frac{3.5\text{ V}}{2\text{ ns}} = 47\text{ mA} \tag{3.2}$$

The peak current in this equation combines nonsymmetrical current usage with the power supply transition current. For low-to-high transitions, this current is added to the quiescent current. For high-to-low transition, the current is subtracted from the quiescent current, since the gate is *sunk* to ground and the capacitive charge from the load must discharge into the output driver's gate, which appears as a short to ground.

For products that are sensitive to input current draw, such as portable electronic devices powered by a battery with a limited time of operation, consideration must be made for all components specified in the Bill of Materials. This includes all second-source components where alternative manufacturers provide a device for a particular function. Different manufacturing processes allow one device to consume more input current than another, either during quiescent operation or logic switching. The measurement, or calculation of the transient output current should be made with all device pins switching simultaneously under maximum capacitive load. This determines the worst case conditions that may exist. Consideration must also be made for usage at elevated temperatures where input current draw may be greater than that specified in its respective data sheet.

Another concern related to radiated EMI emissions is due to the difference between manufacturers of active digital components. Although a digital device may be form, fit, and function compatible, differences exist in the manufacturing process (and design). Not every manufacturer designs its components in the same way; hence, design engineers should not assume identical results from similar components related to functionality or EMC compliance, especially if behavioral models are used for simulation purposes.

3.3 CLOCK SKEW

With increasing performance requirements in high-technology products, greater emphasis on the design of low clock skew circuits is required. Clock skew, the difference in time between simultaneous clock transitions within a network at various points of arrival, is a major component constraint that forms the upper bounds of the system clock frequency. Reduction in system clock skew improves operational performance without having to resort to a higher speed logic device such as ECL or GaAs.

System designers want to utilize as much of the clock cycle as possible without adding unnecessary timing guard bands. Propagation delays of peripheral logic do not scale with frequency. As a result, when the clock period decreases, the designer has less time to perform a specific function with more logic devices to trigger. This is often a difficult task to achieve. A viable option is to use a special clock source that minimizes clock uncertainty.

To illustrate this situation, a 33-MHz component has a clock cycle of $T_{\text{cycle}} = 30$ ns. A 74FCT240, for example, has a high–low uncertainty (t_{plh} to t_{phl}) of approximately 3.3 ns. If a pin-to-pin skew of 1.7 ns exists on the part, along with the propagational delay within the trace, we may have only 25 ns of the clock period available instead of 30 ns. Taking this example to a new level, we see that a 50-MHz system has a penalty of 25%. This allows for a maximum of 10% of the period permitted for clock distribution.

If use of multilevel clock drivers is required, additional clock skew may be added to the circuit. This is exactly what we *want to avoid*. Use of "multi-output," not "multilevel," clock skew buffers is being provided to address this problem. The drawback of these devices is that, although they meet timing requirements in the time domain, a large amount of radiated RF energy will emanate from the device package when observed in the frequency domain. Even with the best design techniques implemented for suppressing of RF energy on a PCB, reducing radiated noise from a component is difficult, if not impossible to eliminate, except through containment and use of a metal case, a grounded heatsink, or overall system shielding.

An important consideration when designing with a clock driver circuit is that the specifications provided are for a fixed, lumped capacitive load. With various devices on the net, the capacitance of the transmission line may be altered from optimal conditions, hence exacerbating the creation of common-mode currents and increasing ground bounce. With this situation, various loads distributed over several inches of a PCB trace can contribute additional delay. The system designer must use caution to minimize total system skew. In other words, changing a clock driver to a low-skew device may not solve all timing problems. In fact, an increase in RF emissions usually occurs.

Skew is divided into three parts: duty cycle skew, output-to-output skew, and part-to-part skew. Depending on the specific application, each component can be of equal or overriding importance.

3.3.1 Duty Cycle Skew

Duty cycle skew is the difference between t_{plh} and t_{phl} related to propagation delay between components. This is illustrated in Fig. 3.2. Because of the difference in t_{plh} and t_{phl}, pulse width distortion of the duty cycle is identified as pulse skew. This skew is critical in applications when both edges, or when the duty cycle of the clock signal, is important. This is generally observed in microprocessor designs [8].

3.3.2 Output-to-Output Skew

Output-to-output skew is the difference between the propagation delay of all outputs of a clock driver. This skew is dependent on the design of the component's output transistors being identical. If the skew between all edges is a critical parameter, we need to add this time parameter to the duty cycle skew to acquire total system skew. Generally, output-to-output skew is smaller than duty cycle skew for TTL and CMOS devices. Because of the near zero-duty cycle skew of differential drivers, (e.g., ECL or LVDS), a single device driver provides a clock signal to a load, or when multiple load devices on the net must be clocked at exactly the same time with respect to each other. This situation is caused by the design of the wafer (die) internal to the component package. Because of the manufacturing process used, this skew will be significantly less than the propagation delay that is specified in the device's data sheet. Figure 3.3 illustrates output-to-output skew [8].

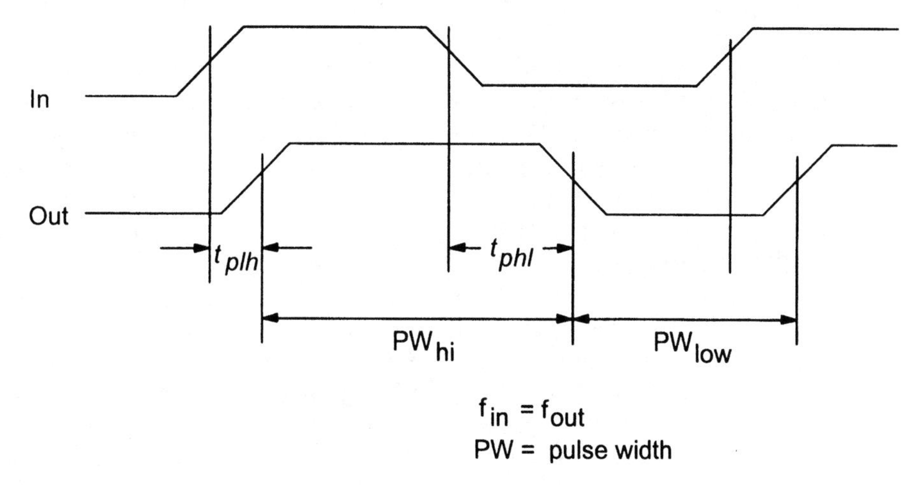

Figure 3.2 Duty cycle skew.

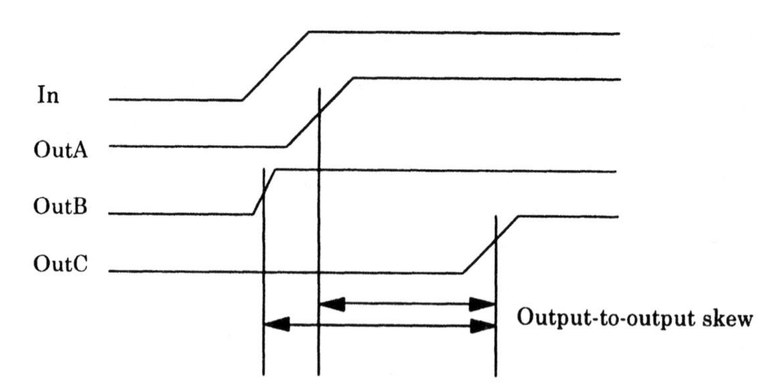

Figure 3.3 Output-to-output skew

3.3.3 Part-to-Part Skew

The part-to-part skew specification is by far the most difficult performance aspect of a device to minimize, owing to the environment in which a device is located in and the manufacturing process being used to manufacture the clock driver. This skew is significantly greater than duty cycle or output-to-output skew and is based on nonvarying environmental conditions. It is advisable to study carefully the data sheet on devices used to ascertain the conditions in which the part is guaranteed to function. If the part-to-part skew specified is different from the propagation delay window for the device, we can assume that constraints must exist for the device's part-to-part skew specification [8].

3.4 COMPONENT PACKAGING

Consideration must always be given to placement of components on a PCB along with their interconnect traces, bus structures, and decoupling capacitors. A design parameter that is generally *not* considered by design engineers (and always overlooked because "that's the way it is") is how digital components are packaged (silicon substrate in its protective case, either plastic or ceramic). Design engineers generally agree that a device is a device and must be used in a design based on desired functionality and cost. The case packaging is a parameter that is generally outside the control of the design engineer. With this situation, why worry about component packaging? Speed is the only important parameter in high-technology designs according to the marketing document or engineering functionality specification. In reality, component packaging plays a major role in the development or suppression of RF currents.

The inductance of individual leads within a component package creates several problems, the greatest concern being that of *lead-length inductance*. This inductance allows several abnormal operating conditions to exist. These concerns are ground bounce and creation of a small loop antenna that may radiate RF currents based on the physical dimensions that exist between source and load. Ground bounce causes glitches to occur in logic input circuitry whenever the device's output switches from one logic state to another. Ground bounce is discussed later in this chapter.

Although it may seem minuscule, the loop area of the die, its bonding wires to a pad, and the component leads to the PCB can become significant contributors to the creation of EMI. This is especially true with very-large-scale integrated (VLSI) components and heavily populated PCBs with high-speed (edge rate) parameters. With multilayer PCB stackups, the trace radiating loops are so small that IC leads can become a large radiating antenna relative to loop area and the frequency generated within the component.

A detailed discussion on how differential-mode currents produce radiated emissions is found in Chapter 2. Differential-mode currents are set up by the existence of a loop between components and a plane on a multi-layer board. For inductance to exist, a loop must be present. To minimize inductance, the loop area must decrease in size. This inductance includes the length of the bond wires internal to components, internal bond leads for capacitors, resistors, and other passive components. A review using Fig. 3.4 examines how loops create radiated emissions.

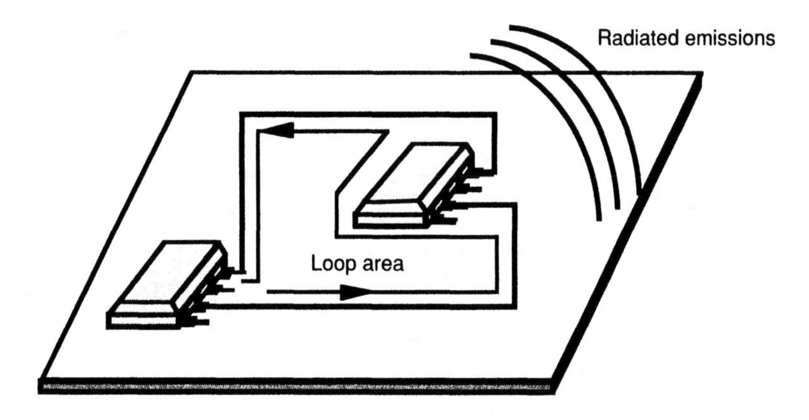

Figure 3.4 Loop area between components.

With lead-length inductance in mind, the worst type of component packaging is the standard TTL Dual-In-Line Package (DIP) where the power and ground pins are at opposite corners. This arrangement is illustrated in Fig. 3.5.

In most PCBs, primary emission sources are established from currents flowing between components. Radiated emissions can be modeled as a small-loop antenna carrying interference current shown in Fig. 3.4. A small loop is one whose dimensions are smaller than a quarter wavelength $(\lambda/4)$ at a particular frequency of interest. For most PCBs, loops exist with small dimensions for frequencies up to several hundred MHz. When a dimension approaches $\lambda/4$, RF currents within the loop will appear out of phase at a distance such that the effect causes the field strength to be reduced at any given point.

Common-mode current is more difficult to control and normally determines the overall emissions performance. Common mode is generally observed from cables affixed to the unit. RF energy is determined by the common-mode potential (usually ground-noise voltage) and is not the same as differential-mode radiated energy. Common-mode current may be modeled as a monotonic antenna driven by ground-noise voltage. For a short monopole antenna of length L over a ground plane, the magnitude of the electric field strength can be measured at a distance r in the far field.

To predict the maximum electric field strength from a loop over a ground plane, Eq. (3.3) is used [4]. Differential-mode radiated emissions is best controlled in the design and layout of the PCB or product.

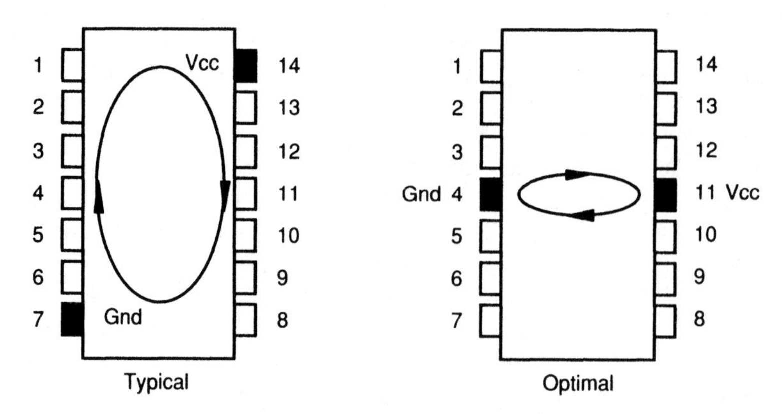

Figure 3. 5 Component packaging related to RF loops—DIP configuration.

$$E = 263 * 10^{-16} (f^2 * A * I_s)\left(\frac{1}{r}\right) \text{ volts per meter} \quad \text{(differential-mode)}$$

$$E = 4\pi * 10^{-7} (I_s * f * l)\left(\frac{1}{r}\right) \text{ volts per meter} \quad \text{(common-mode)} \tag{3.3}$$

where E = effective radiated field (V/m)
A = loop area (cm^2)
f = frequency (MHz)
I_s = the source current (mA)
l = length of the trace or cable (meters)
r = distance from the radiating element to the receiving antenna (meters)

The maximum loop area that will not result in E-field levels to exceed a specific specification level is described by Eq. (3.4).

$$A = \frac{380\,Er}{f^2 I_s} \tag{3.4}$$

In free space (typically described as a minimum distance from the radiating source as wavelength, λ, divided by 2π), radiated energy decreases with inverse proportion to distance between source and antenna. The loop area on the PCB must be known, which is the total area of the circuit between the trace and RF current return path. A convoluted shape may be present, which is at times difficult to determine for a single frequency of interest using Eq. (3.3). The equation must be solved for *each and every loop* (different loop-size areas) and for each frequency of interest if the full profile is to be understood.

Using Eq. (3.3), we can determine if a particular routing topology needs to have special attention as it relates to radiated emissions. This special attention may involve some or all of the following: re-routing the trace stripline, changing routing topology, locating source and load components closer to each other, or providing external shielding of the assembly (containment).

To help minimize loop area, we can select logic components (to the extent possible), with power and ground pins located in the center of the package (not on opposite corners) or physically adjacent to each other. Power pins in the center provide for optimal placement of decoupling capacitors (when these capacitors are placed on the bottom side of the PCB). This configuration also minimizes trace length connections between the device and decoupling capacitor, in addition to minimizing trace length inductance from the power and ground pins *internal* to the silicon wafer (die) of the package. Since a via is required to bring both power and ground to the device (when both power and ground planes are provided in a multilayer PCB stackup assignment), these same vias can also be used for the local decoupling capacitor.

Surface-mount technology (SMT) components have an advantage over through-hole devices by virtue of a smaller loop area related to creation of RF currents. In Fig. 3.5, we notice that a reduction in loop area exists when the power and 0V reference pins are located in the center of the device instead of opposite corners. A similar reduction in loop area also occurs on larger packaged components where the power and ground pins are provided adjacent to each other, as illustrated in Fig. 3.6. Adjacent power and 0V reference interconnect bond wires minimize loop areas for RF currents that may be developed and allow for enhanced flux cancellation (or minimization) between power and ground. If RF currents exist on power input pins due to differential-mode switching currents created by simultaneously switching all pins under maximum capacitive load, a more stable 0V reference system must be available. RF flux will see this alternate return path (0V reference), thus canceling out internally generated RF currents by minimizing differential-mode ground-noise voltage.

With this consideration in mind, use of surface-mount technology (SMT) components is preferred to through-hole in minimizing RF emissions. This characteristic difference is due to shorter lead-length inductance from the die of the component to the circuit trace on the PCB. SMTs by virtue of package size have smaller loop areas. Internal lead-

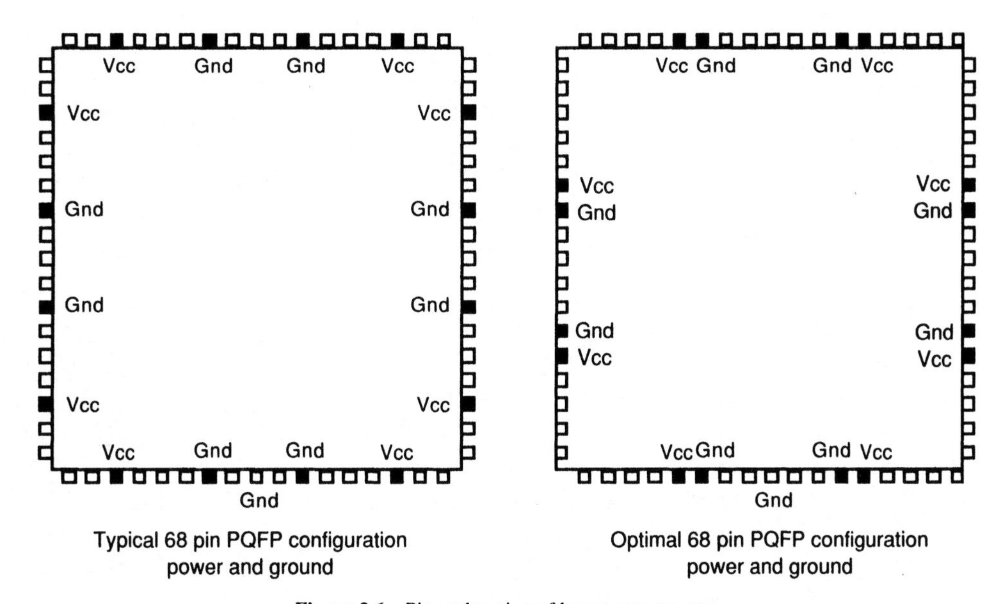

Figure 3.6 Pinout location of larger components.

length inductance also exists. Inductance is a component that generates RF currents. RF currents cause RF emissions, in addition to possibly causing signal integrity problems. Sometimes through-hole devices are installed on sockets. Sockets add greater lead-length inductance and hence create greater amounts of EMI since the loop area is increased.

Another design feature built into components that either promote or demote creation of RF currents is the manner in which IC package leads are bonded to the PCB. An example of two different lead bond configurations is shown in Fig. 3.7.

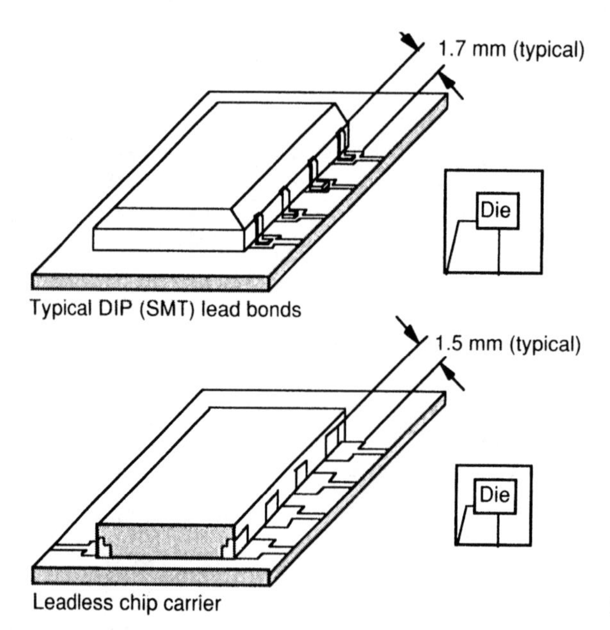

Figure 3.7 Lead bonding to the PCB

Although the loop area appears to be extremely small, bonding wires internal to the package, along with the external interconnect leads, allow for significant RF paths to be developed. Multilayer boards minimize RF currents because the loop area for the traces are small compared to the overall inductive lead-length internal to the device. In other words, bond wires can become significant antennas, especially at high frequencies or with logic devices that operate in the sub-nanosecond range. Issues with bond wires are becoming more common in today's high-technology products. With this situation, the use of DIP packaging provides for the greatest amount of RF field development, especially if a through-hole socket is provided. Table 3.3 presents values of lead-length inductance of various logic packages [2].

SMTs also provide superior performance over DIP components as there is approximately a 40% reduction in package size. This translates to a 64 percent reduction in the lead length inductance in the radiating loop between the die and the mounting pad. In addition, SMTs use less board space with corresponding less trace lengths between components (smaller board size). With the use of SMT components, the decoupling capacitor loop area is also reduced (see Chapter 5).

A variety of packaging configurations exist. Almost all packages when used at high speeds suffer from problems associated with lead-length inductance, lead capacitance, and heat dissipation. The inductance of individual leads within a device package creates a problem identified as *ground bounce*. Ground bounce causes glitches in logic inputs dur-

TABLE 3.3 Lead-Length Inductance of Various Logic Packages

Package Size and Type	Lead-Length Inductance
14 pin DIP	2.0 – 10.2 nH
20 pin DIP	3.4 – 13.7 nH
40 pin DIP	4.4 – 21.7 nH
20 pin PLCC	3.5 – 6.3 nH
28 pin PLCC	3.7 – 7.8 nH
44 pin PLCC	4.3 – 6.1 nH
68 pin PLCC	5.3 – 8.9 nH
14 pin SOIC	2.6 – 3.6 nH
20 pin SOIC	4.9 – 8.5 nH
40 pin TAB	1.2 – 2.5 nH
624 pin CBGA	0.5 – 4.7 nH
Wire bonded to hybrid substrate	1 nH

ing a state transition from a driving source. We will now examine the magnitude of these glitches and their effects.

3.5 GROUND BOUNCE

A major concern associated with the development of RF emissions from a digital device is ground bounce. Ground bounce causes RF noise (differential-mode) to be produced by the simultaneous switching of drivers within an IC package. By examining details within the component, we gain a better understanding of what happens within the PCB. Differential-mode voltages ultimately result in radiated emissions because common-mode currents are accordingly established. To better understand this phenomenon, we will examine a component using a micromodel analysis.

A qualitative relationship exists between ground bounce and emissions at the system level. Designers tend to limit the noise threshold below 500mV for zero-to-peak amplitude of ground bounce glitches. When a ground bounce glitch exceeds a threshold level, emissions increase, along with false triggering of components on the routed net (poor signal quality). Ground bounce is difficult to solve especially when emissions from a product exceed regulatory requirements. Sometimes the unit ceases to function properly. When diagnosing a signal integrity problem, EMI concerns may be eliminated or reduced when the signal integrity problem is solved.

In addition, ground bounce presents a situation in which the ground reference system is not at a constant 0V reference value. Transistors within a component package will not sense an active signal properly if the ground reference subsystem is not stable or is constantly changing.

Ground bounce develops a common-mode potential between the device die and the parent (PCB) image plane (0V return path) and in this mode will couple this potential to all device signals by superimposition. This superimposition can occur on both the power and ground structure. Ground bounce is directly related to the large instantaneous current flow through the power supply inductance and is not due to the output capacitance or inductance of the transmission line being driven. Ground bounce is also dependent on the physical location of the device driver as well as the number of outputs that switch at the same time with respect to a power and ground pad on the die. Bounce is directly related to the dI/dt of the output driver gate (switching speed of the pre-driver within the die circuitry).

Figure 3.8 illustrates an idealized logic circuit internal to a semiconductor. We assume four leads: V_{in}, V_{out}, V_{cc}, and V_{gnd}. The device shown is a totem-pole configuration. In reality, all logic device families exhibit similar ground bounce problems at high speeds of operation. When Switch 2 closes, the load capacitor C is shorted to ground. As the voltage across C falls to 0V, the stored charge flows back to ground, causing a massive current surge within the ground return circuit. This current is identified as $I_{discharge}$.

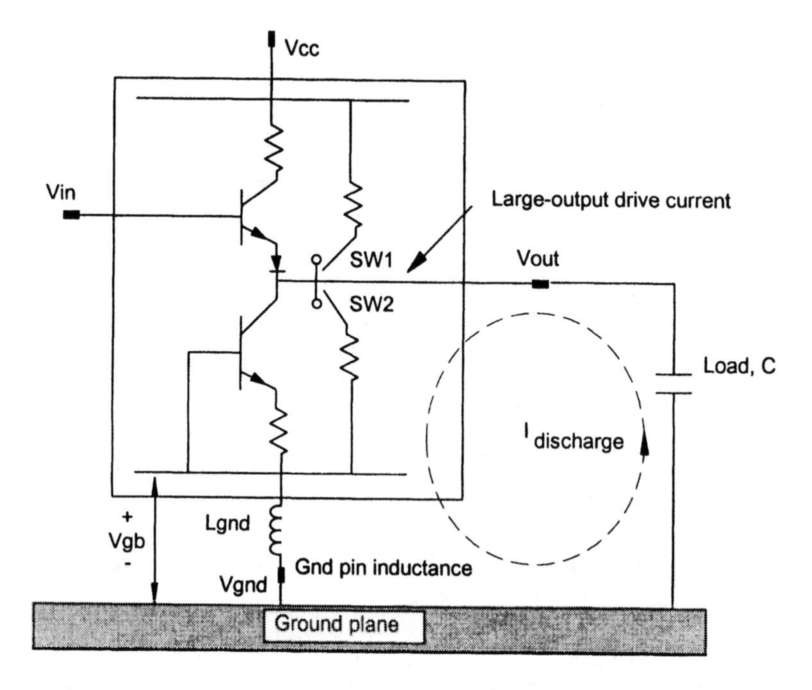

Figure 3.8 Lead inductance within a component.

As the capacitor's current is replenished with voltage and then discharged again, working against the inductance in the ground return pin, L_{gnd}, a voltage V_{gb} is induced between the system ground plane underneath the device and the ground reference internal to the device. The magnitude of this voltage is shown in Eq. (3.5). We identify V_{gb} as the ground bounce voltage.

$$V_{gb} = L_{gnd} \frac{dI_{discharge}}{dt} \tag{3.5}$$

Another explanation of how ground bounce is created has to do with when a gate switches from one logic state to another. Both n and p transistors of the gate are on, and current is sunk between the power and ground planes of the PCB. This current places an additional requirement on the power distribution network which may be insufficient for optimal performance. This explanation of the switching of the two transistors in the gate being in conduction is not the most acceptable one, since the resistance between the power rail (V_{cc}) to the top transistor and from the bottom transistor to ground (Gnd) limits the current from V_{cc} to Gnd significantly. Thus, the primary source of the ground bounce, in the totem-pole configuration of TTL circuits, is the load capacitance discharge to ground through the gate.

Switching elements demand an almost instantaneous change in drive current. The inductance in the lead bonds of the component, trace inductance, and other parasitic inductance causes this instantaneous drive current to occur. The power supply assembly cannot absorb an instantaneous change of current. As a result, a voltage difference is created between the ground and power pins of the component and the lead bond connections. Ground bounce will appear as noise in both the component's power and ground structure. Under this condition, reduced noise margin is observed which may permit false triggering of a voltage-level sensitive trace. From a functional perspective, the noise margin is usually smaller for the low-logic state than for the high-logic state. It is the low-logic state that is of greater concern for system-level functionality.

Usually, the measured ground bounce voltage, V_{gb}, is small compared with the full-swing output signal voltage. Ground bounce does not often affect the transmitted signal. It does, however, interfere with reception of the signal by the load. This is because the receiver compares its input voltage against the internal, local 0V reference. This difference appears as a (+) input connected to a (-) input. Since the internal ground carries the V_{gb} pulse, the actual differential voltage observed at the receiver's input is: $V_{in} - V_{gb}$. This condition is representative of TTL circuits. CMOS compares its input against a weighted average of both power (V_{cc}) and ground. ECL components compare their input against V_{cc}. Although the topology is different between logic families, the concept of ground bounce is the same. If we simultaneously switch N outputs from a component into N corresponding capacitive loads, we have N times as much ground current and pulse V_{gb} grows N times larger.

Fundamentally, information is processed within a digital device by variations of voltages between logic states. Figure 3.9 illustrates a CMOS gate and associated parasitic impedance. In the high-to-low transition state, the load capacitance, C_L, is assumed to be 50 pF. A 5V potential across C_L equates to 250 pCoulombs ($Q = CV$) within the capacitor. This charge must be transferred through the device in order to bring the load to the low state (0V potential). During this high-to-low transition, the charge stored in the capacitor will flow from the load through the ground pin of the device. When this happens, the rate of change of current (dI/dt) develops a voltage drop across the inductance of the ground reference pin. This internal lead and ground return inductance can cause overshoot, undershoot, and even ringing in high-speed logic families.

When ground bounce occurs, the waveform depicted in Fig. 3.10 is observed. The charge that is impressed across the PCB trace results in a common-mode voltage. It is this common-mode voltage that causes RF emissions. Because we are unable to eliminate the transfer of charge between logic state transitions, we must limit the magnitude of the RF current peaks. This is best accomplished by having a very low-impedance path across the power and ground structure of the PCB.

Ground bounce gets worse under the following conditions, as a result of increased current drawn from the power distribution network.

- Capacitive loading is increased.
- Load resistance is decreased.

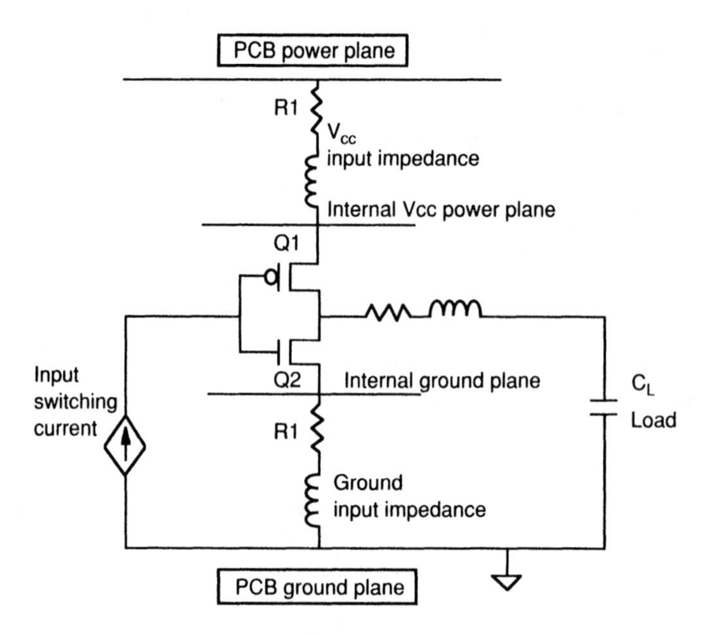

Figure 3.9 Typical CMOS output.

- Lead and trace inductance is increased.
- Multiple gates (devices) switch simultaneously.

To remove ground bounce, several techniques are commonly used. Slowing down the output switching time is the preferred method. Certain components are now being provided with clock skew circuitry to slow down the edge rate, in addition to providing a series resistor internal to the silicon die.

Other manufacturers use multiple ground wire bond leads internal to the device package. This is acceptable if the wires are evenly spaced throughout the device as lead-length inductance is decreased. Spreading the ground connections throughout the components is better than lumping the pins together.

When designing a PCB layout, a separate ground connection should be provided for each ground pin directly to the ground plane. Connecting two ground terminals together and running them through a single trace to a common grounding point (via) defeats the purpose of having independent ground leads.

Other methods to minimize ground bounce include

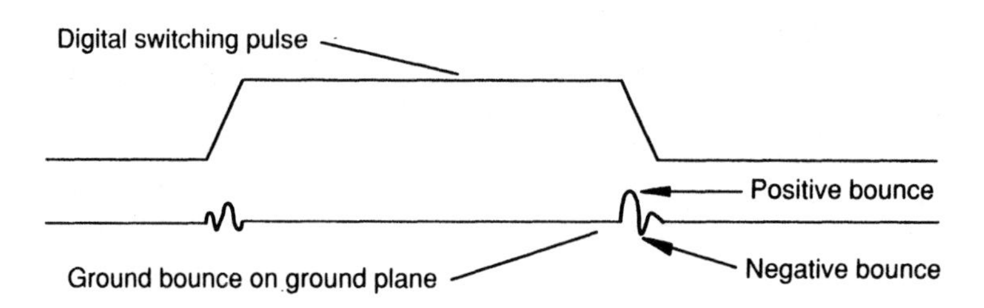

Figure 3.10 Typical ground bounce waveform

1. *Load control*—lower the capacitance and increase resistance.
2. *Layout*—minimize the inductance during layout of the PCB for power and ground, not just the output signal traces.
3. *Component packaging*—use devices with a ground reference pin in the center of the device (4 nH) instead of the corners (15 nH). Surface-mount devices are preferred to through-hole components for this reason.

The design enhancements that a component manufacturer may use to minimize ground bounce for their product include the following list.

1. Decrease trace inductance for both the power/ground pins.
2. Use double bonding wires from the die to the securement pads.
3. Use wider mounting pads as opposed to narrow ones.
4. Decrease the lengths of bond wires within the package.
5. Shorten the height of the pins or lower the profile of the package plastic quad flat pack (PQFPs have less lead-length inductance than pin grid array [PGA] packaging).
6. Provide a high ratio of ground pads to signal pads.
7. Provide an extra ground plane inside the component's package.
8. Provide buried substrate capacitance for application-specific integrated circuits (ASICs).
9. Provide for an on-chip decoupling capacitor internal to the package (built-in power and ground plane between the die and package using the securement glue as the dielectric material).
10. Locate power and ground pins adjacent to each other, preferably in the center of the device so that inductance can cancel magnetic flux lines and minimize development common-mode currents.
11. Provide Low Inductance Capacitor Array (LICA) capacitors internal to Multi-Chip-Module (MCM) packaging.
12. Use low-inductance flip chip packaging.

3.6 LEAD-TO-LEAD CAPACITANCE

A factor that affects the RF emission profile of a component, in addition to lead-length inductance, is stray capacitance between adjacent pins internal to a device. Noise voltages can couple between pins and cause functionality concerns to exist. What really occurs in this situation is crosstalk. If a high RF spectral profile signal is created within a component, and capacitive coupling occurs on an adjacent pin (lead) or trace, it becomes extremely difficult to isolate and implement design enhancements for both signal integrity or regulatory compliance issues. An illustration of capacitive coupling that occurs between component leads is shown in Fig. 3.11.

The percentage of crosstalk that exists between two pins is described by Eq. (3.6). Capacitive crosstalk becomes more pronounced as the rise times become shorter (faster

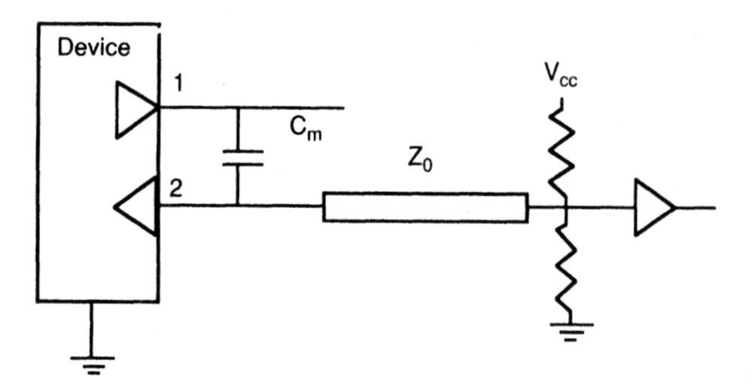

Figure 3.11 Capacitive coupling between component pins.

edge rate). Crosstalk is also made worse when high-impedance transmission lines are provided to the circuit. A high-impedance transmission line adds more capacitance, which contributes to the overall performance of the device. A detailed discussion of crosstalk is presented in Chapter 7, which gives us a better understanding of this equation.

$$\text{Crosstalk} = \frac{ZC_m}{T_{10-90}} \tag{3.6}$$

where C_m = mutual capacitance between lines 1 and 2
Z = the parallel impedance of the trace and terminator $(Z_o \| R_t)$
$T_{10\text{-}90}$ = the 10–90% edge rate of the signal on the output pin

3.7 GROUNDED HEATSINKS

Grounded heatsinks, a new concept in PCB suppression, finds use in specific applications and for certain components. Grounded heatsinks are sometimes required when using VLSI processors with internal clocks in the 75-MHz range and above. These CPU and VLSI components require more extensive high-frequency decoupling and grounding than do most other parts of a PCB.

New technology in wafer fabrication easily allows component densities to exceed 1 million transistors per die. As a result, some components consume 15 watts or more of DC power. Certain components which exceed 15 watts of power and require separate cooling provided by a fan built into their heatsink or by location of the device adjacent to a fan or cooling device. Since these high-power, high-speed processors are being implemented in more designs, special design techniques are now required for EMI suppression and heat removal at the component level.

When we examine the function of a heatsink in the thermodynamic domain, we see that removal of heat generated internal to the processor must occur. Components that dissipate large amounts of heat are usually encapsulated in a ceramic case since ceramic packaging will dissipate more heat than a plastic package. Ceramic cases also cost more. Certain components, due to large junction temperatures between internal gates, generate more heat than the ceramic package can dissipate; hence, a heatsink is required for thermal cooling.

Having briefly discussed the function of heatsinks in the thermodynamic domain, we now examine the metal heatsink in the RF domain. For proper thermal implementation and use of heatsinks, a thermal conductor (silicon compound or mica insulation) is provided. This compound is generally electrically nonconductive. This conductor contains excellent thermal properties for transferring heat from the component to the heatsink. Examining metal heatsinks in the RF domain, we observe the following characteristics, illustrated in Fig. 3.12 and Fig. 3.13.

- Wafer dies operating at high clock speeds generally 75 MHz and higher generate large amounts of common-mode RF current internal within the package.
- Decoupling capacitors remove differential-mode RF current that exists between the power and ground planes and signal pins.
- Certain ceramic packages contain solder pads on top of the package case to provide additional differential-mode power filtering required by the large power consumption in addition to high-frequency decoupling. Decoupling capacitors minimize ground bounce and ground-noise voltage created by the simultaneous switching of all component pins under maximum capacitive load.
- The wafer (or die) internal to the package (Fig. 3.13) is located closer to the top of the case (dimension "X") than the bottom of the package (dimension "Y"). Therefore, height separation from the die to an image plane internal to the PCB is greater than the height of the die to the top of the package case and heatsink. Common-mode RF currents generated internal within the wafer have no place to couple to 0V reference; hence, RF energy is radiated into free space. Differential-mode decoupling capacitors will not remove common-mode noise created within the component.
- Placing a metal heatsink on top of the component provides a 0V reference (image plane) closer to the wafer than the image plane on the PCB. Tighter *common-mode* RF coupling occurs between the die and heatsink than between the die and the first image plane of the PCB.
- Common-mode coupling that occurs to the heatsink now causes this thermodynamically required part to become a *monotonic antenna,* perfect for radiating RF energy into free space.

The net result of using a metal heatsink is the same as placing a monotonic antenna inside the product to radiate clock harmonics throughout the entire frequency spectrum. To deenergize this antenna, the heatsink must be grounded. Although this concept is very simple to understand, it is virtually ignored within the field of PCB design for RF energy suppression.

A VLSI component can be an effective radiator of RF energy. Adding a heatsink adds yet one more design parameter when considered in the frequency domain and not just within the thermodynamic domain. In general, as the size of the heatsink increases, radiation efficiency increases. The maximum amount of radiation will occur at different frequencies depending on the geometry of the heatsink and self-resonant frequency of the assembly if the heatsink is a metallic structure.

Heatsinks must be grounded to the ground planes (or 0V reference) of the PCB by a metal connection on all four sides. Use of a fence (similar to a vertical bus bar) from the PCB to the heatsink will encapsulate the processor. This fence will create a Faraday shield around the processor, thus preventing common-mode noise RF energy internal to the package from radiating into free space or coupling onto nearby components, cables,

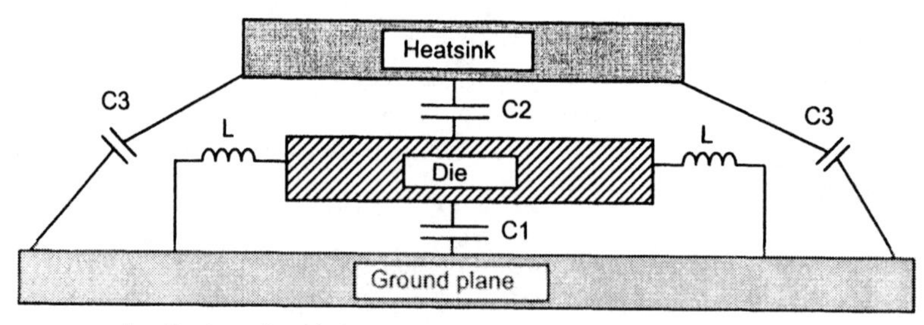

Figure 3.12 Grounded heatsink theory of operation.

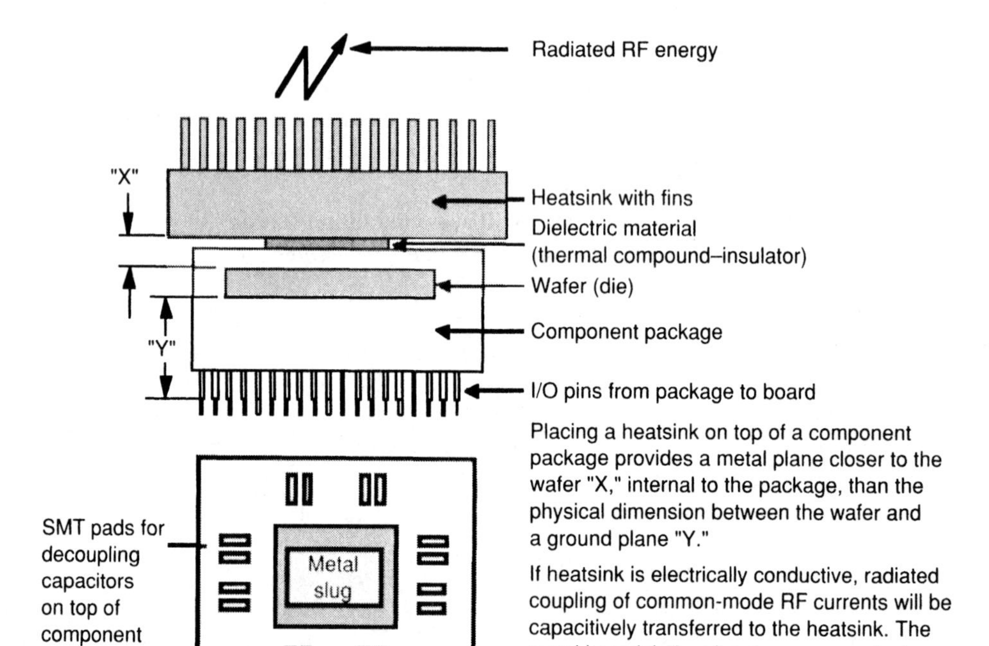

Figure 3.13 Grounded heatsink implementation.

peripherals, or into aperture slots. A technique for providing grounding of the heatsink is shown in Fig. 3.14.

Reduced Instruction Set Computing (RISC) processors or VLSI components generally have a high self-resonant frequency that is a combination of the manufacturing process and internal clock speed, in addition to the impedance present in the power planes during maximum power consumption. As a result, VLSI components radiate RF energy more than many other components if RF suppression techniques were not incorporated by the component manufacturer. Any attempt to remove this self-resonant RF frequency using standard design suppression techniques is almost impossible except through use of the heatsink as a *common-mode decoupling capacitor*.

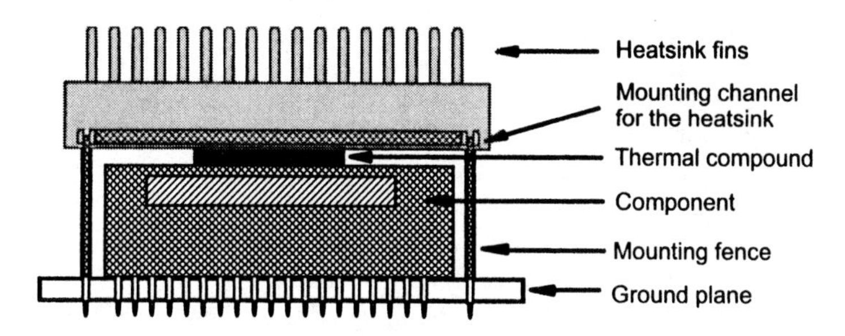

Figure 3.14 Grounding the heatsink.

This heatsink is generally used in conjunction with both differential-mode capacitors located on the top of the components' ceramic package, if provided, in addition to the standard differential-mode capacitors located under the component directly on the PCB. A differential-mode decoupling capacitor connects directly between the power and ground planes to remove switching noise from these planes. A common-mode decoupling capacitor provides an AC shunt to remove CM noise generated internally from the die to the ground reference system.

A grounded heatsink must always be at *ground potential.* The active component is always at *RF voltage potential.* The thermal compound is a dielectric insulator between two large plates. The definition of a capacitor is fulfilled. Thus, a grounded heatsink works as one large *common-mode decoupling capacitor,* while optional discrete capacitors located on top of the device package or directly on the PCB are used for *differential-mode* decoupling. This common-mode capacitor shunts RF currents generated within the processor to ground.

Using a grounded heatsink creates

1. A thermal device to remove heat generated internal to the package.
2. A Faraday shield to prevent RF energy created from the clock circuitry internal to the processor from radiating into free space or corrupting adjacent circuitry.
3. A "common-mode" decoupling capacitor that removes common-mode RF currents generated directly from the die, or the wafer, inside the package, by AC coupling RF energy from the die to ground.

If a grounded heatsink is implemented, the grounding fingers of the fence (spring fingers or other PCB mounting method employed) must be connected to all ground planes, or the 0V reference structure in the PCB on at least 1/4 inch (0.125 cm) centers around the processor. At each and every ground connection, install two sets of parallel decoupling capacitors, alternating between each ground pin of the fence with 0.1 μf in parallel with 0.001 μf, and 0.01 μf in parallel with 100 pF. RF spectral distribution from RISC processors and similar components generally exceed 1-GHz bandwidth. RISC or VLSI processors also require more extensive multipoint grounding around all four sides of the processor than most other types of components. These capacitive values complement the approximate λ/4 mechanical size of typical heatsinks, making them efficient suppressors of EMI spectra.

3.8 POWER FILTERING FOR CLOCK SOURCES

Oscillators are one source of radiated emissions. The output of their periodic waveform is transmitted down a PCB trace to a load. Depending on the layout of the PCB, component placement, trace routing, decoupling, impedance control, and other items related to flux cancellation or minimization, emissions will either exist or be a nonissue related to EMC compliance. In addition, signal integrity concerns must be considered.

In some situations, oscillators (clock generation circuits) will inject RF currents on to a PCB trace. This is in addition to ground bounce that occurs as a result of poor decoupling or power supply immunity. If the oscillator is located within a noisy environment, additional power supply filtering will be required. The amount of this filtering is dependent on how much reduction in jitter must be achieved. Jitter is a small, rapid variation in the waveform owing to mechanical vibrations, fluctuation in supply voltages, and control-system instability. Basically, clock jitter refers to any deviation of a clock's output transition from their ideal operating condition. Trying to determine a precise value for jitter reduction is nearly impossible for the following reasons.

1. Different manufacturers of oscillators have different power requirements, along with a difference in actual edge rates. Although the oscillators may have the same frequency, not all oscillators have the same AC or DC characteristics.
2. Jitter performance is generally not provided in their respective data sheets or application notes. Each manufacturer of oscillators will have a different jitter requirement.
3. RF noise in a system changes when different brands of integrated circuits are used.

To minimize ground bounce and enhance power supply noise reduction, use of a filter circuit is required (see Fig. 3.15). These circuits will achieve a reduction of up to 20 dB in the frequency range above 20 MHz. Use of a two-stage filter will double the attenuation.

It is mandatory to physically locate the filter circuit as close as possible to the power input pin of the oscillator circuit to minimize RF loop currents. Depending on the frequency of the oscillator, a current loop could be present causing radiated emissions. Use of surface-mount devices is preferred over through-hole devices due to less lead-length inductance in the component package.

Two methods can be used to provide power filtering for clock sources. One involves use of an RLC circuit, the other a ferrite bead and capacitor combination, discussed below.

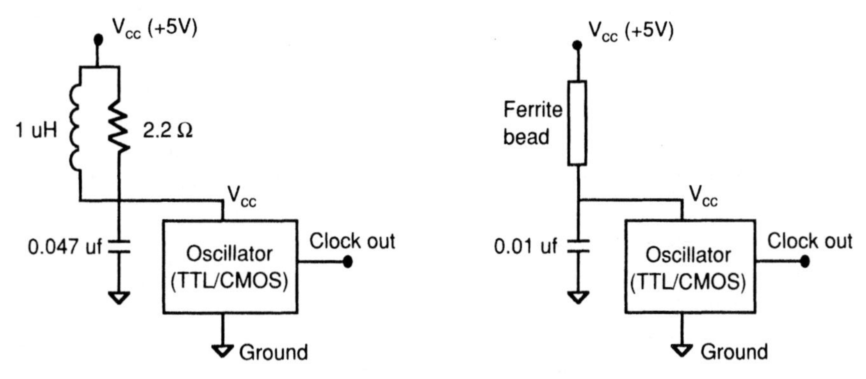

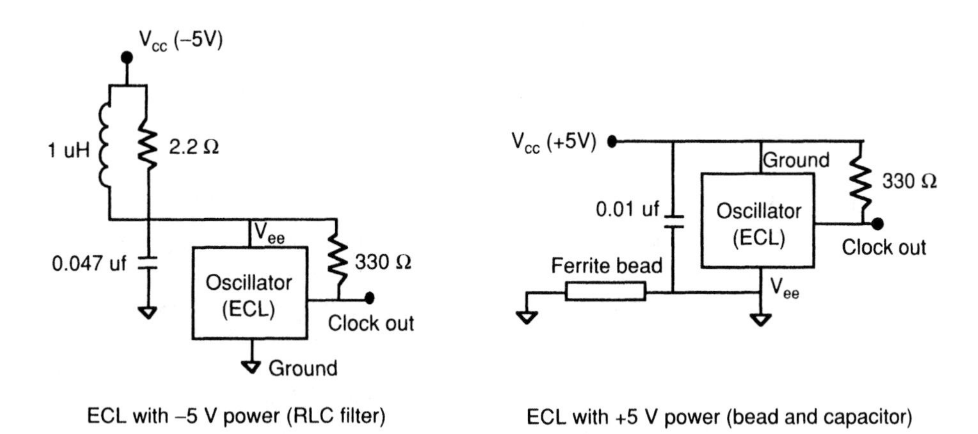

Figure 3.15 Sample filter circuits for oscillators.

A large inductor or capacitor value will enhance the filter's attenuation at lower frequencies if this filter method is used. For any combination of L and C, to achieve 20 dB of attenuation, Eq. (3.7) is used.

$$f_{20dB} = \frac{3.2}{\sqrt{LC}} \tag{3.7}$$

where f is frequency in hertz.

If a resonance occurs in this LC circuit, reduction of the Q of the circuit may be required. This is best accomplished by use of a resistor. This resistor prevents a particular resonance from occurring and is calculated by Eq. (3.8).

$$R = \frac{1}{2}\sqrt{\frac{L}{C}} \text{ (ohms)} \tag{3.8}$$

With either filter, the oscillator is forced to draw transient current from the local decoupling capacitor. The regular decoupling and bulk capacitors located throughout the PCB would be unable to provide the fast-transient current required by the oscillator due to inductance in the power path. This local filter would keep the current loop (power and ground) from the power input circuit small, thus preventing this localized circuit from infecting the rest of the board. The path for the transient current surges would only be from the local filter capacitor to the IC's power pin and out the IC's ground pin back to the negative side of the capacitor, a relatively small loop. This small-current loop description is applicable not only for oscillators, but also for all components that require this type of filtering.

If a ferrite bead-on-lead is provided in place of the inductor, we can eliminate the resistor. This is because the bead-on-lead at DC provides low inductance. Since the oscillator or clock generation circuitry is usually above 10 MHz in today's products, high-frequency RF energy will be prevented from entering the system's main power distribution network and corrupting other functional circuits. The local capacitor provides only decoupling to recharge the power rail for the localized oscillator circuitry. This filter combination minimizes power consumption surges and prevents the possibility of infecting the entire power distribution circuit with RF energy created by the oscillator package.

Caution: Use of some ferrites in power or ground can interact with the output signal's return image in the plane and alter signal quality.

3.9 RADIATED DESIGN CONCERNS FOR INTEGRATED CIRCUITS

Recent advances in integrated circuit (IC) components such as microprocessors, digital signal processors, and application-specific integrated circuits (ASICs) have become significant sources of electromagnetic noise. In recent years, clock rates have increased from 25 and 33 MHz to 200 through 500 MHz. Along with the increase in clock rates, there is a corresponding increase in dynamic power dissipation due to switching currents that exceed 10 watts on a typical VLSI device. Individual circuits, when isolated by themselves, generally do not radiate enough RF energy to exceed mandated regulatory limits. The RF energy that is created is frequently coupled into structures within a product assembly, which will then cause EMI problems to be observed.

These structures and assemblies include cavities created by metallic enclosures, apertures, connected cables, and the like which enhances lower-frequency emissions. Heatsinks are a prime source of radiated RF energy, discussed earlier.

The reason why some, not all, manufacturers of components do not place a high priority on EMC or radiated noise coupling (requiring the end user to accept the responsibility of solving emissions along with ground noise or ground bounce) is based on the following wish-list. These noncompliant, wish-list components are designed for

1. "Infinitely" fast rise times (zero rise times).
2. Unlimited fanout drive (unlimited power output).

With these two items, it becomes even more important to recognize that these conditions exist. Not all manufacturers produce the same product, although they may be form, fit, and function compatible. Noncompliant components must be handled with care using the following techniques.

1. Keeping short lead-lengths (lower the output loop area).
2. Keeping clock signals away from I/O circuitry and lines (prevents coupling).

3. Raising the output resistance of a clock trace with a series impedance (resistor or ferrite bead).

To address these problems of noncompliant components, EMC engineers must advance state-of-the-art principles by implementing EMI suppression techniques for ICs. These techniques must keep up with higher speed designs. Design and cost margins play an important part in determining how a solution will be implemented. Engineers must be able to predict radiated emissions using specialized tools and simulation programs. The main problem that exists with simulation analysis is finding appropriate tools, including the development of behavioral models that reflect IC parameters (actual and parasitic) along with proper voltage and current impulse (surge) characteristics.

Various studies have been performed to determine the difference in characteristic radiated emissions between ICs. These differences include packaging, layout, logic families, and different vendors for the same device. Recorded measurements show differences of up to 10 dB between different versions of the same device. An example of this difference is seen in Table 3.4 [3].

Various efforts by numerous companies involved in simulation and modeling are underway to calculate radiated emissions from components. These emissions are generally common-mode which makes it difficult to model. In contrast, differential-mode currents are easy to simulate and model. Efforts are in process to determine efficient methods of measurement techniques. Until research is available, calculating radiated emissions from ICs will remain a subject for adventurous engineers.

Radiated emissions from components can be reduced through use of the following design techniques:

- Reduction of package size; antenna efficiency (lowers the effective area of radiation).
- Reduction of high-frequency energy created within the die structure.
- Isolation of RF noise produced from the die to any IC pins that connect to external circuitry.

Selected combinations of these techniques must be used in order to reduce radiated emissions. This includes designing the component to have interconnect pads on the substrate adjacent to each other for power and ground to minimize ground bounce. Edge rate control must be implemented on periodic clock signals (series resistance or equivalent technique) to reduce high-frequency RF coupling onto adjacent traces or other metallic structures. Lower impedance bond wires must also be provided.

To minimize radiated emissions from components, see Table 3.5.

TABLE 3.4 Radiated Emissions Between Vendors

Radiated electric field strength (dBμV/m) at 3 m		
Frequency	Vendor A	Vendor B
30 MHz	32.8	42.8
40 MHz	27.5	33.0
50 MHz	23.0	28.0

TABLE 3.5 Design Concerns to Reduce Radiated Emissions from Components

Reduce Radiation Efficiency	Reduce Coupling and Crosstalk	Reduce High-Frequency Switching Energy
Use a smaller package size	Use more ground pins placed strategically around the device	Use drivers with the slowest edge rate acceptable for proper operation
Use a ground plane inside the package	Use small distributed clock drivers instead of a single driver	Isolate areas on the die where necessary which contain clock logic
Use additional ground/power leads near the clock source	Use drivers with the lowest drive voltage possible	Use split power-ground structures to isolate clock noise
Use a shielded package	Use the slowest clock rate possible	Separate areas on the die where isolation is required
Group signals and power leads together	Use current limiting resistors	Isolate signal leads with ground wires
Use low-inductance bond wires in the package	Reduce die trace capacitance Use differential clock drivers	Separate I/O ports from clock pins

3.11 SUMMARY FOR RADIATED EMISSION CONTROL—COMPONENT LEVEL

The following recommendations will help minimize the amount of RF energy that is created from use of certain logic devices, especially digital logic.

- Select devices that consume less input current during logic transition states. Of concern here is the maximum inrush current of all component pins switching simultaneously under maximum capacitive load, not the average or quiescent value.
- Use the slowest logic possible for the function required. Although slower speed devices are becoming more difficult to procure, a best attempt effort is required to prevent use of sub-nanosecond devices for common logic functions.
- Select logic devices with power and ground pins located in the center of the package, with both power and ground pins adjacent to each other.
- Use devices with metal enclosed packaging (oscillators). Ground the metal case or package to the 0V reference with as many low-impedance via connections as possible.
- For devices that contain ceramic packaging and a metal slug on top, provide for a grounded heatsink. Conceptually, this can be incorporated in certain products; however, it may be difficult, if not impossible, to implement.

REFERENCES

[1] Montrose, M. 1996. *Printed Circuit Board Design Techniques for EMC Compliance*. Piscataway, NJ: IEEE Press.

[2] Bakoglu, H. 1990. *Circuit Interconnections and Packaging for VLSI*. Reading, MA: Addison-Wesley, Table 6.2.

[3] Erwin, V., and K. Fisher. 1985. "Radiated EMI of Multiple IC Sources." *Proceedings of the IEEE EMC Symposium.* Piscataway, NJ: IEEE.

[4] Ott, H. 1988. *Noise Reduction Techniques in Electronic Systems.* 2nd ed. New York: John Wiley & Sons.

[5] Goulette, D., and R. Crawhall. 1996. "Quieter Integrated Circuits Ease EMI Compliance." *Nortel Technology.*

[6] Johnson, H. W., and M. Graham. 1993. *High Speed Digital Design.* Englewood Cliffs, NJ: Prentice Hall.

[7] Mardiguian, M. 1992. *Controlling Radiated Emissions by Design.* New York: Van Nostrand Reinhold.

[8] Motorola, Inc. *Low Skew Clock Drivers and Their System Design Considerations* (#AN1091).

[9] Motorola, Inc. 1996. *ECL Clock Distribution Techniques.* (#AN1405).

[10] Williams, Tim. 1996. *EMC for Product Designers.* 2nd ed. Oxford, England: Butterworth-Heinemann.

[11] Brench, C. 1994. "Heatsink Radiation as a Function of Geometry." *Proceedings of the IEEE EMC Symposium.* Piscataway, NJ: IEEE.

[12] Diaz-Olavarrieta, L. 1991. "Ground Bounce in ASIC's: Model and Test Results." *Proceedings of the IEEE EMC Symposium.* Piscataway, NJ: IEEE.

4

Image Planes

4.1 OVERVIEW

In any digital system, especially with high-speed components (fast edge rate), a low-impedance (low-inductance) RF return current path must be present for optimal performance. As examined in Chapter 2, a closed-loop network is required for reasons of functionality. This closed-loop network is required for both time and frequency domain aspects of the circuit. All components and all possible trace or wire interconnects that exist must operate in an environment in which the RF return currents find their way back to their source (low-impedance path).

Since RF currents must return to their source (closed-loop circuit), they will do so using any path possible. We must control all return currents using conductive paths. An alternate conductive return path is better than no path at all. If no conductive path exists, free space becomes the path. Free space is exactly what we do not want as a return path for RF currents, especially with regulatory compliance concerns.

Examining Fig. 4.1, we see that the RF return currents do not have an optimal return path home. Assume that the components are tied to a voltage reference source only by traces to the power supply structure. In the time domain, functionality concerns are met, and the circuit works. The RF return path occurs through the ground wires that provide the 0V reference to the circuit. Why should digital designers worry about EMC issues when the circuit operates per marketing specification and logic signals travel from source to load without functional degradation?

As discussed in Chapter 2, time and frequency domain aspects of a circuit must be considered simultaneously. Return currents (DC voltage reference) occur through the power and 0V reference (ground) structure of the PCB. These return currents exist in both the time and frequency domain for each and every trace. While a single-sided PCB may

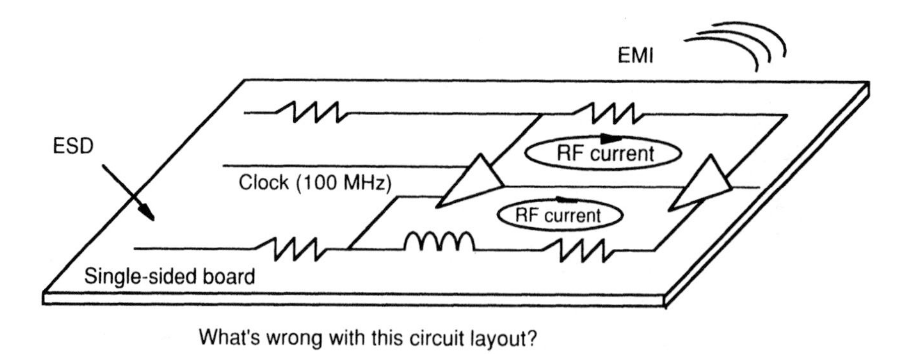

Figure 4.1 Typical PCB design without an RF return current system.

be a cost-effective implementation of a design (which will maximize profits), the probability is high that this simple configuration will not pass various EMC test requirements. Adding a ground plane (two-layer board), or making the assembly a four-layer design will enhance the overall performance of the assembly related to signal integrity and EMC compliance. While cost is being taken out of the PCB by using a single-sided design, alternative methods of EMC compliance may be required, which may include adding an expensive metal cover or metalized plastic enclosure.

In addition to maximizing radiated emissions between two components, the PCB becomes sensitive to ESD events. A high-current pulse, along with its effective radiated field, will see a lower impedance provided by the PCB than that observed by free space. This radiated field becomes impressed into the trace. Component failure may occur by damage to the die internal to the component package. In addition, a functional glitch may also occur, degrading the performance of the product.

High-technology CMOS products are sensitive to ESD events. This sensitivity requires special handling during the assembly cycle, yet protection against ESD on the PCB is frequently not considered, especially if cost has to be added to the board.

A good 0V reference (ground) system is the foundation of any digital PCB. If the 0V reference system is poor, it becomes difficult to fix an EMI problem when one develops. Indeed, a poor 0V reference implementation may be the actual cause of the problem. The only remedy is to redesign the board or start the design over from scratch. Adding in two more layers to a double-sided PCB, for example, a power and ground plane, requires only minimal work, yet will achieve improvement between 10 to 20 dB on radiated emissions as documented in numerous EMC publications, textbooks, and technical papers (not identified) herein. (For a sample list of publications, see References at end of this chapter and the Bibliography.)

EMI test failures can negate the cost-effectiveness of using less expensive double-sided boards. An extra round of EMC tests (emissions and immunity), redesign and relayout, and a prototype build for functionality testing, tying up engineering resources and adding many weeks of delay to the schedule, can easily result in costs exceeding tens of thousands of dollars. An incremental cost in adding two more layers may be cheaper than trying to maintain a double-sided structure. Depending on the number of boards to be produced, a cost saving may occur using multilayer boards. In one experience, $50,000 was spent in additional engineering resources to optimize a double-sided board—all in an effort to save $10,000 in production costs. Management must consider these financial values before making a decision to remain with an ill-considered design.

Taking this concern to the next level, we may find an indefinite number of parallel return paths. This is because many interconnects occur between components, along with connection into a power distribution network. Since a large number of return paths may be present, we can take advantage of this feature and convert "infinity" to one ($\infty \rightarrow 1$). This number one (1) is identified as an image plane. We generally refer to the 0V reference structure of the PCB as a ground plane. In ECL systems, the power plane is referred to as the 0V reference. In reality, any copper laminate in a multilayer stackup provides a return path with minimal impedance for RF return currents. Since RF return currents flow on the copper laminate in the first level using skin effect, the voltage potential (e.g., +5V, +12V, etc.) is not a major concern, except under certain operating conditions. The disadvantage of using a multilayer board lies in cost. For many applications, use of a multilayer design is not economically possible. For this situation, an alternate and effective return path for RF return currents must be established, perhaps through use of a gridded ground system, ground traces, or other creative means, as described later.

Discussion of multilayer boards is predominant in this chapter because technology is evolving at a rapid rate. There is practically no such thing as a slow-speed logic device anymore. Manufacturers of components are constantly improving their yield production using a "die shrink" process. To accomplish die shrink and increase yields, in addition to making their circuit desirable for use in a competitive marketplace, an increase in operating speed becomes mandatory. In addition, the lithography line widths within the die becomes smaller, with a corresponding increase in speed and faster edge rate.

Most "component vendors" concern themselves only with profit, not with EMC compliance. It is generally not a priority issue for these component vendors to recognize that their customers (users) are required, by law in many cases, to have their end product comply with emissions and immunity requirements. According to many component vendors, components do not cause EMI. Their components are always used with other components on a PCB; hence, many vendors consider themselves exempt from regulatory compliance concerns. A faster edge rate device will still work in slower speed products. Why, then, should semiconductor vendors worry about retooling their equipment to build a slower speed device when a faster device can be made available for less money and be pin-for-pin compatible?

A *brief* discussion of single- and double-sided boards will be presented for completeness. High-technology products require use of multilayer stackups.

4.2 5/5 RULE

The 5/5 rule[1] indicates when use of multilayer boards becomes necessary. The rule states that when clock speeds in excess of 5 MHz, or when rise times faster than 5 ns exist, a multilayer board should be used as the "crossover" point, beyond which faster edges and higher frequencies proportionally increase the need for multilayer boards. With proper design and layout techniques, the 5/5 rule can be changed to use faster clock and edge rates, *only* if the designer is aware of the problems that can exist when using these faster edge rate devices and high clock speeds. The designer must have extensive experience in designing high-technology products using a simpler board stackup assignment. This is an extremely difficult task to accomplish with minimal cost [3] .

[1]The 5/5 rule and definition were first used by Daryl Gerke and Bill Kimmel [3].

4.3 HOW IMAGE PLANES WORK

In Chapter 2, we examined the need for flux cancellation or minimization. Image planes provide flux cancellation or minimization by allowing RF return currents to image back along its source path differentially. Here the term *differentially* describes the phase relationship between the signal and its return image. A detailed discussion of common-mode and differential-mode currents is found in Chapter 2. When an RF return path is placed in close proximity to a wire or trace, magnetic lines of flux which are opposite in polarity cancel each other out. We now examine the physics of this discussion.

When current travels through a PCB trace, an electromagnetic field is generated by magnetic lines of flux created within the transmission path. Maxwell's equations describe the development of an electric field from magnetic lines of flux, and vice versa. Depending on the length of the routed trace, radiated emissions may be created. Traces and copper planes have a finite amount of inductance. This inductance inhibits current buildup and charge whenever a voltage is applied to the trace or transmission line.

Research has shown [6] that if a two-wire transmission line is slightly unbalanced, the trace will radiate as an asymmetrical dipole antenna. This unbalanced structure will create common-mode radiated emissions at levels much greater than the differential-mode radiation that exists within the closed-loop circuit, detailed in Chapter 2.

Before examining how an image plane works within a PCB, the following briefly summarizes the difference between various types of inductance within the board structure [2]. These are

Partial inductance: the inductance that exists in a wire or PCB trace.

Self partial inductance: the inductance from one wire segment relative to an infinite segment.

Mutual partial inductance: the effects that one inductive segment has on a second inductive segment.

4.3.1 Inductance

At any frequency, a conductive element such as a wire or PCB trace exhibits inductance. The distributed inductance, capacitance, and resistance of traces, vias, and planes on a PCB must be considered at the same time as lumped parameters of all circuit components. The most difficult parameter to investigate or quantify is inductance. Unlike capacitance and resistance, inductance is a dynamic property of a closed-loop current path.

Inductance is defined as the ratio of total magnetic flux that couples (passes through) a closed-loop path to the amplitude of the current that produces the magnetic flux. Inductance is described by Eq. (4.1).

$$L_{ij} = \frac{\psi_{ij}}{I_i} \text{ henries} \tag{4.1}$$

where ψ = magnetic flux and I is the current in the loop structure. If the wire is configured in a closed-loop circuit, the inductance is a function of loop geometry as well as the shape and dimensions of the wire itself.

The inductance of a wire or PCB trace is frequently overlooked when designing a PCB. Inductance is always associated with a closed-loop circuit. To describe the effects of inductance on a loop circuit, we must examine the effects of *partial inductance* and *mutual partial inductance*.

4.3.2 Partial Inductance

Partial inductance is defined as the internal inductance of a conductor due to magnetic flux that is present within the conductor [2]. But this definition is not entirely true.

Inductance is defined only for closed-loop circuits. To simplify the need to study partial inductance, we investigate separate sections of a current loop. This approach allows investigation of the overall effect that a transmission path has in a circuit. To lower the overall inductance of the circuit, or circuit geometry, it is first necessary to reduce the inductance of the section that has the greatest amount of inductance. Reduction may occur by shortening a routed trace length, removing vias, increasing the width of the conductor, or other methods, including trace reorientation. Partial inductance is useful for estimating the voltage drop across part of a circuit due to the inductance of that particular section. Care must be taken since the voltage drop or potential difference is not uniquely defined in the presence of time-varying fields.

The total partial inductance of a closed-loop segment is the sum of all sections. This is shown by

$$L_{\text{total}} = L_{\text{partial segment1}} + L_{\text{partial segment2}} + \ldots + L_{\text{partial segment } n} = \sum_{i=1}^{n} Li \tag{4.2}$$

With Eq. (4.2), the static current within each segment is identical. L_{total} is the total flux of current in the loop. With this information, we can define partial inductance for a particular segment as the ratio of flux coupling to the current within a particular segment, Eq. (4.3).

$$L_{\text{partial segment}}\ i = \frac{\psi_i}{I} = \frac{\text{flux due to segment "}i\text{" that couples the loop}}{\text{amplitude of the current in segment } i} \tag{4.3}$$

The concept of partial inductance is for a single loop. Obviously, different loops will have different values of partial inductance.

The total internal inductance of a conductor will decrease, but the internal impedance still increases with the square root of the frequency owing to skin effect. Because of skin effect, the inductance that exists within the center portion of the conductor plays a minor role in the overall inductive performance of the conductor. The parameter of interest is the partial inductance, which is frequency-independent, not the total inductance of the trace. With knowledge of frequency independent inductance, the mutual inductance between two parallel conductors (or a trace over image plane) can be determined by the general equation, Eq. (4.4). A more detailed presentation of partial and mutual partial inductance is provided in Eq. (4.5).

4.3.3 Mutual Partial Inductance

Mutual partial inductance [2,6,9] is the key element that allows an image plane to provide for flux cancellation. Flux cancellation occurs by allowing magnetic lines of flux to link and find an optimal return path for RF currents.

Self partial inductance applies to a given segment of a loop independent of the location or orientation to any other loop segment. Given a current within a wire or trace, a nominal rectangular loop is defined, bounded by the wire segment on one side and infinity on the other side. These two perpendicular wire segments extend from the ends of the segments into infinity. This is illustrated by Fig. 4.2. Since self partial inductance is present between a wire segment and an infinite structure, we can develop the concept of *mutual partial inductance* [9] .

Consider an isolated conductor (or trace), length L, carrying current I. The *self partial inductance* of the conductor, L_p, is the "ratio of net magnetic flux generated by a current I passing through the loop (or between a conductor and through infinity, beyond the trace), divided by the current I within the wire segment" [2].

Self partial inductance is, of course, theoretically independent of the proximity to adjacent conductors. Closely spaced conductors, however, can alter the self partial inductance of one or both of the conductors. This is because one conductor will interact with the other conductor and cause current distributions over the entire length of the conductor to deviate from a uniform condition. This typically occurs when the ratio of wire separation to radius is less than approximately 5:1. A separation radius of 4:1 for two identical wires means that a third wire may fit between the two original wires if the wire radius is identical [2].

Between two conductors, *mutual partial inductance* exists. Mutual partial inductance, M_p, is based on the distance spacing between parallel traces or wire segments. The distance, s, is the ratio of "magnetic flux due to current in the first conductor that passes between the second conductor and into infinity" to "the current in the first conductor that produced it." Mutual partial inductance is observed in Fig. 4.2 with the electrical schematic detailed in Fig. 4.3. The voltage developed across the conductors from this configuration is described by Eq. (4.4) [2] .

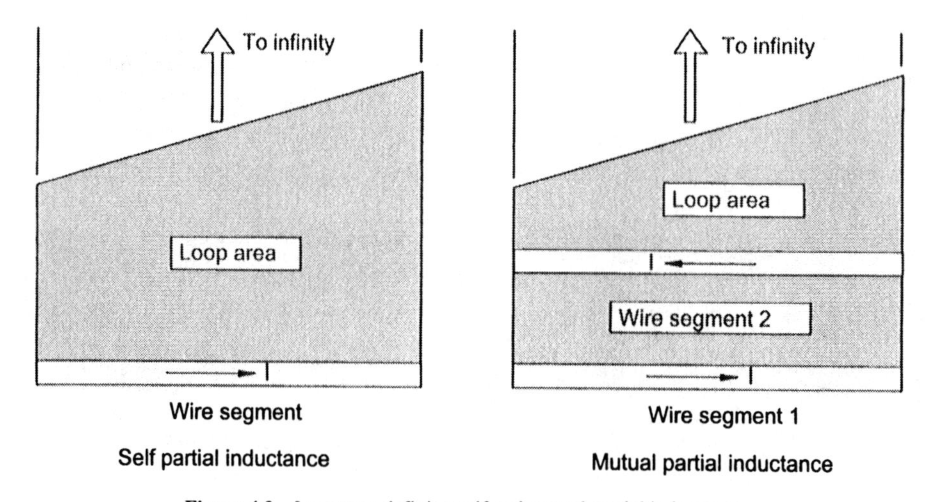

Figure 4.2 Loop area defining self and mutual partial inductance.

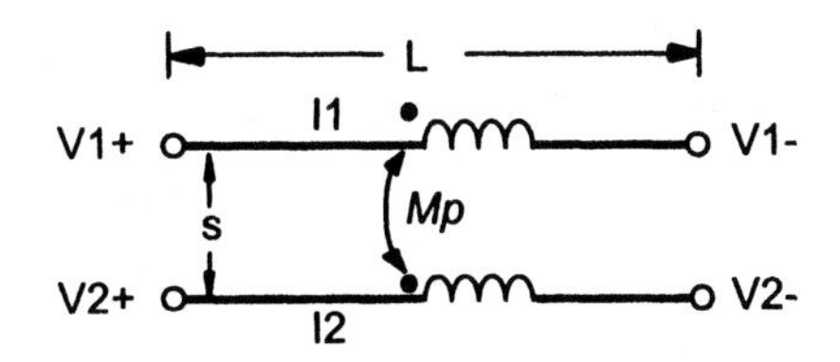

Figure 4.3 Mutual partial inductance between two conductors.

$$V_1 = L_{p1}\frac{dI_1}{dt} + M_p\frac{dI_2}{dt}$$
$$V_2 = M_p\frac{dI_1}{dt} + L_{p2}\frac{dI_2}{dt} \tag{4.4}$$

With the concept of mutual partial inductance, consider the two traces in Fig. 4.3 are now carrying a signal of interest, for example, clock. The trace identified as V_1 is the signal path, and the trace identified as V_2 is the RF current return path. Assume two conductors constitute a signal path and its associated return so that $I_1 = I$ and $I_2 = I_1 = -I$. If there is no mutual coupling between two conductors, the circuit cannot function, for a closed-loop circuit will not exist (Chapter 2). The voltage drop within the circuit of Fig. 4.3 becomes

$$V_1 = (L_{p1} - M_p)\frac{dI}{dt}$$
$$V_2 = -(L_{p2} - M_p)\frac{dI}{dt} \tag{4.5}$$

According to Eq. (4.5), in order to reduce the voltage drop across a conductor, we must *maximize* the mutual partial inductance between that conductor and its associated conductor within the same circuit. The easiest way to maximize mutual partial inductance is to provide a path for RF return current as close as possible to the signal trace. The most optimal design technique is use of an RF return plane located adjacent to the signal trace with the smallest distance spacing that is manufacturable. An alternative way to maximize mutual partial inductance for single- and double-sided PCBs is to provide an RF return path (trace) adjacent to the signal trace with a distance spacing that is as small as possible.

To view the effects of both partial and mutual partial inductance, consider two traces or a trace over a plane. Partial inductance will always exist in a conductor (by default), and inductance will equate to an antenna at a specific resonant frequency. Mutual partial inductance minimizes the effects of partial inductance. By locating two conductors close together, the individual partial inductance becomes minimized, which is a desired design requirement for EMI compliance within the boundary of an "image" between the conductors.

To optimize mutual partial inductance, the currents in the two conductors must be equal in magnitude and opposite in direction. This is why image planes (and ground traces) work as well as they do. Because mutual partial inductance exists between two parallel wires, a certain amount of inductance will be present. Table 4.1 provides details on the mutual partial inductance between two parallel wires with various spacings [2].

Since mutual partial inductance was examined for signal traces, how does this inductance relate to power and ground planes separated by a dielectric material? The mutual

TABLE 4.1 Mutual Partial Inductance Between Two Parallel Wires

	Common Length		
Conductor Separation	1 inch	10 inches	20 inches
1/2 in. (1.25 cm)	3.23 nH	137.9 nH	344.9 nH
1/4 in. (0.63 cm)	6.12 nH	172.4 nH	414.7 nH
1/8 in. (0.32 cm)	9.32 nH	207.3 nH	484.8 nH
1/16 in. (0.16 cm)	12.7 nH	242.2 nH	555.0 nH

partial inductance between planes is maximized when the distance spacing is minimized. In addition to minimizing mutual partial inductance, interplane capacitance is increased, which is desirable for reasons detailed in Chapter 5. When we maximize the mutual partial inductance between the power and ground planes, RF signal currents that are observed within the power distribution plane are canceled out by equal and opposite RF return currents.

4.3.4 Image Plane Implementation and Concept

A solid plane can produce common-mode radiation. Figure 4.4 illustrates what an image plane structure looks like within a PCB assembly along with mutual partial inductance. In Fig. 4.4, the majority of the RF currents within the signal trace will return on the plane located directly below the signal trace. Within this return "image" structure, the RF return current will encounter a finite impedance (inductance). This return current produces a voltage gradient, which is referred to as ground-noise voltage. Ground-noise voltage will cause a portion of the signal current to flow through the distributed capacitance of the ground plane [6].

Common-mode currents, I_{cm}, are typically several orders of magnitude less than differential-mode currents, I_{dm}. However, common-mode currents (I_1 and I_{cm}) produce

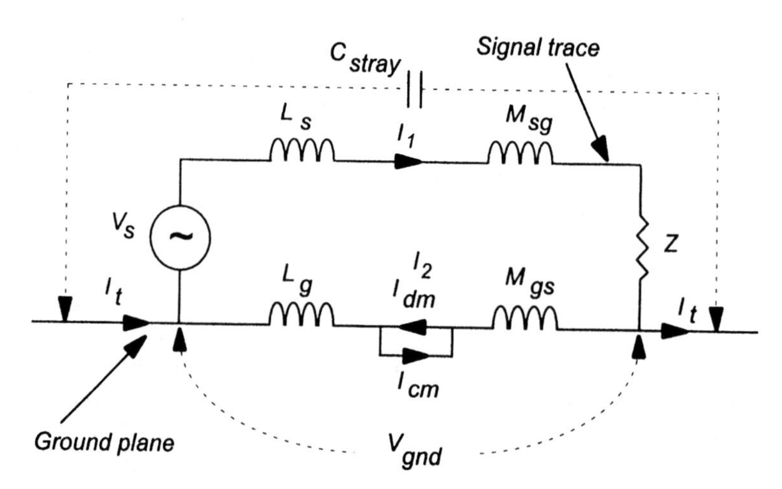

The currents in the return path may also be identified as I2.

Figure 4.4 Schematic representation of a ground plane within a PCB.

higher emissions than those created by differential-mode currents (I_1 and I_{dm}). This is because common-mode RF current fields are additive, whereas differential-mode fields tend to cancel [6,7,8].

To reduce ground-noise voltage, it is necessary to increase the mutual partial inductance between the trace and its nearest image plane. Doing so provides an enhanced return path for signal return current to mirror image back to its source. We calculate ground-noise voltage V_{gnd} using Eq. (4.6):

$$V_{gnd} = L_g \frac{dI_2}{dt} - M_{gs} \frac{dI_1}{dt} \tag{4.6}$$

where in Fig. 4.4 and Eq. (4.6)

L_s = partial self-inductance of the signal trace
M_{sg} = partial mutual inductance between signal trace and ground plane
L_g = partial self-inductance of the ground plane
M_{gs} = partial mutual inductance between ground plane and signal trace
C_{stray} = distributed stray capacitance of the ground plane
V_{gnd} = ground plane noise voltage

To reduce I_t currents, shown in Fig. 4.4, ground-noise voltage (V_{gnd}) must be reduced. This is best accomplished by reducing the distance spacing between the signal trace and ground plane. In most cases, there is a limitation on ground-noise reduction since the spacing between a signal plane and image plane must be at a specific, finite distance to maintain a constant impedance of the board for functionality reasons. Hence, there are limits to making the distance separation between the two planes any closer than physically manufacturable. Ground-noise voltage can also be reduced by providing an additional path for RF currents to flow through. This additional return path includes ground traces.

Since mutual partial inductance minimizes the creation of radiated RF currents, let's examine how differential-mode, I_{dm}, and common-mode, I_{cm}, currents are affected. Use of image planes significantly reduces these currents, as illustrated in Fig. 4.5. As discussed in Chapter 2, and as will be reaffirmed later in this chapter, differential-mode RF currents are canceled out when equal and opposite currents exist within the signal trace and RF return current path. If the cancellation of currents is not 100%, the amount of current that is left over becomes common mode. It is this common-mode current that functions as an excitation source and develops the majority of EMI that propagates from a product. This is because the leftover RF return current in the return path is added to the primary current in the signal path. To minimize common-mode currents, we must maximize the mutual partial inductance between signal trace and image plane to "capture the flux," hence canceling unwanted RF energy.

When an RF return plane or path is provided within a PCB assembly, optimal performance results when the return path is connected to a reference source. This reference source must be connected to the reference pins of components physically located at both the source and load ends of the transmission line [11]. For TTL and CMOS, the power and ground pins inside a component die (wafer) are connected to a reference source, power and ground. Certain geometrical factors impact these connections. Only when the RF return path is connected to the power and ground pins of a component will a real

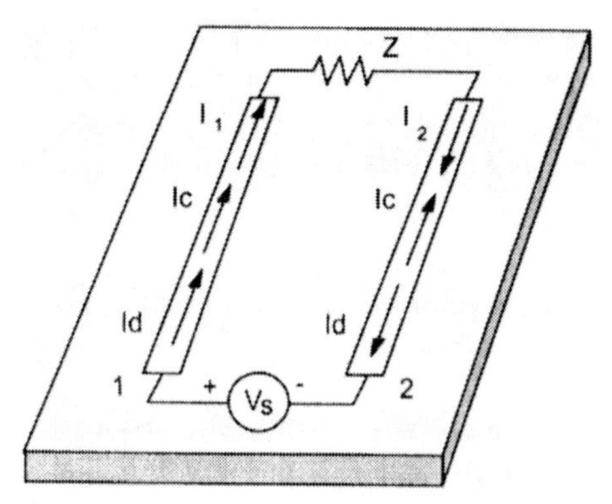

Circuit model-2 traces over a plane

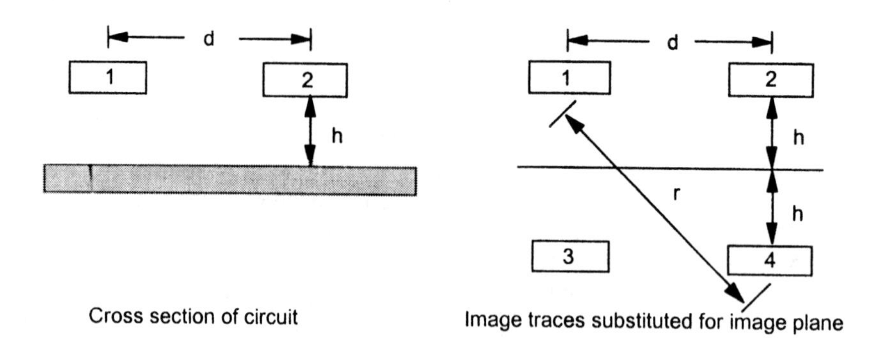

Cross section of circuit

Image traces substituted for image plane

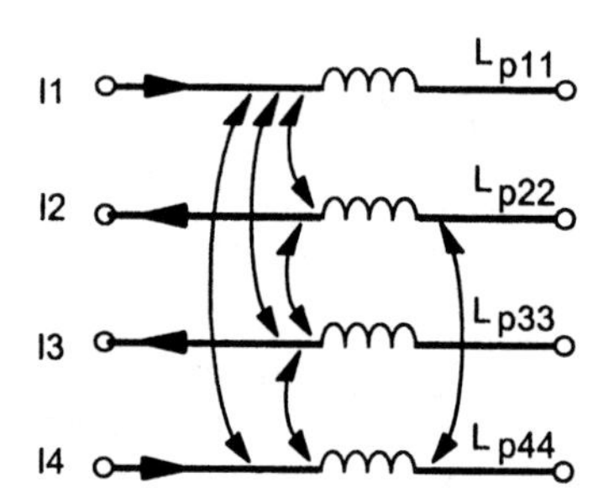

Partial inductance model of the circuit
I3 is the image of I1, while I4 is the image of I2

Figure 4.5 Use of image plane related to partial inductance. (*Source: Introduction to Electromagnetic Compatibility*, Clayton Paul © 1992. Reprinted by permission of John Wiley & Sons, Inc.)

image plane exist. An isolated solid sheet of copper foil (a layer within the PCB structure), that is not connected to any reference source, will not work as an image plane under any condition. How is the return current going to get back to the source if the return path is broken?

An image plane containing differential-mode voltage and currents will produce common-mode currents. Depending on the distance spacing between the trace and image plane, differential-mode currents will be reduced by increased mutual partial inductance. How much differential-mode current travels in the planes is dependent on the minimized distance separation between the two conductive surfaces.

Image planes function because digital components are connected to a power and ground plane structure. The ground connection internal to the device, connected to the ground plane, provides for the reference to exist. This connection internal to the component package is what makes image planes work.

When the image plane is removed, a phantom image return path is created between a trace and plane. The RF image associated with these currents will cancel out along with a reduction of radiated energy because each trace pair (the original current and its image) is closely spaced. For an image plane to perform as desired, the plane should be infinite in size and not contain disruptions, slots, or cuts [2].

4.4 GROUND AND SIGNAL LOOPS (NOT EDDY CURRENTS)

Loops are a major contributor to the propagation of RF energy. RF current will attempt to return to its source through any path or medium: components, wire harnesses, ground planes, adjacent traces, and so forth. RF current is always created between a source and load due to the return path, where there is a voltage potential difference between these two points. Path inductance, however, causes magnetic coupling of RF currents to occur between a source and victim circuit, thus increasing RF losses in the path.

One of the most important design considerations for EMI suppression on a PCB is ground or signal return loop control. An analysis must be made for each and every ground stitch connection (mechanical securement between the PCB and chassis ground) related to RF currents generated from RF noisy electrical circuits. High-speed logic components and oscillators should always be located as close as possible to a ground stitch connection to minimize the formation of loops in the form of eddy currents to the chassis ground. This design requirement will now be examined in detail.

An example of loops that could occur in a computer with adapter cards and single-point grounding is shown in Fig. 4.6. As observed, an excessive signal-return loop area is present. Each loop will create a distinct electromagnetic field and spectra. RF currents will create an electromagnetic radiated field at a unique frequency, depending on the physical size of the loop. Containment measures must now be used to keep these RF currents from coupling to other circuits or radiating to the external environment as EMI. Internally generated RF loop currents are to be avoided.

To expand on the concept of loop area shown in Fig. 4.6, we have Fig. 4.7 which shows the loop area between two components.

To reiterate the importance of minimizing loops within a PCB structure, we examine the effects a loop has in creating EMI. This concept is important in understanding how

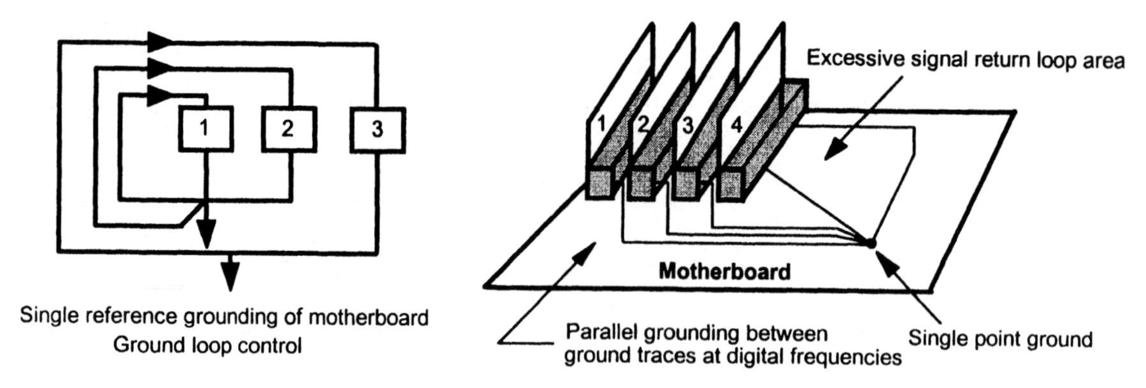

Figure 4.6 Ground loops within a PCB assembly.

RF energy is created within a PCB. (For a discussion of how loops create EMI within components, refer back to Chapter 3, Section 3.4.)

With RF energy concentrated within a loop structure, how can this energy be removed if a return path is not provided for the RF current? A ground connection to chassis ground or a 0V reference source assists in removing this undesirable accumulation of RF current, also identified as loop area control.

4.4.1 Loop Area Control

Figure 4.8 illustrates how various loop areas are created within a PCB structure using both a single- and double-sided assembly and a multilayer stackup. The electromagnetic field induced into a loop structure by a magnetic field can be represented as a voltage source within that loop. This voltage source is proportional to the total area of the loop. To minimize magnetic field coupling, we must minimize the loop area. The electric field pickup reception is also dependent on the loop area forming the receive antenna.

When an electric field is present, a current source is created between a two-conductor system (power and ground). Electric fields do not couple line-to-line but rather line-to-ground, including common-mode currents. Therefore, the only loop valid for this mode of coupling is the conductor to chassis coupling. Of course, the H-field that accompanies the E-field also couples into wiring loops (line-to-line and line-to-ground).

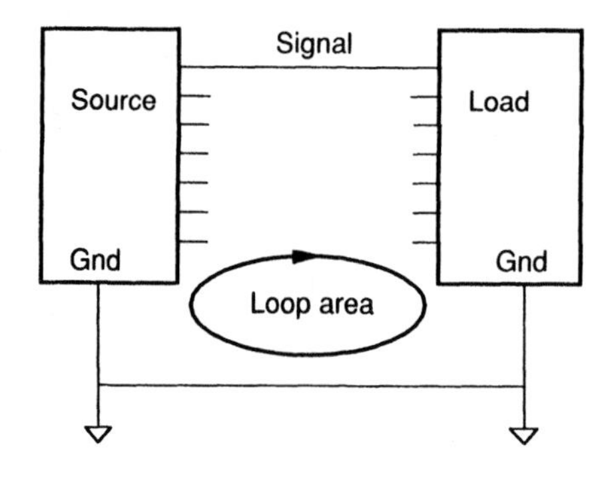

Figure 4.7 Loop area between components.

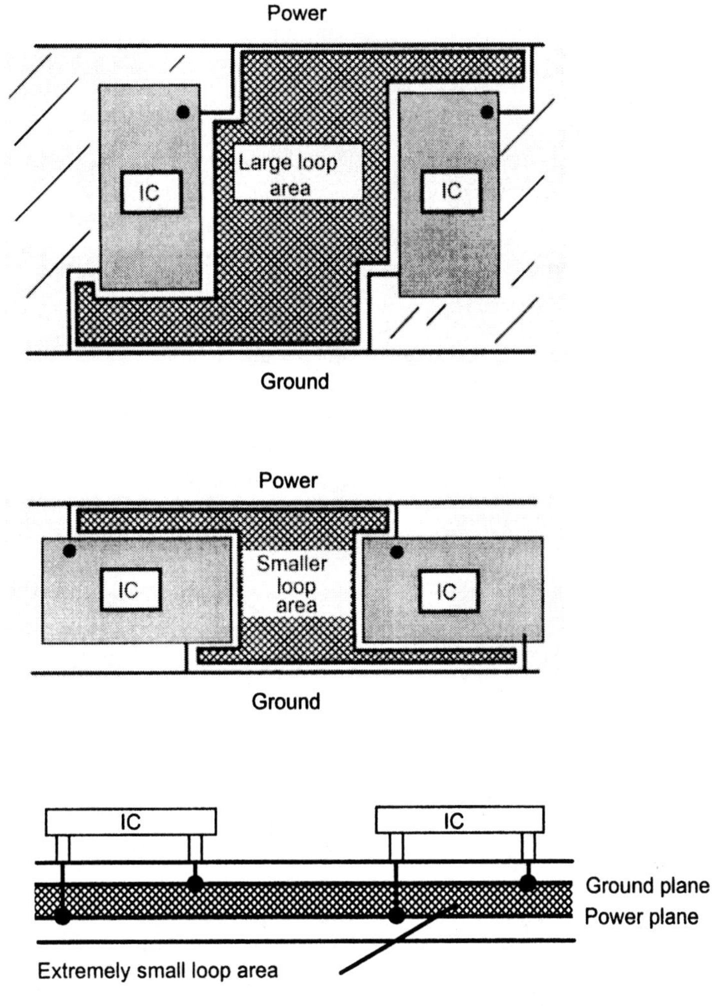

Figure 4.8 Loop areas that exist within a PCB. (*Source: Introduction to Electromagnetic Compatibility,* Clayton Paul © 1992. Reprinted by permission of John Wiley & Sons, Inc.)

It is generally overlooked during a PCB layout that loop areas may be created between the power and 0V reference structure. Figure 4.8 illustrates a poor layout in the top drawing that may be performed by a PCB designer because it is easy to create using computer CAD software. This software permits easy routing of busses between components located next to each other. With a large loop area on the PCB, susceptibility to the pickup of ESD-induced (or other) fields can occur. A multilayer stackup minimizes the potential of ESD disruption, in addition to minimizing creation of a magnetic field that will be radiated into free space.

Using power and ground planes helps reduce the inductance of the power distribution system. Lowering the characteristic impedance of the power distribution system reduces the voltage drop across the board. With less voltage drop, ground bounce potential is minimized. In addition to lowering the characteristic impedance of the structure, we increase the capacitance between the two parallel planes. This capacitance reduces the effects of any induced voltages. (Decoupling is discussed in Chapter 5.)

Large loop areas may be created when signal lines travel between components. This is seen in Fig. 4.9. Signal lines are generally forgotten when analyzing why a PCB has radiated emission problems. Although we may have high signal integrity (time domain), EMI still exists (frequency domain) because signal loop areas create more problems than those of the power distribution system, especially from the viewpoint of ESD. This is because an ESD event may be injected directly into the loop and into the input pins of components. To mitigate the consequences of a harmful disruption from an ESD event, reducing loop area is the easiest technique to use. A power and ground plane distribution network provides a low-impedance path that allows transfer of the ESD energy into a 0V return reference plane. After all, loops are loops, and if they can emit fields, then they also can receive fields.

In addition to reducing ground-noise voltage, an image plane prevents RF ground loops from being developed because RF currents tightly couple themselves to their source trace without having to find an alternate return path home. When loop control is maximized, flux cancellation is enhanced. *This is one of the most important concepts of sup-*

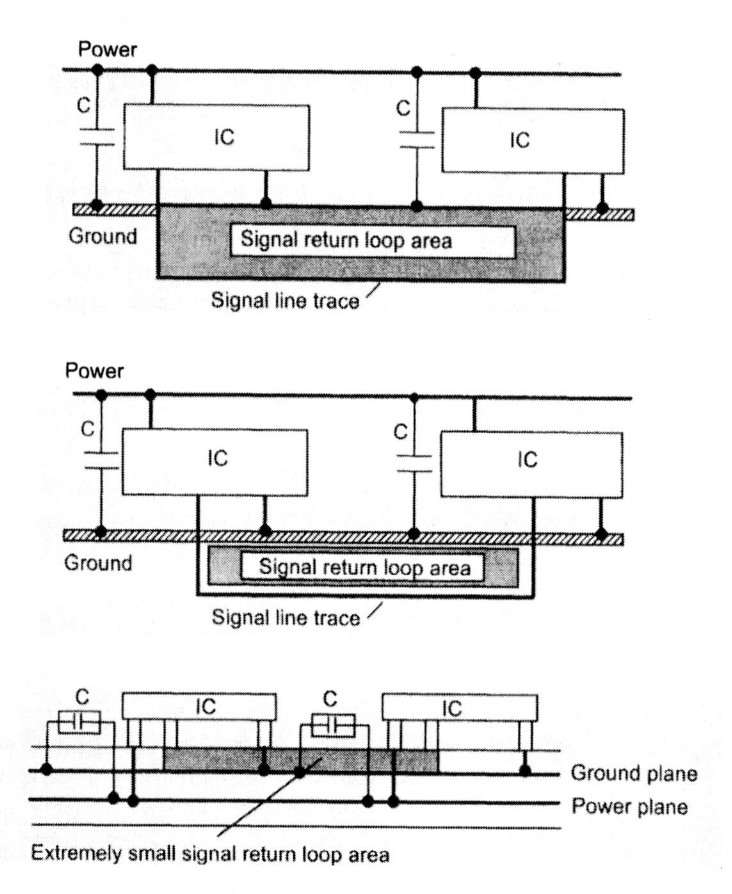

Figure 4.9 Reduction in loop areas that exist within a PCB. (*Source: Introduction to Electromagnetic Compatibility*, Clayton Paul © 1992. Reprinted by permission of John Wiley & Sons, Inc.)

pression of RF currents at the PCB level. Proper placement of an image plane adjacent to each and every signal plane removes common-mode RF currents created by signal traces coupling to its return path. Image planes carry large amounts of RF currents that must be sourced to ground or 0V reference potential. To help remove excess RF potentials and uncontrolled eddy currents, all ground and chassis planes, (if 0V reference is used), can be connected to chassis ground through a low-impedance ground stitch connection [1,6,9,10,11].

We now examine optimal spacings for creating a low-impedance ground stitch connection to remove RF currents into the 0V reference or return structure.

4.5 ASPECT RATIO—DISTANCE BETWEEN GROUND CONNECTIONS

Aspect ratio is a term commonly used in the television industry to refer to the ratio of frame width to frame height. The term also refers to the ratio of a longer dimension to a shorter one. With these definitions, how does aspect ratio relate to EMC? When providing ground stitch connections in a PCB using multipoint grounding to a metallic structure, we must concern ourselves with the distance spacing in all directions of the ground stitch location.

RF currents that exist within the power and ground plane structure will tend to couple to other components, cables, peripherals, or other electronic items within the assembly. This undesirable coupling may cause improper operation, functional signal degradation, or EMI. When using multipoint grounding to a metal chassis, and providing a third wire ground connection to the AC mains, RF ground loops become a major design concern. This configuration is typical with personal computers. (An example of a single-point ground connection for a personal computer was shown in Fig. 4.6).

Because the edge rate of components is becoming faster, multipoint grounding is becoming a mandatory requirement, especially when I/O interconnects are provided in the design. Once an interconnect cable is attached to a connector, the unit at the other end of the interconnect may provide an RF path to a third wire AC ground mains connection (if provided) to its respective power source (e.g., the negative terminal of a battery) or simply through distributive radiation RF impedance to earth or through free space. A large ground loop on the I/O interconnect can cause undesirable levels of radiated common-mode energy. How can we minimize loops that may occur within a PCB structure? The easiest way is to design the board with many ground stitch locations to chassis ground, if chassis ground is provided. The question that now exists is, how far apart do we make the ground connections from each other, assuming the design has the option of specifying this design requirement?

The distance spacing between ground stitch locations should not exceed $\lambda/20$ of the highest frequency of concern, not just the primary frequency (including harmonics). If many high-bandwidth components are used, multiple ground stitch locations are typically provided. If the unit is a slow edge rate device, connections to chassis ground may be minimized, or the distance between ground locations increased.

For example, $\lambda/20$ of a 64-MHz oscillator is 23.4 cm (9.2 in.). If the straight-line distance between any two ground stitch locations to a 0V reference (in either the x- and/or y-axis) is greater than 9.2 inches, then a potential efficient RF loop exists. This loop could

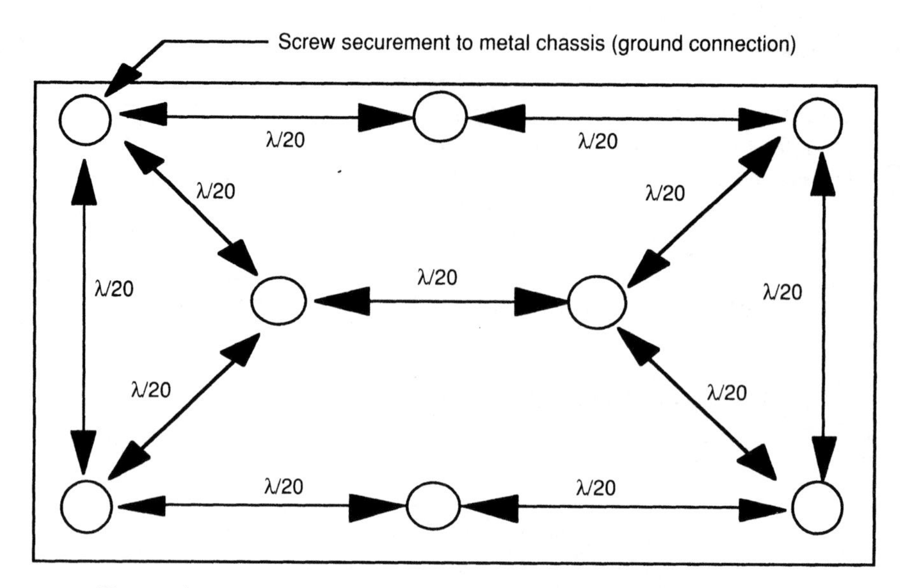

Figure 4.10 Aspect ratio.

be the source of RF energy propagation, which could cause noncompliance with international EMI emission limits. Unless other design measures are implemented, suppression of RF currents caused by poor loop control is not possible and containment measures (e.g., sheet metal) must be implemented. Sheet metal is an expensive Band-Aid that might not even work for RF containment. An example of *aspect ratio* is given in Fig. 4.10 [1].

Proper placement of components is critical in any PCB layout. Most designs incorporate functional subsections or areas (by logical function). Grouping each functional area adjacent to other subsections minimizes signal trace lengths and reflections, and makes trace routing easier along with maintaining signal integrity. Vias should be avoided where possible, for vias increase the inductance of the trace by approximately 1 to 3 nH each. Figure 4.11 illustrates the functional grouping of subsections (or areas) using a stand-alone CPU-motherboard as an example.

Extensive use of chassis ground stitch connections is also observed in Fig. 4.11. High-frequency designs (fast edge rates) develop very high spectral frequency profiles and require new methodologies for bonding ground plane(s) to chassis ground. Use of these multipoint grounding points effectively partitions common-mode eddy currents emanating from various segments on the design from coupling into other segments. Products with clocks above 50 MHz generally require frequent ground stitch connections to chassis ground to minimize the effects of common-mode currents and ground loops present between functional sections. At least four ground points surround each subsection. These ground points illustrate best case implementation of aspect ratio. Note that a chassis bond connection (screw or equivalent) is located on both ends of the DC power connector (Item P) used for powering external peripheral devices. RF noise generated on either the PCB or peripheral power subsystem must be AC shunted to chassis ground by parallel bypass capacitors. These capacitors minimize power-supply-generated RF currents from coupling into signal or data lines. Removal of RF currents on the power connector will

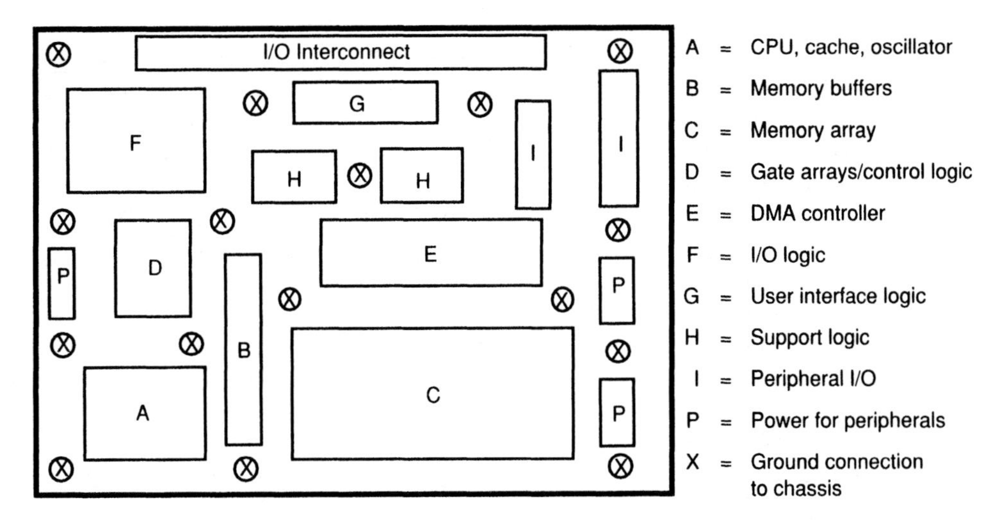

Figure 4.11 Multipoint grounding—implementation of aspect ratio.

optimize signal quality for data transfer between the motherboard and external peripheral devices in addition to reducing emissions [1,10].

Most PCBs can be arranged to consist of functional subsections or areas. A typical personal computer contains the following: CPU, memory, ASICs, I/O, bus interface, system controllers, PCI/IDE bus, SCSI bus, peripheral interface (fixed and floppy disk drives), and other components. Associated with each subsection are different bandwidths of RF energy. Different logic families generate RF energy across the frequency spectrum. The higher the frequency component of the signal, the greater the bandwidth of RF spectral energy. RF energy is generated from the higher frequency components and the time-variant edges of digital and analog signals. Clock signals are the greatest contributors to the generation of RF energy. This is because clocks are periodic signals providing coherent spectral distribution (50% duty cycle) and generally have fast edge rates.

To prevent coupling between different bandwidth areas, functional partitioning is used. Partitioning refers to the physical separation between functional sections. Partitioning is product specific and may be achieved using separate PCBs, isolation, topology layout variations, or other creative means.

Proper partitioning allows for optimal functionality, ease of routing traces, and minimization of trace lengths. It also permits smaller loops to exist while optimizing signal quality. The design engineer will specify which components are associated with each functional subsection. Use the information provided by the component manufacturer to optimize component placement prior to routing any traces.

4.6 IMAGE PLANES

An image plane is a layer of copper (voltage plane, ground plane, or chassis plane) internal to a PCB physically adjacent to a circuit or signal plane. Image planes are used to provide a low-impedance path for RF signal currents to return to their source (flux return), thus completing the RF current return path and reducing EMI emissions. The term *image plane* was popularized by the German, Ott, and Paul [7], and is now used as industry standard terminology.

RF currents must return to their source one way or another. This return path may be a mirror image of its original trace route, through another trace located in the near vicinity, a power plane, a ground plane, or a chassis plane. RF currents will capacitively (or by mutual inductance) couple themselves to a conductive medium (e.g., low-impedance path such as the copper that makes up a trace or plane). If this coupling is not 100%, common-mode RF currents can be propagated between traces and their nearest image plane. An image plane internal to the PCB reduces ground-noise voltage in addition to allowing RF currents to return to their source (mirror image) in a tightly coupled (nearly 100%) manner. Tight coupling provides for flux cancellation, which is another reason for use of a solid plane. Solid planes also prevent common-mode RF current from being generated in the PCB by those traces rich in RF energy.

Figure 4.12 illustrates the concept and use of an image plane and what happens when tight coupling does not exist between the signal trace and 0V reference (ground) plane. The voltage developed across a return conductor is referred to as ground drop. The lower the value of ground drop between two points on a PCB return structure, the lower the radiated emissions from the PCB.

One concern related to image planes involves the concept of skin effect. *Skin effect* refers to current flow that resides in the first skin depth of the material at high frequencies. Current does not and cannot flow in the center of traces and wires, and is predominately observed on the outer surface of the conductive media. Different materials have different skin depth values. The skin depth of copper is extremely small, above 30 MHz. Typically, this is observed at $6.6 * 10^{-6}$ (0.0017 mm) of an inch at 100 MHz. RF current present on a ground plane cannot penetrate 1 oz. 0.0014″ (0.036 mm) thick copper. As a result, both common-mode and differential-mode currents flow only on the top (skin) layer of the plane. No significant current flows internal to the image plane or on its bottom. Placing an additional image plane beneath this ground plane would not provide additional EMI reduction. If the second plane is at voltage potential (the primary plane at ground potential), a decoupling capacitor will be created. These two planes can now be used as both a decoupling capacitor and dual image planes but with some concern regarding flux cancellation (see Section 2.1) [6].

With regard to image plane theory, the material presented herein is based on a finite-sized plane, typical of all PCBs. Image planes cannot be relied on for reducing cur-

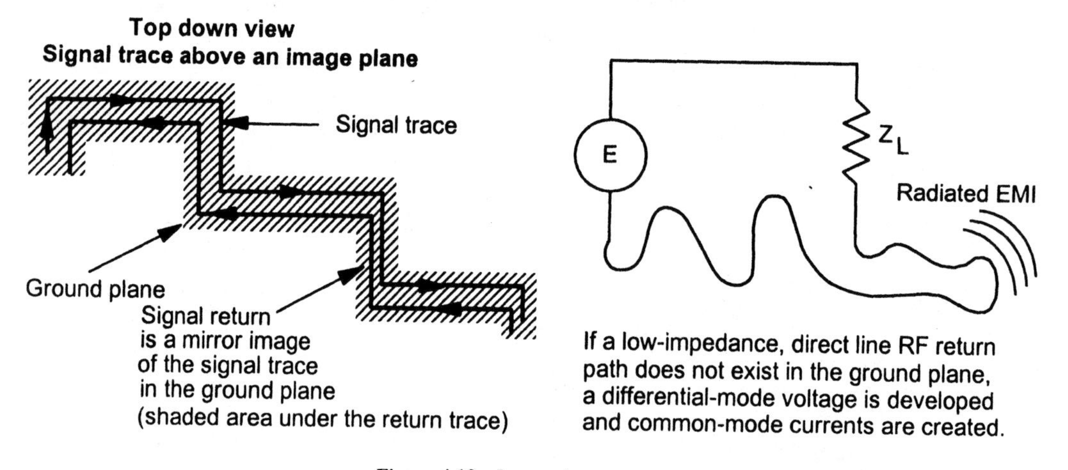

Figure 4.12 Image plane concept.

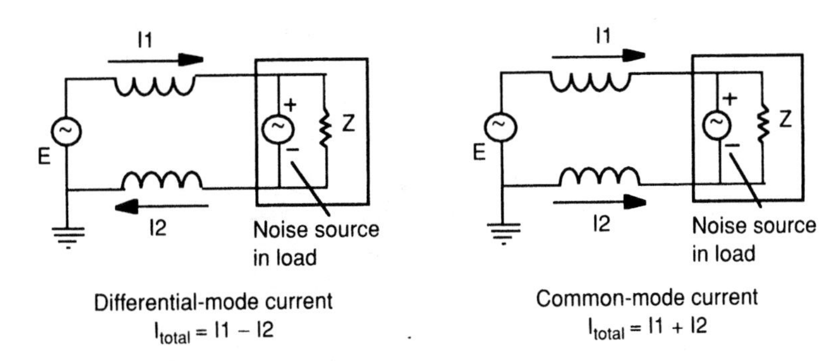

Figure 4.13 Common-mode and differential-mode currents.

rents on I/O cables because approximating finite-sized conductive planes is not always valid. When I/O cables are provided, the dimensions of the configuration and source impedance are important parameters to remember [12].

An example of common-mode and differential-mode currents is shown in Fig. 4.13. The measured *E*-field of the differential-mode current will be the difference of I1 and I2. This difference is negligible because of a 180 degree phase difference. The measured *E*-field due to common-mode current is the sum of I1 and I2, which could be substantial due to the summing effect. Common-mode currents are always much smaller than differential-mode currents.

If "three" internal signal planes (stripline configuration) are physically adjacent to each other in a multilayer board stackup, the middle signal plane, (e.g., the one not adjacent to a reference plane), will couple its RF currents to the other two signal planes, thus causing RF energy to be transferred (by mutual inductance and capacitive coupling) to the other two planes. After this first level of coupling, a second level of coupling occurs to the real image or RF return plane. This coupling can cause significant crosstalk to occur, which may include nonfunctionality. Flux cancellation performance is sometimes enhanced when the signal routing layer is adjacent to a ground plane, but not to a power plane, as described throughout this chapter.

4.7 IMAGE PLANE VIOLATIONS

For an image plane to be effective, all signal traces must be located adjacent to a solid plane and must not cross an isolated area of copper. Exceptions can occur using special trace routing techniques. If a signal trace, or even a power trace (e.g., +12 V trace in a +5 V power plane) is routed within a solid plane, this solid plane becomes fragmented (split) into smaller parts. Provisions have now been made for a ground or RF signal return loop to be developed for RF return currents that are observed on the adjacent layer across this violation. This RF loop occurs by not allowing RF current present in a signal trace to seek a straight-line, low-impedance path back to its source.

Figure 4.14 illustrates a violation of the image plane concept. These planes can now no longer function as a solid 0V reference to remove common-mode RF currents. The losses across the plane segmentations may actually produce RF fields. Vias placed in an image plane do not degrade the imaging capabilities of the plane, except where ground slots are provided as discussed next.

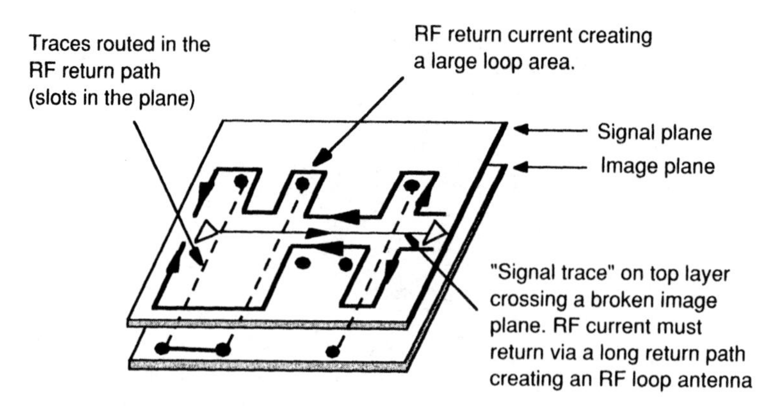

Figure 4.14 Image plane violation with traces.

Another area of concern that lies with ground plane discontinuities is the use of through-hole components. Excessive use of through-holes in a power or ground plane creates the Swiss Cheese Syndrome [5]. The copper area in the plane is reduced because many holes overlap (oversized through-holes), leaving large areas of discontinuities. This effect is observed in Fig. 4.15. The return current flows on the image plane around the through-hole pattern, while the signal trace is on a direct line route across the discontinuity. As seen in Fig. 4.15 [1], the return currents in the ground plane must travel around slots or holes. As a result, extra trace length is present for return currents that must flow around these slots in the image plane. This extra trace length adds more inductance in the signal return trace, $E = L(dI/dt)$. With additional inductance in the return path, there is reduced differential-mode coupling between signal trace and the RF current return path (less flux cancellation). For through-hole components that have a space between pins (nonoversized holes), optimal reduction of signal and return current is achieved through less inductance in the signal return path and the existence of the solid plane.

If a signal trace is routed "around" the through-hole discontinuities (not shown in the left side of Fig. 4.15), a constant image plane (RF return path) would be maintained along the entire signal route. The same is true for the right side of Fig. 4.15. There are no ground plane discontinuities and hence, shorter trace length. The longer trace route on the left side of the figure adds more trace length inductance $E = L(dI/dt)$. This length can cause reflections that affect signal integrity and functionality, and may also create a loop for RF current. Problems arise when the signal trace travels through the middle of slotted holes in the PCB (in an attempt to minimize trace length routing) when a solid plane does not exist in this oversized through-hole area. When routing traces between through-hole components, use of the *3-W Rule* (defined later in this chapter) must be maintained between the trace and through-hole clearance area.

Generally, a slot in a printed circuit board with through-hole components will not cause RF problems for the majority of signal traces that route between the through-hole device leads. However, it can cause electromagnetic fields to be developed around the holes. For high-speed, high-threat[2] signals, alternative methods of routing traces between

[2]High-threat refers to high-bandwidth, RF spectral components that propagates as an electromagnetic field down a transition line or trace. These signals include clocks, video, address lines, analog circuits, and the like. All of these highly sensitive circuits may either radiate RF energy, or be susceptible to an externally induced field disturbance, requiring a mandatory, low-impedance RF return path to complete the closed-loop circuit.

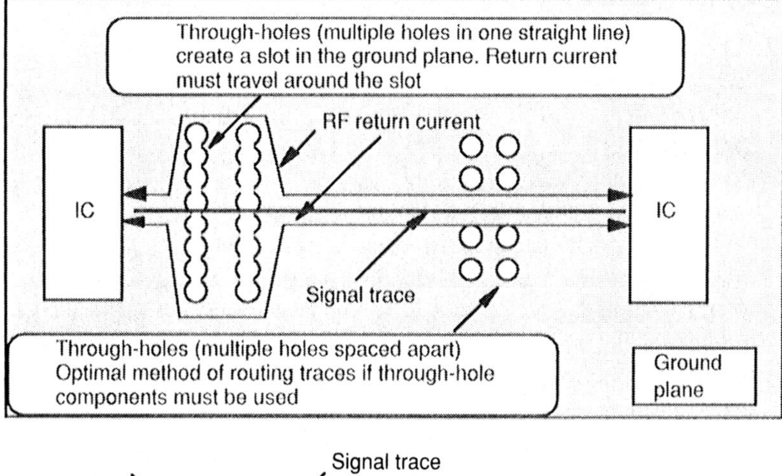

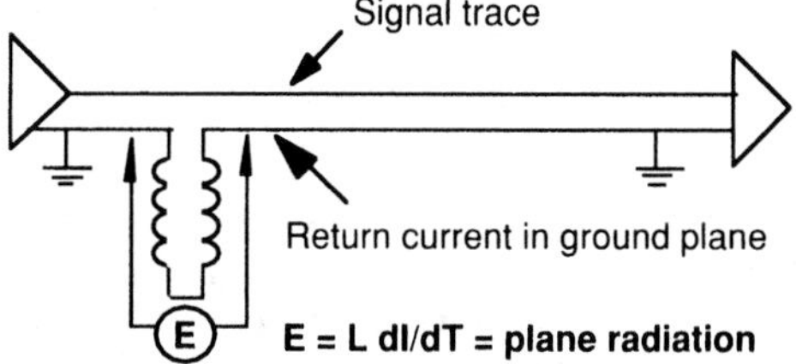

Equivalent circuit showing inductance in the return paths. This inductance is approximately 1 nH/cm.

Figure 4.15 Ground loops when using through-hole components (slots in the plane).

through-hole component leads must be devised. For those applications where a trace must traverse across a slot or partition within the PCB assembly, Fig. 4.16 provides a design technique which allows RF return current to jump the slots using capacitors.

Capacitors provide an AC shunt for RF currents to traverse across a moat or slot. A significant performance improvement of up to 20 dB has been observed during functional testing. The capacitor must be chosen for optimal performance based on the self-resonant frequency of the component trace signal. It is cautioned, however, that this technique may result in reactance-based phase shifts in the current relationships between the traces and their images, impacting the magnitude of flux cancellation or minimization.

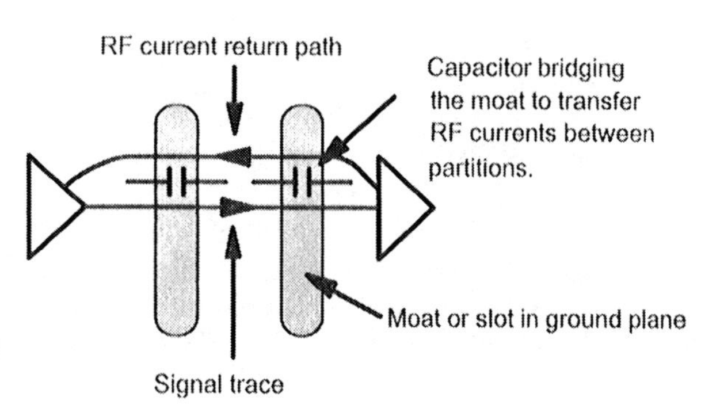

Figure 4.16 Crossing a moat and allowing RF return current to pass.

4.8 LAYER JUMPING—USE OF VIAS

When routing clock or high-threat signals, it is common practice to via the trace to a routing plane (e.g., x-axis) and then via this same trace to another plane (e.g., y-axis) from source to load. It is generally assumed that if each and every trace is routed adjacent to an RF return path, there will be tight coupling of common-mode RF currents along the entire trace route. In reality, this assumption is partially incorrect.

As a signal trace jumps from one layer to another, RF return current should follow the trace route. When a trace is routed internal to a PCB between two planar structures, commonly identified as the power and ground planes, or two planes with the same potential, the return current is shared between these two planes. The only time the return current can jump between the two planes is at a location where decoupling capacitors are positioned. If both planes are at the same potential (e.g., 0V reference) the RF return current jump will occur at a via connecting both planes to a device or component assigned to that via.

When a jump is made from a horizontal to a vertical layer, the RF return current *cannot* fully make this jump. This is because a discontinuity is placed in the trace route by the via. The return current must now find an alternate low-inductance (impedance) path to complete its route. This alternate path may not exist in a position that is immediately adjacent to the location of the via used for the jump. As a result, RF currents on the signal trace can couple to other circuits and pose problems as both crosstalk and EMI. Use of vias in a trace route will always create a concern in any high-speed, high-technology product.

To minimize development of EMI and crosstalk due to layer jumping, the following design techniques have been found to be effective:

1. Route all clock and high-threat signal traces on only one routing layer as the initial approach concept. This means that both x- and y-axis routes are in the same plane. (*Note:* This technique is likely to be rejected by the PCB designer as being unacceptable because it makes autorouting of the board nearly impossible.)
2. Verify that a solid RF return path is adjacent to the routing layer, with no discontinuities in the route created by use of vias or jumping the trace to another routing plane.

If a via must be used for routing a sensitive trace (high-threat or clock signal) between the horizontal and vertical routing layer, the designer should incorporate ground vias at "each and every" via location where the signal axis jumps are executed. The ground via is always at 0V potential.

A ground via is a via that is placed directly adjacent to each signal route via from a horizontal to a vertical routing plane. Ground vias can be used only when there are more than one 0V reference planes internal to the PCB. This via is connected to all ground planes (0V reference) in the board that serves as the RF return path for the signal jump currents. This via essentially ties the 0V reference planes together adjacent and parallel to this signal trace location. When using two ground vias per signal trace via, a continuous RF return path will now exist for RF return current throughout its entire trace route. This

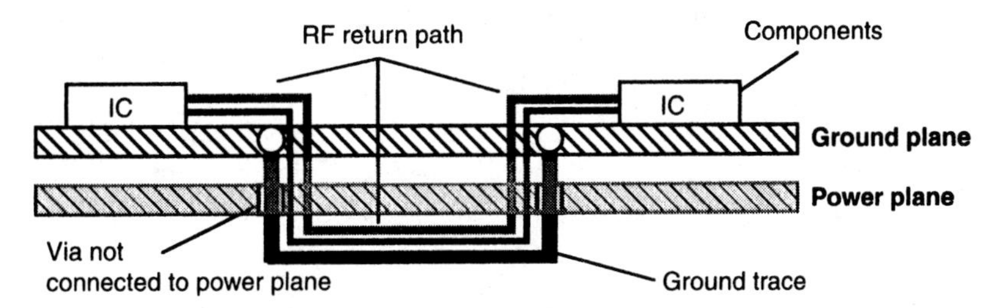

Figure 4.17 Routing a ground trace to assure a compete RF return path exists.

ground via will maintain a constant RF return path (through use of image planes) located 100% adjacent to a signal route.[3]

What happens when only one 0V reference (ground) plane is provided and the alternate plane is at voltage potential as commonly found with a four-layer PCB stackup assignment. To maintain a constant return path for RF currents, the 0V reference plane should be allowed to act as the primary return path. The signal trace must be routed against this 0V reference plane. When the trace must route against the power plane, use of a *ground trace* is required, with the vias at both ends of the ground trace routed parallel to the signal trace tied to the 0V reference plane. Using this configuration, we can now maintain a constant RF return path (see Fig. 4.17).

How can we minimize use of ground vias when layer jumping is mandatory. In a properly designed PCB, the first routed traces will be clock signals, "manually routed." Since much freedom is permitted in routing the first few traces by the PCB designer (e.g.,

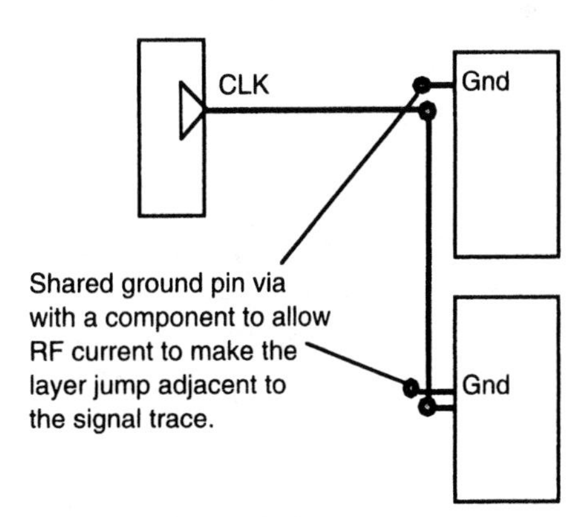

Figure 4.18 How to route the first trace within a PCB.

[3]Use of ground vias was first identified and presented to industry by W. Michael King. Ground vias are also described in [1] and [13].

all clocks and high-threat signals), the designer can route the board using the shortest trace distance routing possible (shortest Manhattan length), making the layer jump adjacent to the *ground pin via* of any component. This layer jump will co-share this component's ground via. The ground via being referenced will perform the function of providing 0V reference to a component while allowing RF return current to make a layer jump as detailed in Fig. 4.18.

4.9 SPLIT PLANES

When multilayer PCB assemblies are used, the power and ground planes are sometimes split on the same plane. An example is separation of analog circuitry from digital logic, isolation of I/O interconnects (detailed under Partitioning in Section 4.10), separation of voltage reference areas (e.g., +5V section from a –48V partition), component isolation, and the need to force RF return currents to travel a designated route through the PCB. This designated RF return path can be likened to a road map. We travel only on the roads provided in a predefined manner. Why not cause RF return currents to do the same thing, travel a predefined road or path?

One of the PCB designer's primary design and layout concerns is to guarantee that overlaps on a split plane do not occur. If an overlap on a plane is present, a finite-sized capacitor will be created between the overlapping plane segments as seen in Fig. 4.19. This finite-sized capacitor, C1, will allow RF energy (which is an AC waveform) to traverse from one plane (e.g., a noisy plane) to a separate, quiet, or isolated plane. The DC voltage potentials of the planes remain intact due to passing the DC voltage from one isolated area to another through filters.

If additional high-frequency isolation is required, we can isolate one plane (power) or both potentials (power and ground) with ferrite bead-on-leads, not inductors. We must be careful with this technique. If both planes contain high-frequency RF noise, it is usu-

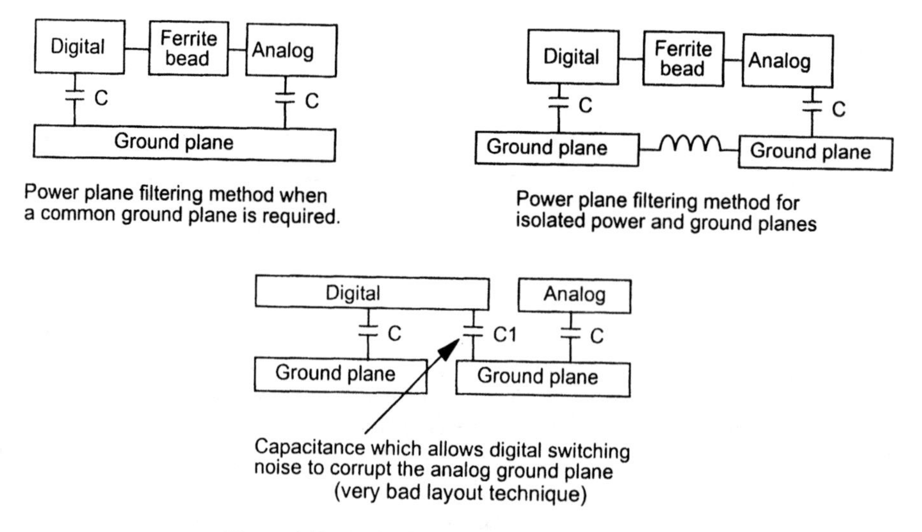

Figure 4.19 Variations on split plane configurations.

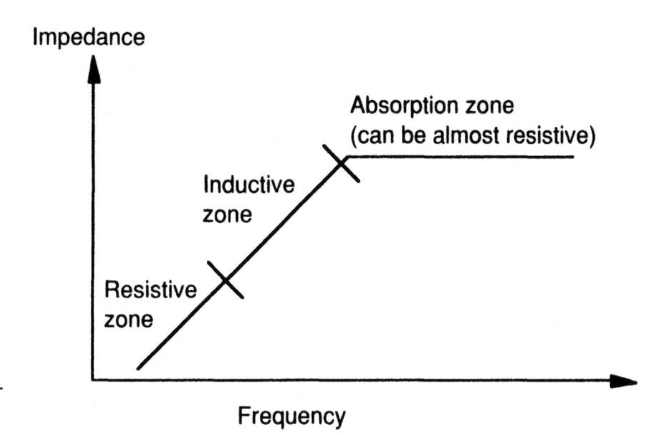

Figure 4.20 Ferrite material performance characteristics.

ally better to isolate both planes. If a common digital-to-analog ground reference is required, and analog power is needed for device operation, the ferrite bead-on-lead should be placed only across the power plane partition.

When a split plane[4] occurs, the common ground plane must be located "directly" under the discrete filter components. All signal traces must then be routed adjacent to this solid ground plane under the filter in an area identified as a bridge. Bridges are discussed later in this chapter. The advantage of this design technique is to maintain the integrity of the 0V reference (image) plane necessary for high-frequency EMI control and to provide an optimal RF return path from load to source.

The reason not to use inductors is easy to visualize and is best shown in Fig. 4.20. Ferrite material has practically zero impedance (or DC resistance) at DC voltage or signal transition levels, including very low-frequency signals. It is essentially transparent to DC voltage and acts as a small inductor or resistor having little effect at low frequencies. At higher frequencies, RF currents are created within the power distribution structure, and the resistive characteristics of the ferrite material dominate, providing a high impedance to the circuit. The high impedance of the material is present until the ferromagnetic properties reaches a predefined operating frequency, where the ferromagnetic material ceases to function as desired. Basically, a ferrite component is a large RF resistor that keeps RF energy from traveling between two isolated locations. An inductor, on the other hand, has a large inductive value with an inductive reactance, $j\omega L$. Inductive reactance is exactly what we *do not want* within a transmission path. Parasitic capacitance will exist between the two terminals of the inductor, plus the capacitance between the inductor windings and 0V reference. With an L and C component present within the device, a resonant circuit is created. Depending on the values of L and C, we may be allowing RF currents, at a particular frequency, to pass between the isolated areas. Once the RF currents pass through the circuit, these RF currents are now allowed to cause harmful disruption to functional circuits, which were supposed to be operated from clean filtered power.

If an isolated plane contains only low-frequency circuits (analog) and another isolated plane has high-frequency (digital) switching currents, it sometimes becomes mandatory to isolate both the power and ground planes between these two areas with ferrites de-

[4]A split plane refers to a solid copper structure that has been segmented into two or more partitions. An example of this split is easily seen in Fig. 4.19. The top left circuit has a continuous ground plane. The other two circuits have been partitioned into separate functional ground planes: analog and digital. One split is connected by a ferrite bead, the other totally isolated.

pending on the device's function and the manufacturer's requirements for power and/or plane isolation. This isolation technique is required only if no high-frequency energy can be allowed to pass between the two areas. If both areas contain only low-frequency components and there are no high-frequency RF energy threats (high-edge rate switching noise), ferrite components are not required. A single-point connection is permissible between the two planes.

4.10 PARTITIONING

Designing I/O circuits involves two basic areas of concern: functional subsystems, and quiet areas. Each is briefly discussed separately below, with more detail presented herein.

4.10.1 Functional Subsystems

Each I/O should be considered as a different subsection on a PCB, for each may be unique in its particular application. To prevent RF coupling between subsystems, partitioning may be required. A functional subsystem is a group of components along with their respective support circuitry. Locating components close to each other minimizes trace length routing and optimizes functional performance. Every hardware and PCB designer generally tries to group components together, but, for various reasons, it is sometimes impractical to do so. I/O subsystems must still be treated differently during layout than any other section of the PCB. This is generally done through layout partitioning.

Layout partitioning enhances signal quality and functional integrity by preventing high-bandwidth emitters (e.g., backplane interconnect, video devices, data interfaces, Ethernet controllers, Small Computer System Interface [SCSI] devices, and central processing units [CPUs]) from corrupting serial, parallel, video, audio, asynchronous/synchronous ports, floppy controller, front panel console displays, local area and wide area networks controllers, and so on. Each I/O subsystem must be conceived, designed, and treated as if the subsystems were separate PCBs.

4.10.2 Quiet Areas

Quiet areas are sections that are physically isolated from digital circuitry, analog circuitry, and power and ground planes. This isolation prevents noise sources located elsewhere on the PCB from corrupting susceptible circuits. An example is power plane noise from the digital section entering the power pins of analog devices (analog section), audio components (audio section), I/O filters, interconnects, and so on, detailed in Fig. 4.21 [1].

Each and every I/O port (or section) must have a partitioned (quiet) ground/power plane. Lower-frequency I/O ports may be bypassed with high-frequency capacitors (usually 470 pF to 1000 pF) located near the connectors.

Trace routing on the PCB must still be controlled to avoid recoupling RF currents into the cable shield. A clean (quiet) ground must be located at the point where cables leave the system. Both power and ground planes must be treated equally, for both planes act as a path for RF return currents. RF return currents from switching devices to I/O con-

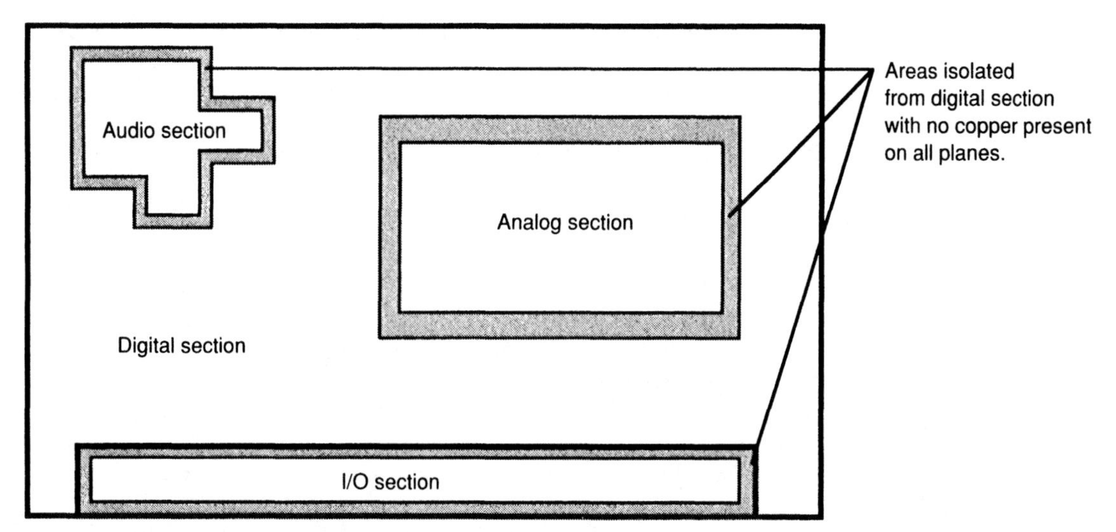

Figure 4.21 Quiet areas.

trol circuitry can inject high-bandwidth switching RF noise into the I/O cables and interconnects.

To implement a quiet area, use of a partition is required. This quiet area may be

1. 100% isolated with I/O signals entering and exiting through an isolation transformer;
2. data line filtered;
3. filtered through a high-impedance common-mode inductor; or
4. protected by a ferrite bead-on-lead component.

The main objective of partitioning is to separate *dirty power and ground planes and other functional areas from clean or quiet zones and areas.*

4.11 ISOLATION AND PARTITIONING (MOATING)

Isolation and partitioning refers to the physical separation of components, circuits, and power planes from other functional devices, areas, and subsystems. Allowing RF currents to propagate to different parts of the board by radiated or conductive means can cause problems not only in terms of EMI compliance, but also with regard to functionality.

Isolation is created by an absence of copper on *all* planes of the board through use of a moat. Absence of copper is created using a wide separation, typically 0.050 inch (50 mils) minimum from one section to another. In other words, an isolated area is an island on the board, similar to a castle with a moat. Only those traces required for operation or interconnect can travel to this isolated area. The moat serves as a "keep out" zone for signals and traces that are unrelated to the moated area or its interface. Two methods exist to connect traces, and power and ground planes to this island. Method 1 uses isolation transformers, optical isolators, or common-mode data line filters to cross the moat. Method 2 uses a bridge in the moat. Isolation is also used to separate high-frequency bandwidth components from lower bandwidth circuits, in addition to maintaining low-EMI bandwidth I/O in terms of the RF spectrum propagating from I/O interconnects.

4.11.1 Method 1: Isolation

Method 1 involves use of an isolation transformer or optical isolator. An I/O area must be 100% isolated from the rest of the PCB. Only at the metal I/O connector is RF bonding to chassis ground performed, and then only through a low-impedance, high-quality securement path to ground. We want to keep chassis ground outside this isolated area. The use of bypass capacitors from shield ground (or braid) of the I/O cable to chassis ground is sometimes needed in place of a direct connection when required by the interface specification. *Shield ground* (or drain wire) refers to a discrete pin or wire in the interface connector that connects the internal drain wire of the external I/O cable to its mylar foil shield, also located internal to the cable.

Pigtails should not be used under any condition to connect the shell of the BNC connector to chassis ground or to any other ground system. Measurements are well documented showing a 40- to 50-dB difference between a pigtail and a 360° connection of the cable shield to the BNC connector shell in the 15- to 200-MHz region for RF emissions. In addition to improvement in reducing RF emissions, a greater level of ESD immunity is provided due to less lead inductance that is presented to the ESD event. For most applications, the recommendation is to connect the cable shield to the BNC connector shell in a 360° fashion. This backshell then mates with a bulkhead panel containing a solid metallic contact with chassis ground.

Common-mode data line filters may be used in conjunction with isolation transformers to extend common-mode rejection. Common-mode data line filters (usually toroidal in construction) may be used for both analog and digital signal applications. These filters minimize common-mode RF currents carried on the signal traces to the I/O section or cable. If power and ground are required in the isolated area (e.g., +5 VDC for a keyboard or mouse), the moat should be crossed with a ferrite bead-on-lead for the power trace and a single solid trace three times the width of the power trace for a return. Use of a common-mode torroid in the power and ground connections is also an acceptable method. The secondary short-circuit fuse (required for product safety) can be located on either side of the ferrite bead, if required. Sometimes, capacitive decoupling is required to remove digital noise from filtered I/O power. This optional decoupling capacitor can be located with one terminal of the capacitor to the filtered side of the ferrite bead (output side) and the other terminal to the isolated ground plane. The power filtering components can be located across the moat at the far outside edge of the board. Both power and ground trace should be routed adjacent to each other to minimize RF ground loops that can be developed between these two traces if located on opposite sides of the moat. This is shown in Fig. 4.22 [1].

4.11.2 Method 2: Bridging

Method 2 uses a bridge between a control section and an isolated area. A bridge is a break in the moat at only one location where signal traces, power, and ground cross the moat. This is illustrated in Fig. 4.23 [1]. Violation of the moat by any trace not associated

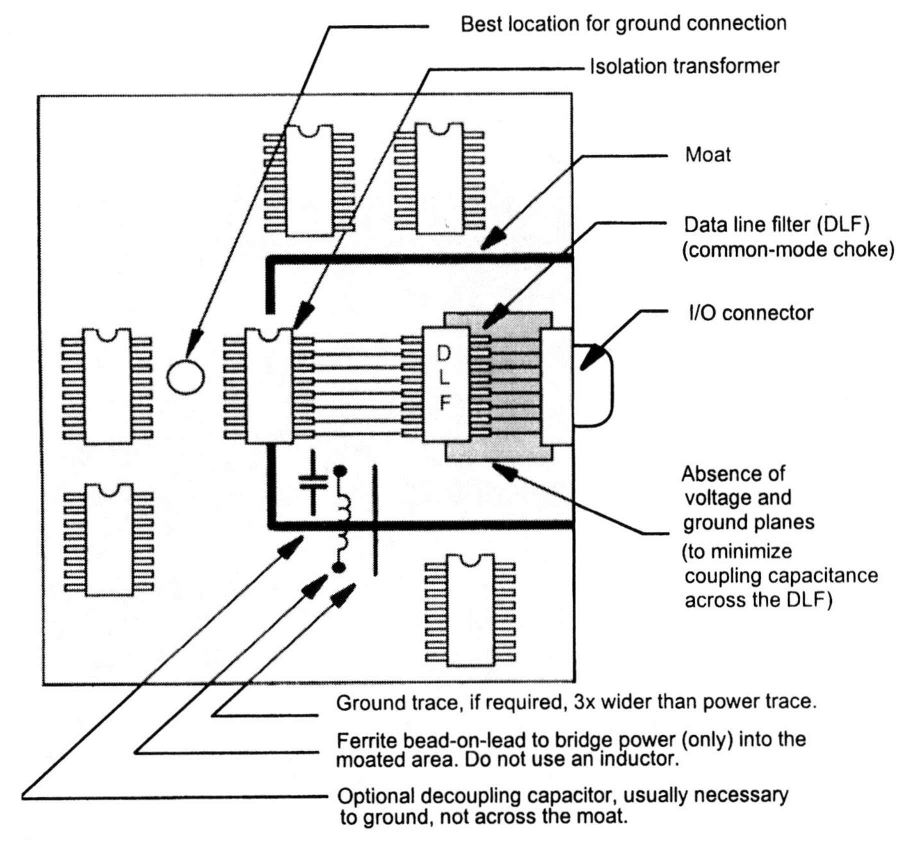

Figure 4.22 Using isolation in moating—Method 1.

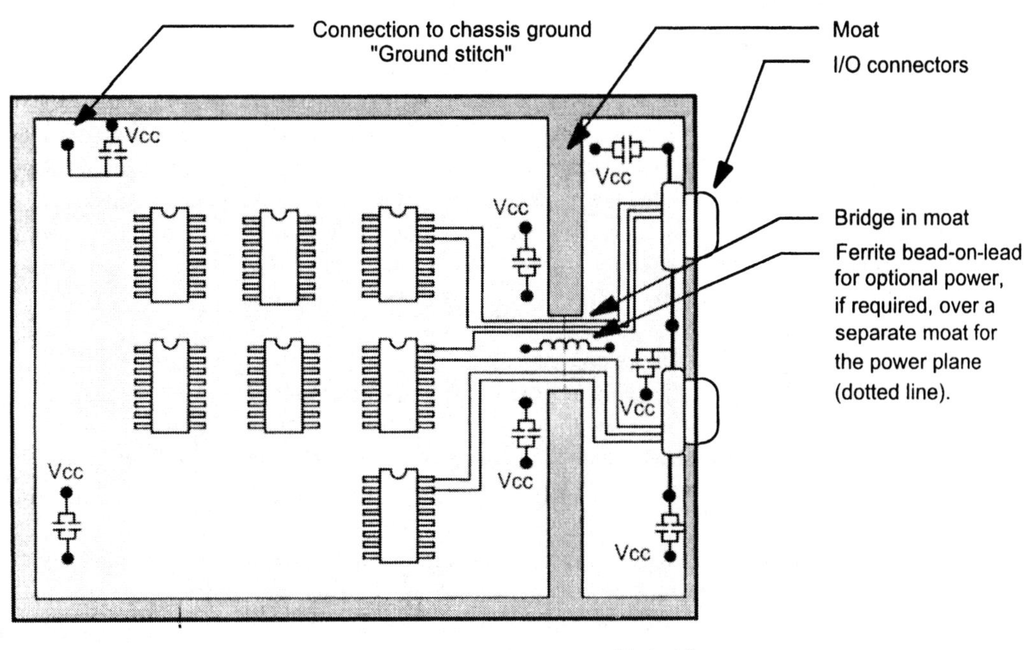

Figure 4.23 Bridging a moat—Method 2.

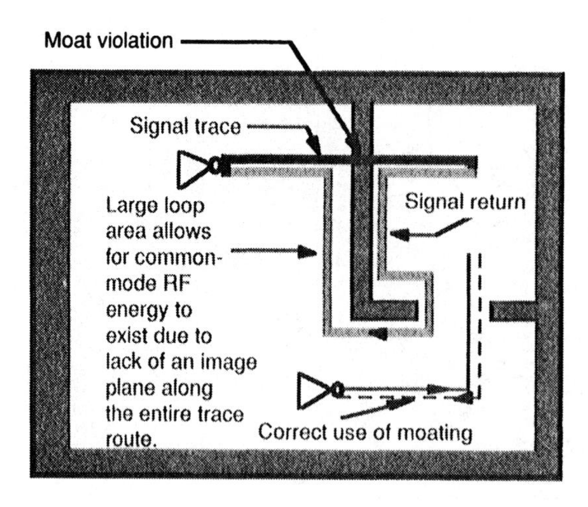

Figure 4.24 Violating the concept of moating.

with the I/O circuit can cause both emissions and immunity problems. RF loop currents will be created, detailed in Fig. 4.24 [1]. RF currents must image back along their trace route. Common-mode noise will be generated between the two separated areas. Unlike Method 1, power and ground planes are directly connected between the two areas; hence, this method forms a partition.

The advantage of using a bridge is similar to the castle concept with a moat. Only the signals that have a passport to cross the bridge will be allowed to pass. With the requirement and need for RF return currents to image back along the trace route, optimal flux cancellation (minimization) will occur. This one image return path is the only path that can be allowed to exist.

Sometimes, only the power plane is isolated, and the ground plane is fully connected through the bridge. This technique is common for circuits where a common ground plane is required, or separately filtered, where regulated power is needed. In this case, a ferrite bead-on-lead is typically used to bridge the moat for the filtered power only. This bead must be located in the bridge area and not over the moat. If analog or digital power is not required in the isolated area, this now unused power plane can be redefined as a second 0V (ground) plane referenced to the main ground plane. When a split plane partition is provided, one should guarantee that the traces that cross through the bridge do so along a solid 0V reference (ground) plane, and not against the split power plane.

When using bridging, grounding both ends of the bridge to chassis or frame ground is highly recommended if multipoint grounding is provided in the chassis and system-level design. Grounding the entrance to the bridge performs two functions:

1. It removes high-frequency common-mode RF components in the power distribution network (ground-noise voltage) from coupling into the partitioned area.
2. It helps remove eddy currents (for improved ground loop control) that may be present in the chassis or card cage. A much lower impedance path to ground is provided for RF currents that would otherwise find their way to chassis ground through other paths, such as RF currents in an I/O cable.

Grounding both ends of the bridge also increases electrostatic discharge immunity. If a high-energy pulse is injected into the I/O connector, this energy may travel to the main control area and cause permanent damage. This energy pulse must be sunk to chassis ground through a very low-impedance path.

Another reason to ground both sides of a bridge is to remove RF ground-noise voltage created by voltage gradients that appear between the partitioned area and main control section. If the RF common-mode noise contains high-frequency RF energy, decoupling capacitors for the RF energy (AC waveform) should be provided at each chassis ground stitch connection.

Figure 4.24 illustrates how traces are routed when using both digital and analog partitions. Since digital power plane switching noise may be injected into the analog section, isolation or filtering may be required. All traces that travel from the digital to analog section must be routed through the bridge. For analog power, a ferrite bead-on-lead should be used to cross the moat. A voltage regulator may also be required. The moat for analog power is usually 100% complete around the entire partition.

Certain analog components require analog ground to be referenced to digital ground but only through a bridge as shown in Fig. 4.25 [1]. Many analog-to-digital and digital-to-analog devices connect their analog ground (AGND) and digital ground (DGND) (indicated on the pin designation) together within the device package. When such is the application of a partition that is internal to the component, only one ground connection between analog and digital ground is required during PCB layout. AGND and DGND should be moated away from each other only when the circuit devices themselves provide separate AGND to DGND isolation *inside* the device package. It is important that the designer consult the recommendation made by the device manufacturer on how to properly isolate or connect AGND and DGND during layout.

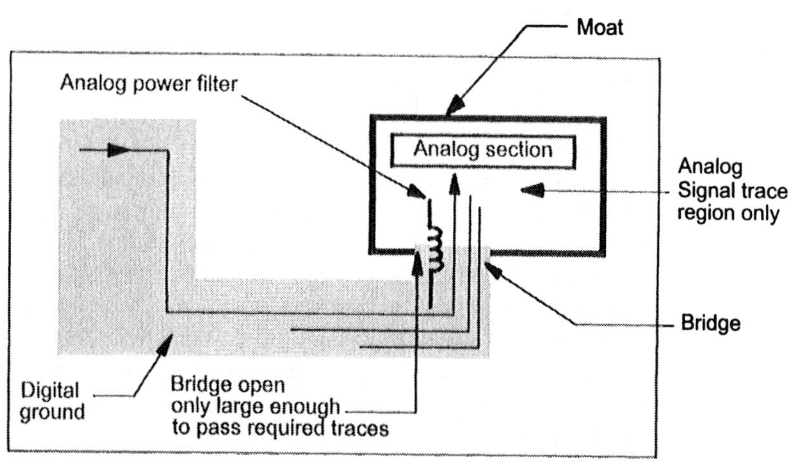

Note: All signal traces must pass through this region only (bridge).
No signals are to pass over a plane void region (moat).
For the analog section, the power plane is 100% moated.
If a bridge is used for ground, both digital and analog ground will be at the same potential.

Figure 4.25 Concept of digital and analog partitioning.

4.12 INTERCONNECTS AND RF RETURN CURRENTS

When designing a product with interconnects, either internal or external, a decision must be made concerning how to create a single system that is compatible with all operational subassemblies. It is preferable to have a single system (PCB) rather than several smaller PCBs interconnected by cables assemblies that are not referenced to each other. Having a common reference will limit the voltage drops that are developed between various 0V reference structures (grounds). It is easier to limit this voltage drop when all ground references are located on the same PCB assembly than if they are placed on separate PCBs and interconnected by cables that are highly inductive by the nature of their physical construction.

The impedance between two subsystems can be reduced through use of an image plane or a low-impedance ground grid structure. All ground structures must be connected together in as many locations as is physically possible. The more ground connections, be it through vias, cable connectors, ground stitch connections, and the like, will allow RF return currents to be controlled from the assembly. With less RF return current within the interconnect structure, less RF emissions will be present.

When using cables to interconnect PCBs or peripheral devices to another PCB, creating a low-impedance path to ground becomes difficult because it is difficult to intersperse ground wires, or ground pins in an optimal position within the connector. This is especially true if the connector pinout assignment is random without consideration for the RF return current path during the design cycle. Use of a predefined bus structure may present functionality concerns because the number of signal traces provided may require the majority of available pin connections, which may also significantly far exceed an equal number of RF returns or ground pins. The summation of all RF return currents in a single ground return pin, with many I/O or signal pins, may cause the circuit to be nonfunctional related to EMI (radiated emissions) because of excessive current and ground bounce across the single ground pin.

A typical I/O configuration is shown in Fig. 4.26. Notice that because of a poor pinout assignment, large RF loop currents will occur between power and ground, or between a signal trace (with or without a periodic clock signal) and a 0V RF current return path. High-frequency RF voltages may be developed between the main PCB and interconnect area. These high-frequency voltages can create common-mode currents flowing between the assemblies. These currents accentuate both radiated and conducted emissions. If all components are located on the same PCB assembly, common-mode currents between interconnects must not exist.

For the routing configuration identified as "poor pinout configuration," Fig. 4.26, the clock trace shown is positioned in a poor location within the interconnect assembly; it is not surrounded by a 0V reference (ground). The RF return currents that are created on the clock trace, in addition to the switching currents that exist within the power distribution network, must travel to the opposite side of the connector in its attempt to return to its source (complete the circuit). With a large loop area, a magnetic field is created. This magnetic field creates an electric field that may be observed during compliance testing.

Under the enhanced pinout configuration, optimal pinout design for RF return currents exists, thus minimizing loop areas. The clock trace is routed in a stripline configura-

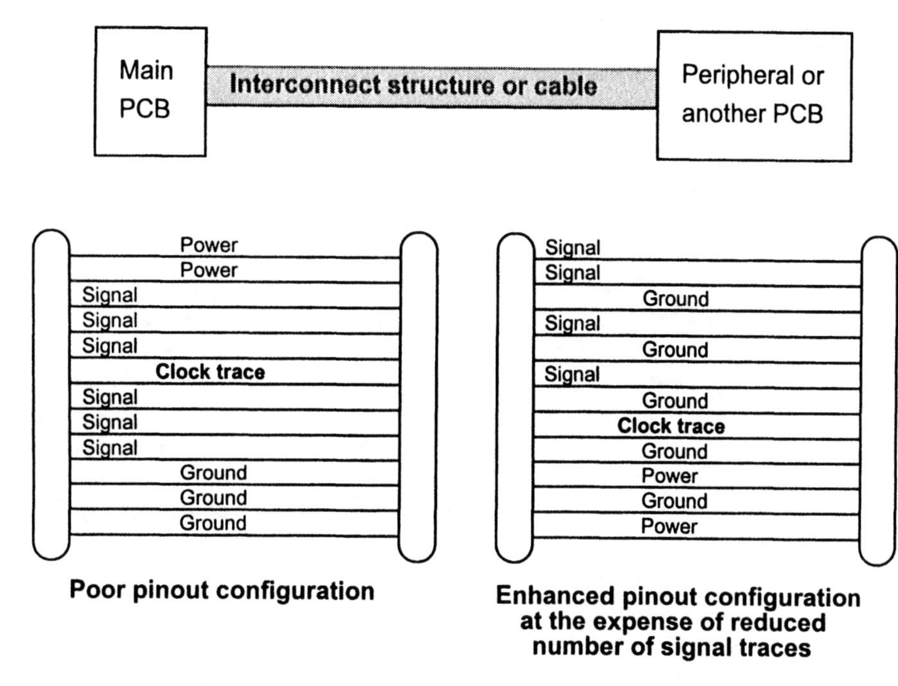

Figure 4.26 Pinout configuration typical of interconnects.

tion. Routing a clock trace stripline within an interconnect provides enhanced system performance and prevents development of RF currents losses. We also have a much lower impedance power distribution system to minimize ground bounce and the transference of high-frequency, high-bandwidth RF energy. This RF energy may be created with certain components injected into the power distribution system. This energy may also be transferred to other components sensitive to RF noise corruption by the power distribution network.

High-speed signals that travel between interconnects should be buffered at the entry point to reduce drive-capability and fan-out concerns, capacitive loading effects, and ground bounce when a signal crosses the interconnect barrier. A buffer reduces RF currents in the interconnect by a factor of four, in addition to reducing common-mode currents. An additional benefit of a buffer allows the source driver to consume less power when driving a loaded trace, especially if a high-impedance interconnect exists which already causes a large resistance-based drop to be developed.

When many nonperiodic signals are present within an interconnect structure, several signal traces may be contained or interspersed between 0V reference traces. The purpose of this configuration is to force the return currents on the interconnect to return to their source by the path of least impedance. In general, the concept is to design the PCB layout to force the RF return currents to travel the way we want them to travel within the PCB assembly. It becomes the designer's job to direct RF currents on the board in an optimal, low-impedance manner. It is not desirable to allow the RF return currents to travel any way they can within an assembly, for a random RF current return path will tend to maximize radiated emissions.

4.13 LAYOUT CONCERNS FOR SINGLE- AND DOUBLE-SIDED BOARDS

Special concerns exist for single- and double-sided PCB assemblies. With high-speed, high-technology products, use of single- and double-sided assemblies presents additional concerns relating to EMC compliance. These concerns are difficult to implement using specialized or advanced layout design techniques. For lower technology designs that are sensitive to cost, the use of single- or double-sided assemblies is frequently desirable. Consideration must be made for the RF return current to complete its return home to the source in an optimal, low-impedance manner.

One should think in terms of using transmission lines for both signal and power. Power and return lines must be routed parallel to each other as they are distributed to the component devices. Dedicated return traces should also be provided for high-threat traces, clocks, and so on to minimize loop structures that radiate and pick up electromagnetic energy. In double-sided boards, loop area control is the key to signal quality and EMI performance.

It is important to note, especially for EMC compliance, that there is *no such thing as a double-sided PCB*, although it physically exists. When analyzing how a double-sided PCB functions, related to EMC compliance, it should be noted that for a typical PCB the standard thickness of 0.062 inch (1.6 mm) is provided for the core material. The distance spacing between the top layer with components and a bottom layer with a ground plane or 0V reference structure is often assumed to provide an image plane for RF return currents created on the top layer. In reality, the distant spacing between the signal trace and image plane is so great that flux cancellation cannot occur efficiently. Flux cancellation cannot occur efficiently because of the lack of mutual partial inductance between the trace and return plane. The field distribution from a signal trace can be small, while the distance separation between trace and plane is extremely large.

The proper way to describe a double-sided PCB is to think of the board as two single-sided designs. We must route both the top and bottom layers of the PCB using design rules and techniques appropriate for single-sided designs.

For example, if the width of the trace is 0.008 inch (2 mm), the field distribution at a distance from the trace approaches 0.008 inch (2 mm). If a reference plane is greater than 0.008 inch (2 mm), then flux cancellation occurs with less efficiency and the RF return current can travel partly through free space. This distance spacing on a double-sided board is typically 0.062 inch (1.6 mm), which is much greater than 0.008 inch (2 mm). This is illustrated in Fig. 4.27.

What is the implementation of a return path for RF currents on single- or double-sided PCBs? We must remember that double-sided PCBs must be considered as two single-sided PCBs. This is difficult to achieve with full level of success. Examples are shown below. To allow for return currents, we must use ground traces (guard trace) or a gridded system at 0V potential. A ground trace or gridded system provides an alternate return path for RF currents. This alternate return path allows RF current to return to the source in a low-impedance manner, which is not an optimal implementation since a full return plane does not exist. For single-sided boards, ground traces are the primary design technique that allows RF currents to return to their source, thus controlling loop areas and, with that EMI.

For both single- and double-sided PCBs, plenty of local filtering or decoupling must occur for every device. Additional small high-frequency filtering to the critical signal

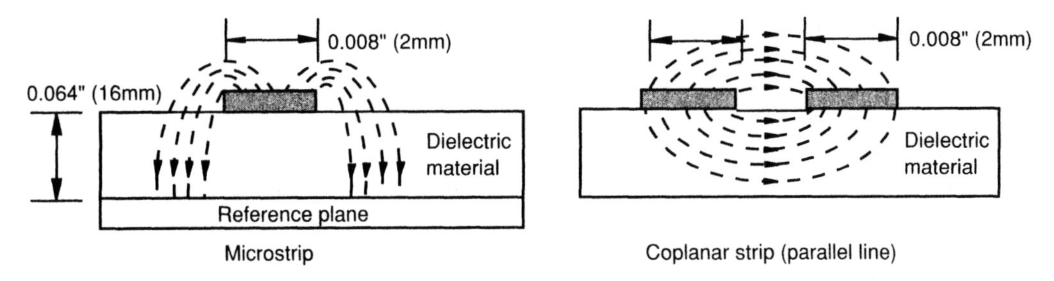

Figure 4.27 Field distribution for microstrip and coplanar strips. (*Source: Introduction to Electromagnetic Compatibility*, Clayton Paul © 1992 Reprinted by permission of John Wiley & Sons, Inc.)

lines must also occur directly at the component. We do not have the benefits of a ground plane; hence, different design techniques must be implemented.

4.13.1 Single-sided PCBs

For single-sided PCBs, there is only one conceptual design technique that provides for RF return currents. This technique is to use a ground trace (guard trace) that is placed as physically close to the high-threat signal trace as possible. This is shown in Fig. 4.28. The power and ground return traces must also be routed parallel to each other with decou-

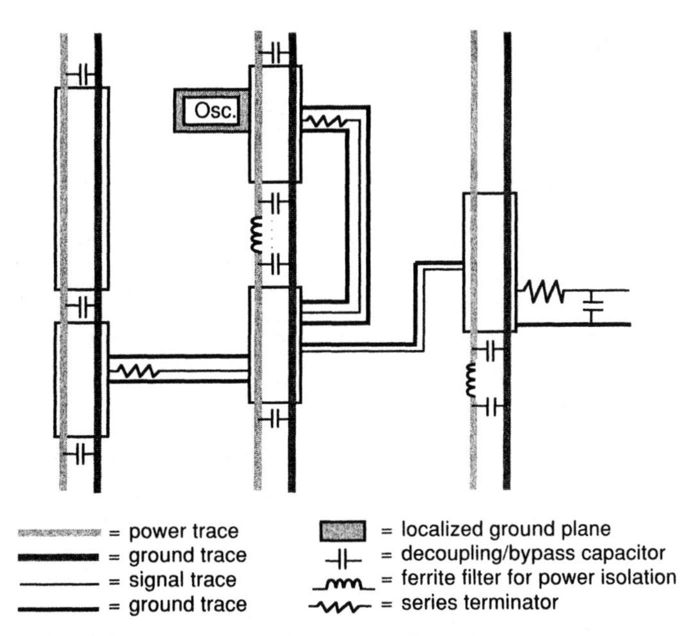

Figure 4.28 Single-sided routing for RF return currents.

pling capacitors provided for each and every component that injects switching energy into the power distribution system.

When a gridded power and ground layout methodology is provided, care must be taken to guarantee that the grids are tied together in as many places as possible. If a grid system is not used, RF loop currents from components may not find a low-impedance RF return path by any reliable means, thus exacerbating emission. By routing power and return traces together in parallel runs, a low-impedance, small loop area transmission line structure can be created, depending on how the parallel runs are implemented during layout. Signal traces referenced to the 0V structure can still create significant current loops if the distance spacing between the trace and 0V reference is excessive.

A problem with single-sided PCBs centers on how traces are routed between components when a power and ground grid exists. In almost every application, it becomes impractical to fully grid a single-sided board. The most optimal layout technique is to use ground fill to substitute as an alternate return path for loop area control and reduced impedance for RF return currents to travel home. This ground fill must be connected to the 0V reference point in as many places as possible.

4.13.2 Double-sided PCBs

There are two types of implementation for providing an alternate return path for RF currents.

1. Symmetrically placed components (e.g., memory arrays)
2. Asymmetrically placed components

4.13.3 Symmetrically Placed Components

There is one primary implementation technique for providing a low-impedance path for RF return currents for two-layer boards related to EMC compliance. The first is for older technology (slow speed components). These designs usually consist of Dual-In-Line Packages (DIPs) placed in a straight row or matrix configuration. Very few products currently use this technique or technology.

Routing horizontal traces on the solder side and vertical traces on the circuit side is the most commonly used technique for double-sided boards. This becomes a design rule that is usually not violated when using symmetrically placed components. The power trace is routed on the top (or bottom) layer, while the ground trace is routed on the opposite layer. All interconnects are made using plated through-holes. For areas that are not being used for either power, ground, or signal traces, ground fill must be used to aid in providing a low-impedance path to ground for RF return currents.

To summarize Fig. 4.29:

- Layer the power and ground in a grid style with the total loop area formed by each grid square not exceeding 1.5 square inches (3.8 square cm), although faster edge-times may demand smaller grids.
- Run power and circuit traces orthogonally to each other, power on one layer, ground on the other layer.
- Locate decoupling capacitors between the power and ground traces at all connectors and at each IC.

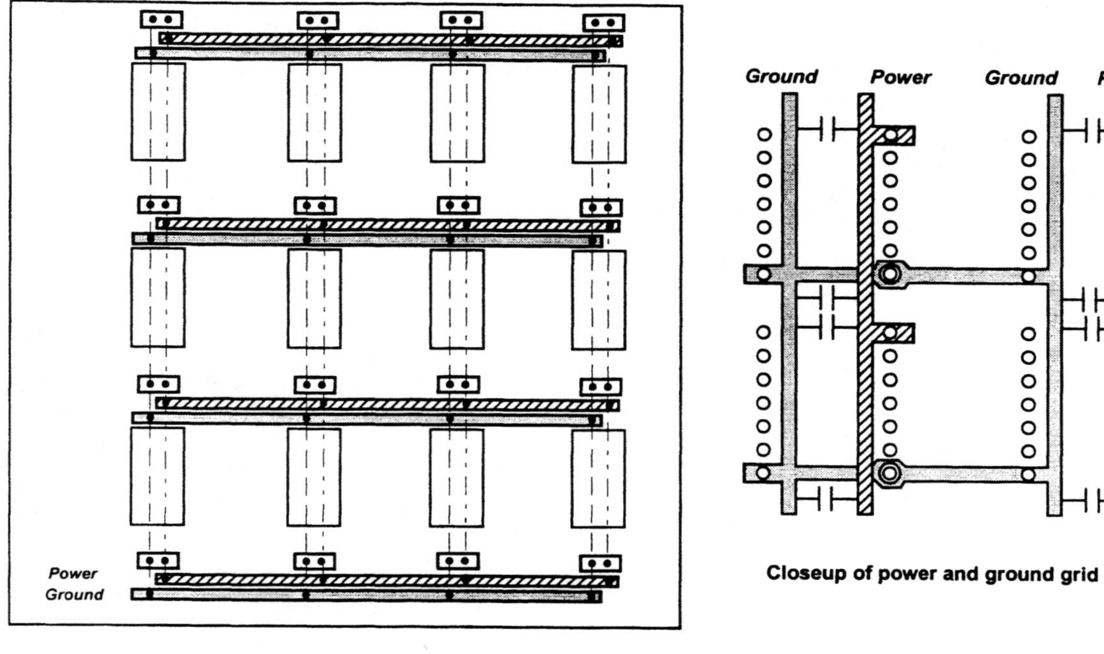

1. Top of board has all vertical traces.
2. Bottom of board has all horizontal traces.
3. Feedthroughs where power and ground traces intersect.
4. Decoupling capacitors between power and ground at connectors and at each IC.
5. Signal lines follow vertical/horizontal pattern.

Figure 4.29 Two-layer PCB with power and ground grid structure.

A power and ground grid system works because the grid structure provides a common return path for RF currents when an image plane is not present.

4.13.4 Asymmetrically Placed Components

Asymmetrically placed components are found in many current designs. This layout design is commonly used in low-frequency analog systems—less than 1 kHz and nearly all low-speed, older technology products.

- Route all power traces in a *radial fashion* from the power supply to all components on the same routing layer. Minimize the total length of all traces.
- Route all ground and power traces adjacent (parallel) to each other. This minimizes loop currents that may be created by high-frequency switching noise (internal to the components) from corrupting other circuits and control signals. Ideally, the only time these traces should be separated by a distance not greater than the width of any individual trace is when they must separate for connection to the decoupling capacitor. Signal flow should parallel these ground paths.
- Prevent loop currents by not tying different branches of a tree to another branch.

In examining Fig. 4.30, observe that at low frequencies, parasitic L and C generally do not cause problems as they do in high-frequency application. For this situation, single-point grounding is possible.

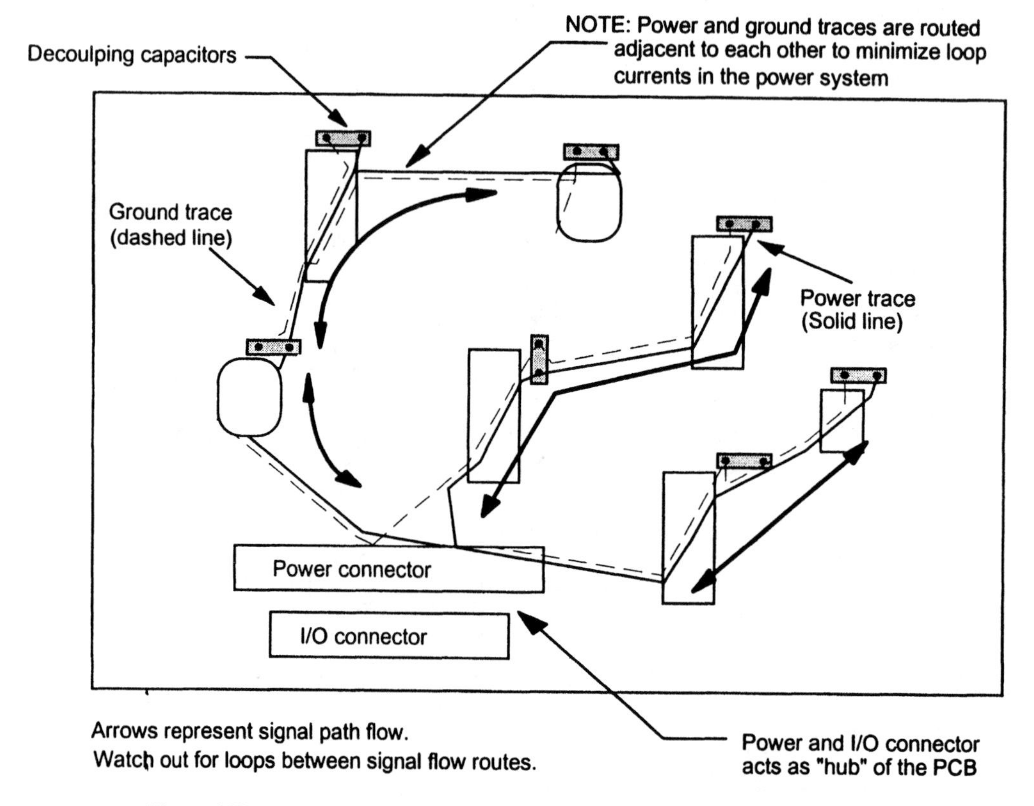

Figure 4.30 Two layer board with radial structure for power routing and flow migration.

In Fig. 4.30, the following is noted. For high-frequency applications, control the surface impedance *(Z)* of all signal traces and their return current path. When used in a low-frequency application, instead of impedance control, topology layout is a primary concern. Loop currents can be prevented from being created by not having components tied together.

4.14 GRIDDED GROUND SYSTEM

A gridded ground system is an effective method of reducing trace inductance and allows for an RF current return path to exist. This grid system can be incorporated within the design layout and is usually found on only single- or double-sided PCBs. When a multilayer structure is provided, a gridded ground system is not sufficient to provide significant control of RF currents as the image planes are more efficient for flux cancellation. A gridded ground system contains both horizontal and vertical ground paths on the PCB, shown in Fig. 4.31. A grid size spacing of 0.5 inch (1.27 cm) is typical, although larger spacing is acceptable depending on the edge rate of the signals and the complexity of the component layout. A generally accepted criterion for grid spacing is one-twentieth of a wavelength, based on the highest frequency that the grid is expected to handle. The main objective is to limit trace inductance by limiting the space between the grids and to make the ground grid and interconnecting elements as "fat" as possible.

A good rule of thumb is to use a grid size whose spacing allows a grid to exist between every IC on the board. This spacing provides an alternate RF return path when a ground plane cannot be implemented. This grid could exist on single-sided boards (extremely difficult, if not impossible); however, it is more optimal to implement a grid on a double-sided board. When using a double-sided stackup assignment, the *x*-axis traces are

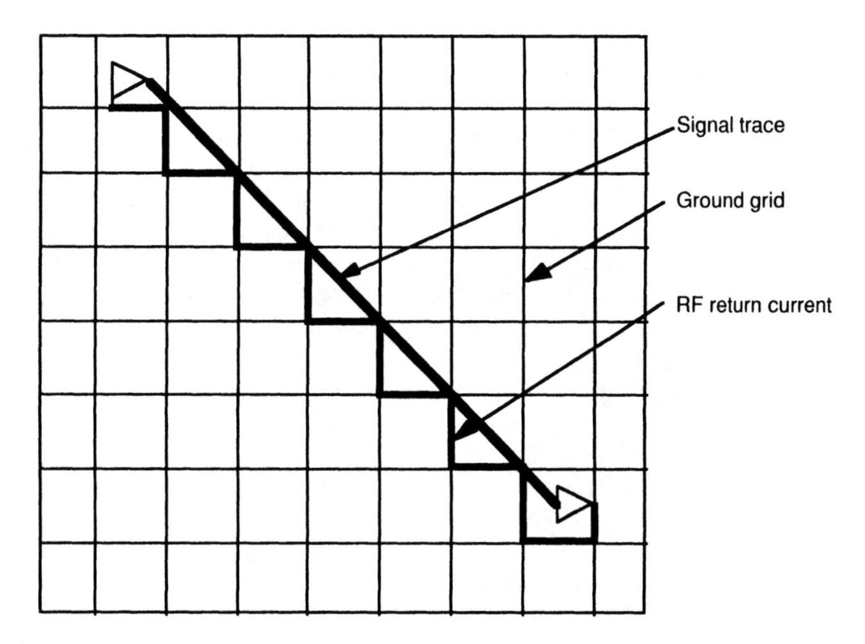

Figure 4.31 Gridded ground structure.

usually routed on the top layer, while *y*-axis traces are located on the bottom side. These traces are connected together by vias wherever they jump layers. This layer jumping allows ample room for necessary signal routing and interconnects. The ground grid on one side of the PCB is connected to the ground grid on the opposite side of the board using as many vias as possible.

The voltage developed across a return conductor is referred to as ground drop or ground bounce. The lower the value of ground drop between two points on a PCB return structure, the lower the radiated emissions from the PCB.

In studying Fig. 4.31, we notice that if the grid spacing was smaller than that provided, the RF return current would mirror image back closer to the signal trace with enhanced mutual partial inductance. Because there is no straight-line path, a convoluted loop area is created which enhances creation of RF energy.

To optimize the design and layout of a PCB using a gridded ground structure, it is imperative to design the grid structure before component placement occurs or signal traces are routed. It becomes difficult to implement a gridded structure after the board is routed. This grid adds no per-unit cost to the product. For single- and double- sided boards, this grid structure may be the only noise suppression technique possible.

A common question regarding the grid structure is, "How wide do I make the traces"? The optimal answer is, "as wide as possible." In reality, the grid can be made with a narrow conductor, since the impedance of the traces are added together in parallel, thus creating a total low-impedance return path. This impedance will still be higher compared to that of an image plane. The only design concern is to guarantee that the width of the traces can handle the 0V return current (from the power supply, not RF return current). Note that a grid developed from narrow traces is preferred to not having a grid at all.

4.15 LOCALIZED GROUND PLANES

The following layout techniques allows for the capture of RF flux generated internal to components and oscillators. This design concept is called a localized ground plane, and it forms a part of the partition concept.

Oscillators, crystals, and all clock support circuitry (e.g., buffers, drivers, etc.) can be located over a single localized ground plane. This localized ground plane is on the component (top) layer of the PCB and ties directly into the main internal ground planes of the PCB through both the oscillator ground pin and a minimum of two additional ground vias. This ground plane should also be positioned next to and connected to a ground stitch location. An example of this localized ground plane is shown in Fig. 4.32 [1].

The following are the main reasons for placing a localized ground plane under the clock generation area:

- Circuitry inside the oscillator demands RF currents. If the oscillator package is a metal can, the DC power pin is relied on for both DC voltage reference and a path for RF currents to be sourced (or sunk) to ground from the oscillator circuitry. Depending on the type of oscillator chosen (CMOS, TTL, ECL, etc.), RF currents created internal to the package can become so excessive that the ground pin is unable to efficiently source this large $L dI/dt$ current with low loss (L from

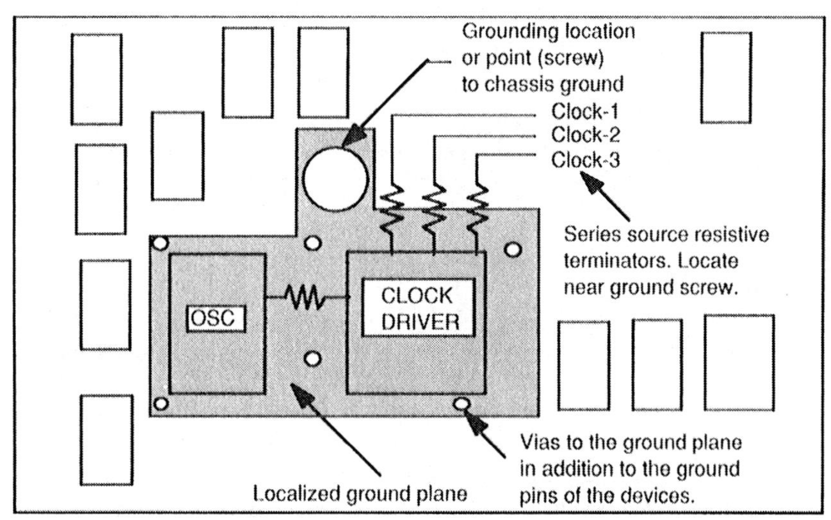

Note 1: Do not run any traces on layer 1 through the localized ground plane.
Note 2: If two microstrip layers exist, do not route any traces on layer two of the localized ground plane (route keep-out area).
Note 3: The localized ground plane is a solid copper plane without solder mask bonded to the main ground plane(s) by vias "and" bonded to the ground stitch location by a screw or equivalent method.

Figure 4.32 Localized ground plane.

the pin lead) to ground. As a result, the metal case becomes a monotonic antenna. The nearest image or ground plane (internal to the PCB) is sometimes two or more layers away and is thus inefficient as a radiated coupling path for RF currents to ground.

- If the oscillator is a surface-mount device, the situation mentioned above is made worse because SMT packages are often plastic. RF currents created internal to the package can radiate to free space and couple to other components. The high impedance of the PCB material, relative to the impedance of the ground pin of the oscillator, prevents RF currents to be sourced to ground. SMT packages will always radiate more RF energy than a metalized case.
- Placing a localized ground plane under the oscillator and clock circuits provides an image plane that captures common-mode RF currents generated internal to the oscillator and related circuitry, thus minimizing RF emissions. This localized ground plane is also at RF hot potential. To contain differential-mode RF current that is also sourced to the localized ground plane, multiple connections to all system ground planes must be provided. Vias from the localized plane, on layer 1, to the ground planes internal to the board will provide this lower impedance path to ground. To enhance performance of this localized ground plane, clock generation circuits should also be located adjacent to a chassis ground (stitch) connection. Connect this localized ground plane to the plated through-hole, 360° connection, preferably not using a wagon wheel configuration. Ensure a low-impedance RF bonding connection to ground exist. Connection through traces to a ground stitch location can defeat a low-impedance connection. While thermal relief "wagon

wheel" connections usually are acceptable, they also degrade the performance of the low-impedance connection.

- When using a localized ground plane, *"do not run traces through this plane"!* This violates the functionality of an image plane. If a trace travels through a localized ground plane, the potential for small ground loops or discontinuities exists. These ground loops can generate problems in the higher frequency range. Why install a plane when you defeat its functional use by running traces through it severing its continuity?
- Support logic circuitry (clock drivers, buffers, etc.) must be located adjacent to the oscillator. Extend this localized ground plane to include this support circuitry. Generally, an oscillator drives a clock buffer. This buffer is usually a super-high-speed, fast edge rate device. Because of the functional characteristics of this driver, RF currents will be created at harmonics of the primary clock frequency. With a large voltage swing and drive current injected onto the signal trace, both common-mode and differential-mode RF currents will exist. These currents can cause functionality problems and possible noncompliance to EMC requirements.

4.15.1 Digital-to-Analog Partitioning

Concerns exist for proper partitioning of digital-to-analog circuits, components, and functional subsections. Because of the application of the circuit partition, and how the component manufacturer designed the silicon substrate, a common ground reference structure may or may not be required. If the vendor designed its component for filtered analog power using a common digital/analog ground, then it is only necessary to filter the power plane or power pin.

If the component has designed into the silicon a separate partition for a distinct digital ground and distinct analog ground, the component itself may be partitioned in the layout on the PCB depending on the transfer characteristics across the device's silicon partition. This partition is achieved through use of moating between the component pins. All analog discrete components must reside within the analog section as detailed in Fig. 4.33.

Within Fig. 4.33, we have two configurations, both with a localized ground plane and moating within the multilayer stackup. We can extrapolate this localized ground plane to be both a formal power and ground plane within a multilayer stackup assembly. The only difference between the localized ground plane and the internal ground plane is that the localized plane is on the top first layer directly under the component or oscillator.

While constructing the PCB during component placement and the partitioning implementation stage, we may sometimes create a very convoluted shape that zigzags between the pins of a component if the component manufacturer did not provide an optimal pinout configuration that allows for ease of digital-to-analog partitioning.

For both design applications, the analog power input to the device is filtered with a ferrite bead-on-lead and capacitors. The filtered side of the bead is located within the "quiet" analog plane or localized ground plane. The output of the analog device, if required, is filtered with an appropriate device. This device may be another ferrite device or an inductor, based on functional application and use.

If partitioning is performed on the power and ground plane structure, the moat must occur on all planes present within the board stackup assignment. It becomes critical that overlapping planes do not occur as was detailed in Fig. 4.19. It is imperative to prevent

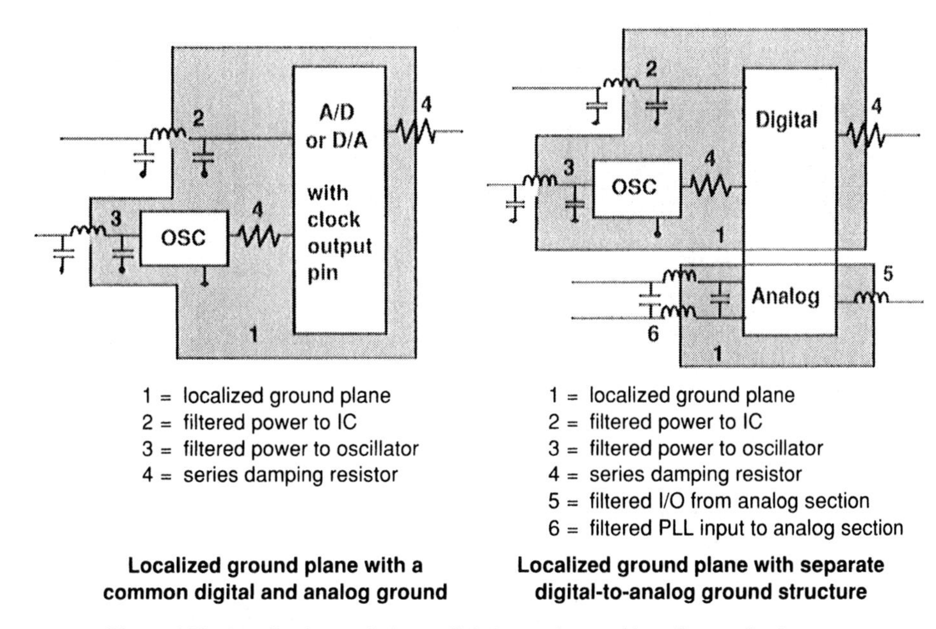

Figure 4.33 Localized ground plane—digital-to-analog partition. (*Source: Designers Guide to Electromagnetic Compatibility*, EDN. © 1994. Cahners Publishing Co. Reprinted with permission.)

capacitive coupling between noisy digital planes and quiet analog planes, thus circumventing possible resonances from developing common-mode noise.

4.16 SUMMARY

An image plane is a term commonly used to identify a return path for RF currents to complete their journey home. This plane consists of a solid copper sheet laminated within a multilayer PCB stackup assignment. An image plane provides a low-impedance RF transmission path for magnetic lines of flux to mirror image themselves against their source transmission line. The closer the distance spacing between source and return path, the more enhanced flux cancellation becomes. Higher density PCB stackups provide approximately six to eight dB of RF suppression per image plane pair due to enhanced flux cancellation.

Benefits of Multilayer Boards

- One or more planes can be dedicated exclusively to power and ground. The principal benefit is due to the presence of the first solid plane.
- A well-decoupled power distribution system exists.
- Circuit loop areas are reduced, thereby reducing differential-mode radiated emissions and susceptibility. Reduction of differential-mode currents will keep common-mode RF energy from being created.

- The signal and power return path (ground) will have minimal impedance levels.
- Characteristic impedance of traces is maintained throughout a trace route.
- Crosstalk will be minimized between adjacent traces.

REFERENCES

[1] Montrose, M. I. 1996. *Printed Circuit Board Design Techniques for EMC Compliance*. Piscataway, NJ: IEEE Press.

[2] Paul, C. R. 1992. *Introduction to Electromagnetic Compatibility*. New York: John Wiley & Sons, Inc.

[3] Gerke, D., and W. Kimmel. 1994. "The Designers Guide to Electromagnetic Compatibility." EDN (January 20).

[4] Ott, H. 1988. *Noise Reduction Techniques in Electronic Systems*. 2nd ed. New York: John Wiley & Sons.

[5] Mardiguian, M. 1992. *Controlling Radiated Emissions by Design*. New York: Van Nostrand Reinhold.

[6] Dockey, R. W., and R. F. German. 1993. "New Techniques for Reducing Printed Circuit Board Common-Mode Radiation." IEEE International Symposium on Electromagnetic Compatibility, pp. 334–339.

[7] German, R. F., H. Ott, and C. R. Paul. 1990. "Effect of an Image Plane on Printed Circuit Board Radiation." IEEE International Symposium on Electromagnetic Compatibility, pp. 284–291.

[8] Paul, C. R., K. White, and J. Fessler. 1992. "Effect of Image Plane Dimensions on Radiated Emissions." IEEE International Symposium on Electromagnetic Compatibility, pp. 106–111.

[9] Hubing, T. H., T. P. Van Doren, & J. L. Drewniak. 1994. "Identifying and Quantifying Printed Circuit Board Inductance." IEEE International Symposium on Electromagnetic Compatibility, pp. 205–208.

[10] Montrose, M. I. 1991. "Overview of Design Techniques for Printed Circuit Board Layout Used in High Technology Products." IEEE International Symposium on Electromagnetic Compatibility, pp. 284–291.

[11] Montrose, M. I. 1996. "Analysis on the Effectiveness of Image Planes Within a Printed Circuit Board." IEEE International Symposium on Electromagnetic Compatibility, pp. 326–332.

[12] Hsu, T. 1991. "The Validity of Using Image Plane Theory to Predict Printed Circuit Board Radiation." IEEE International Symposium on Electromagnetic Compatibility, pp. 58–60.

[13] Johnson, H. W., and M. Graham. 1993. *High Speed Digital Design*. Englewood Cliffs, NJ: Prentice Hall.

5

Bypassing and Decoupling

Bypassing and decoupling refers to preventing energy transference from one circuit to another in addition to enhancing the quality of the power distribution system. Three circuit areas are of primary concern: power and ground planes, components, and internal power connections.

Decoupling is a means of overcoming physical and time constraints caused by digital circuitry switching logic states. Digital logic usually involves two possible states, "0" or "1." Some conceptual devices may not be binary but ternary. The setting and detection of these two states is achieved with switches internal to the component that determines whether the device is to be at logic LOW or logic HIGH. There is a finite time period for the device to make this determination. Within this window, a margin of protection is provided to guarantee against false triggering. Moving the logic state near the trigger level creates a degree of uncertainty. If we add high-frequency noise, the degree of uncertainty increases and false triggering may occur.

Decoupling is also required to provide sufficient dynamic voltage and current for proper operation of components during clock or data transitions when all component signal pins switch simultaneously under maximum capacitive load. Decoupling is accomplished by ensuring a low-impedance power source is present in both circuit traces and power planes. Because decoupling capacitors have an increasingly low impedance at high frequencies up to the point of self-resonance, high-frequency noise is effectively diverted from the signal trace, while low-frequency RF energy remains relatively unaffected. Optimal implementation is achieved by using bulk, bypass, and decoupling capacitors. All capacitor values must be calculated for a specific function. In addition, we must properly select the dielectric material of the capacitor and not leave it to random choice from past usage or experience.

Three common uses of capacitors follow. Of course, a capacitor may also be used in other applications such as timing, wave shaping, integration, and filtering.

Decoupling. Removes RF energy injected into the power distribution network from high-frequency components consuming power at the speed the device is switching at. Decoupling capacitors also provides a localized source of DC power for devices and components, and is particularly useful in reducing peak current surges propagated across the board.

Bypassing. Diverts unwanted common-mode RF energy from components or cables. This is essential in creating an AC shunt to remove undesired energy from entering susceptible areas in addition to providing other functions of filtering (bandwidth limiting).

Bulk. Used to maintain constant DC voltage and current to components when all signal pins switch simultaneously under maximum capacitive load. It also prevents power dropout due to dI/dt current surges generated by components.

An ideal capacitor has no losses in its conductive plates and dielectric. Current is always present between the two parallel plates. Because of this current, an element of inductance is associated with the parallel plate configuration. Because one plate is charging while its adjacent counterpart is discharging, a mutual coupling factor is added to the overall inductance of the capacitor.

5.1 REVIEW OF RESONANCE

All capacitors consist of an *LCR* circuit where L = inductance related to lead length, R = resistance in the leads, and C = capacitance. A schematic representation of a capacitor is shown in Fig. 5.1. At a calculable frequency, the series combination of L and C becomes resonant, providing very low impedance and effective RF shunting at resonance. At frequencies above self-resonance, the impedance of the capacitor becomes increasingly inductive and bypassing or decoupling becomes less effective. Hence, bypassing and decoupling are affected by the lead-length inductance of the capacitor (including surface mount, radial, or axial styles), the trace length between the capacitor and a components, feed-through pads, and so forth.

Before discussing bypassing and decoupling of circuits on a PCB, a review of resonance is provided. Resonance occurs in a circuit when the reactive value difference between the inductive and capacitive vector is zero. This is equivalent to saying that the circuit is purely resistive in its response to AC voltage. Three types of resonance are common:

- Series resonance
- Parallel resonance
- Parallel C—series RL resonance

Resonant circuits are frequency selective since they pass more or less RF current at certain frequencies than at others. A series *LCR* circuit will pass the selected frequency (as measured across C) if R is high and the source resistance is low. If R is low and the source resistance is high, the circuit will reject the chosen frequency. A parallel resonant circuit placed in series with the load will reject the chosen frequency.

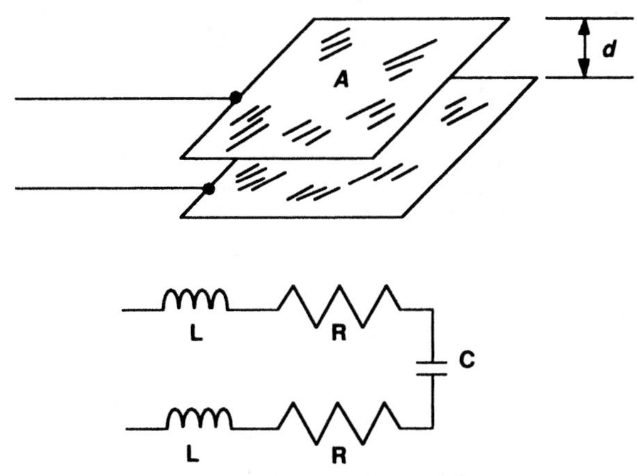

Figure 5.1 Physical characteristics of a capacitor with leads.

5.1.1 Series Resonance

The overall impedance of a series RLC circuit is $Z = \sqrt{R^2 + (X_L - X_c)^2}$. If an RLC circuit is to behave resistively, the value can be calculated as shown in Fig. 5.2 where ω *(2πf)* is known as the *resonant-angular frequency.*

With a series RLC circuit at resonance,

- Impedance is at minimum.
- Impedance equals resistance.
- The phase angle difference is zero.
- Current is at maximum.
- Power transfer (IV) is at maximum.

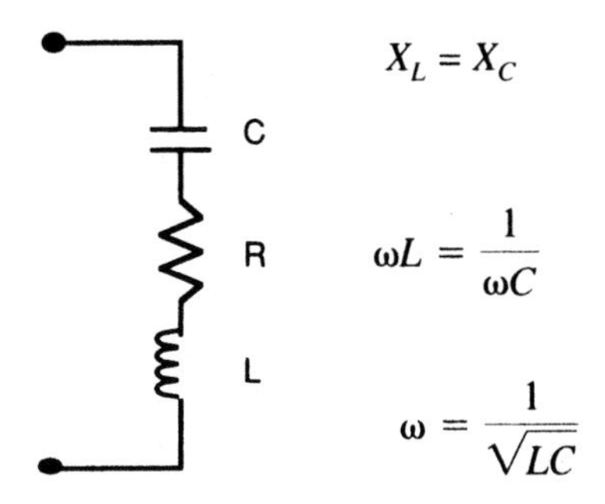

$$X_L = X_C$$

$$\omega L = \frac{1}{\omega C}$$

$$\omega = \frac{1}{\sqrt{LC}}$$

Figure 5.2 Series resonance.

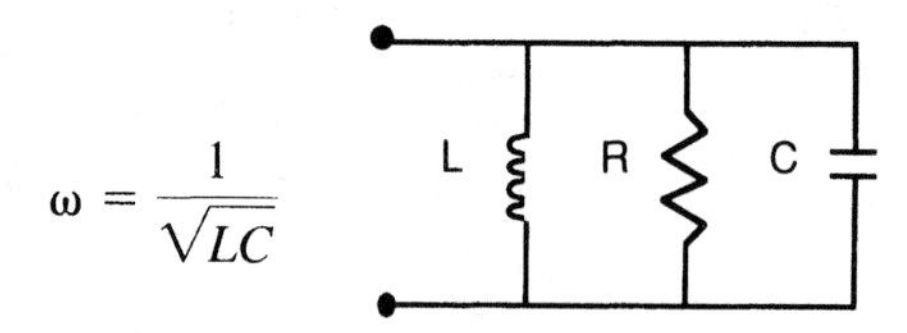

Figure 5.3 Parallel resonance.

5.1.2 Parallel Resonance

A parallel RLC circuit behaves as shown in Fig. 5.3. The resonant frequency is the same as for a series RLC circuit.

With a parallel RLC circuit at resonance,

- Impedance is at maximum.
- Impedance equals resistance.
- The phase angle difference is zero.
- Current is at minimum.
- Power transfer (IV) is at minimum.

5.1.3 Parallel C—Series RL Resonance (Antiresonant Circuit)

Practical resonant circuits generally consist of an inductor and variable capacitor in parallel. Since the inductor will possess some resistance, the equivalent circuit is shown in Fig. 5.4. The resistance in the inductive branch may be a discrete element or the internal resistance of a nonideal inductor.

At resonance, the capacitor and inductor trade the same stored energy on alternate half cycles. When the capacitor discharges, the inductor charges, and vice versa. At the antiresonant frequency, the tank circuit presents a high impedance to the primary circuit current, even though the current within the tank is high. Power is dissipated only in the resistive portion of the network.

The antiresonant circuit is equivalent to a parallel RLC circuit whose resistance is Q^2R.

$$w = \sqrt{\frac{1}{LC} - \left(\frac{R}{L}\right)^2} \approx \frac{1}{\sqrt{LC}} \quad [R << \omega_o L]$$

$$Q = \frac{1}{\omega_o CR} = \frac{X_C}{R} = \frac{X_L}{R}$$

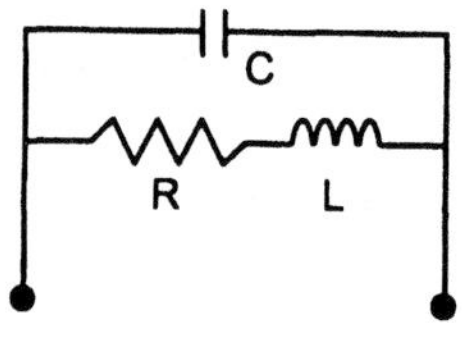

Figure 5.4 Parallel C—series RL resonance.

5.2 PHYSICAL CHARACTERISTICS

5.2.1 Impedance

The equivalent circuit of a capacitor was shown in Fig. 5.1. The impedance of this capacitor is expressed by

$$|Z| = \sqrt{R_s^2 + \left(2\pi fL - \frac{1}{2\pi fC}\right)^2} \tag{5.1}$$

where Z = impedance (Ω)
R_s = Equivalent Series Resistance—ESR (Ω)
L = Equivalent Series Inductance—ESL (H)
C = capacitance (F)
f = frequency (Hz)

From this equation, $|Z|$ exhibits its minimum value at a resonant frequency f_o such that

$$f_o = \frac{1}{2\pi\sqrt{LC}} \tag{5.2}$$

In reality, the impedance equation (Eq. 5.1) reflects hidden parasitics that are present when we take into account ESL and ESR.

Equivalent Series Resistance (ESR) is a term referring to resistive losses in a capacitor. This loss consists of the distributed plate resistance of the metal electrodes, the contact resistance between internal electrodes, and the external termination points. Note that skin effect at high frequencies increases this resistive value in the leads of the component. Thus, the high-frequency "ESR" is higher in equivalence than DC "ESR."

Equivalent Series Inductance (ESL) is the loss element that must be overcome as current flow is constricted within a device package. The tighter the restriction, the higher the current density and the higher the ESL. The ratio of width to length must be taken into consideration to minimize this parasitic element.

Examining Eq. (5.1), we have a variation of the same equation with ESR and ESL, shown in Eq. (5.3).

$$|Z| = \sqrt{(ESR)^2 + (X_{ESL} - X_C)^2}$$

where $X_{ESL} = 2\pi f\,(\text{ESL})$ (5.3)

$$X_c = \frac{1}{2\pi fC}$$

For certain types of capacitors with regard to dielectric material, the capacitance value varies with temperature and DC bias. Equivalent Series Resistance varies with temperature, DC bias, and frequency, while Equivalent Series Inductance remains fairly unchanged.

For an ideal planar capacitor where current uniformly enters from one side and exits from another side, inductance will be practically zero. For those cases, Z will approach R_s at high frequencies and will not exhibit an inherent resonance, which is exactly what a power and ground plane structure within a PCB does. This is best illustrated by Fig. 5.5.

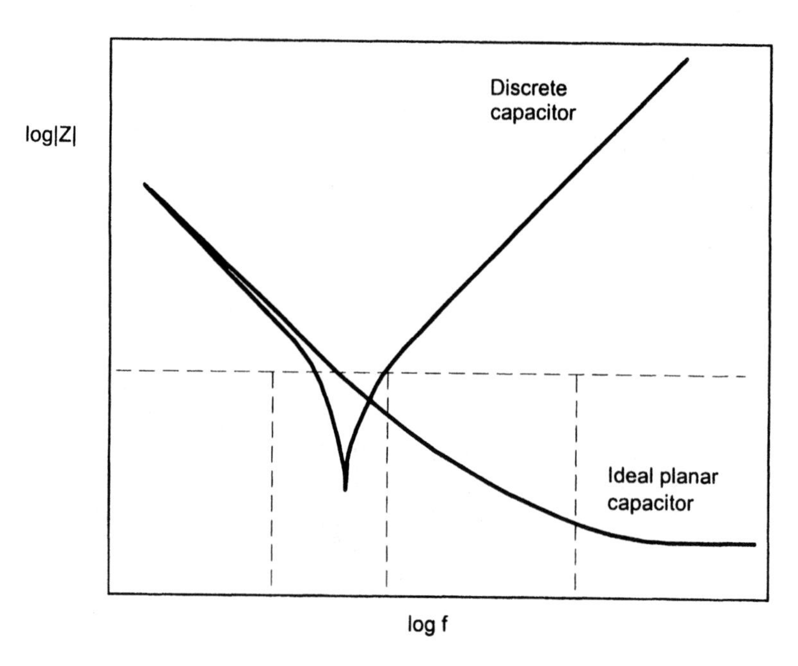

Figure 5.5 Theoretical impedance frequency response of ideal planar capacitors.

The impedance of an "ideal" capacitor decreases with frequency at a rate of –20 dB/decade. Because a capacitor has inductance in its leads, this inductance prevents the capacitor from behaving as desired, described by Eq. (5.2).

It should be noted that long power traces in two-sided boards that are not laid out for idealized flux cancellation are in effect, extensions of the lead lengths of the capacitor, and this fact seriously alters the self-resonance of the power distribution system.

Above self-resonance, the impedance of the capacitor becomes inductive and increases at +20 dB/decade as detailed in Fig. 5.6. Above the self-resonant frequency, the capacitor ceases to function as a capacitor. The magnitude of ESR is extremely small and, as such, does not significantly affect the self-resonant frequency of the capacitor.

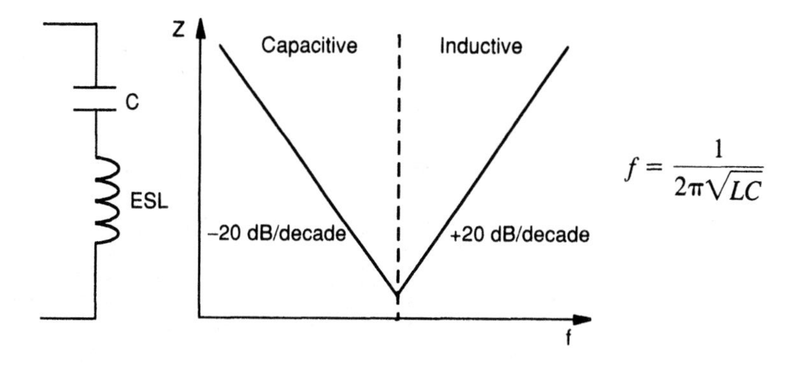

Figure 5.6 Effects of lead-length inductance within a capacitor.

The effectiveness of a capacitor in reducing power distribution noise at a particular frequency of interest is illustrated by Eq. (5.4)

$$\Delta V(f) = |Z(f)| \cdot \Delta I(f) \tag{5.4}$$

where ΔV is the allowed power supply sag; ΔI is the current supplied to the device; and f is the frequency of interest. To optimize the power distribution system by ensuring that noise does not exceed a desired tolerance limit, $/Z/$ must be less than $\Delta V/\Delta I$ for the required current supply. The maximum $/Z/$ should be estimated from the maximum ΔI required. If $\Delta I = 1A$, and $\Delta V = 3.3V$, the impedance of the capacitor must be less than 0.3 Ω.

In order for an ideal capacitor to work as desired, the device should have a high C in order to provide a low impedance at a desired frequency and a low L so that the impedance will not increase at higher frequencies. In addition, the capacitor must have a low R_s to obtain the least possible impedance. For this reason, power and ground planes structures are optimal in providing low-impedance decoupling within a PCB over discrete components.

5.2.2 Energy Storage

Decoupling capacitors ideally should be able to supply all the current necessary during a state transition of a logic device. This is described by Eq. (5.5). Use of decoupling capacitors on two-layer boards also reduces power supply ripple.

$$C = \frac{\Delta I}{\Delta V/\Delta t}$$

$$\text{that is,} \quad \frac{20\ ma}{100\ mv/5\ ns} = 0.001\ \mu f \quad \text{or} \quad 1000\ pf \tag{5.5}$$

where ΔI = current transient

ΔV = allowable power supply voltage change (ripple)

Δt = switching time

Note that for ΔV, EMI requirements are usually more demanding than chip supply needs.

The response of a decoupling capacitor is based on a sudden change in demand for current. It is useful to interpret the frequency domain impedance response in terms of the capacitor's ability to supply current. This charge transfer ability is also for the time domain function that the capacitor is generally selected for. The low-frequency impedance between the power and ground planes indicates how much voltage on the board will change when experiencing a relatively slow transient. This response is an indication of the time-average voltage swing experienced during a faster transient. With low impedance, more current is available to the components under a sudden change in voltage. High-frequency impedance is an indication of how much current the board can initially supply in response to a fast transient. Boards with the lowest impedance above 100 MHz can supply the greatest amount of current (for a given voltage change) during the first few nanoseconds of a sudden transient.

5.2.3 Resonance

When selecting bypass and decoupling capacitors, calculate the charge and discharge frequency of the capacitor based on logic family and clock speed used (self-resonant frequency). One must select a capacitance value based on the reactance that the capacitor presents to the circuit. A capacitor is capacitive up to its self-resonant frequency. Above self-resonance, the capacitor becomes inductive, which minimizes RF decoupling. Table 5.1 illustrates the self-resonant frequency of two types of ceramic capacitors, one with standard 0.25-inch leads and the other surface mount. The self-resonant frequency of SMT capacitors is always higher, although this benefit can be obviated by connection inductance. This higher self-resonant frequency is due to lower lead-length inductance provided by the smaller case package size and lack of long radial or axial lead lengths.

In performing SPICE testing or analysis on various package-size SMT capacitors, all with the same capacitive value, the self-resonant frequency changed by only a few MHz between package sizes, while keeping all other measurement constants unchanged. SMT package sizes of 1210, 0805, and 0603 are common in today's products using various types of dielectric material. Only the lead inductance is different between packaging with capacitance valve remaining constant. The dielectric material did not play a significant part in changing the self-resonant frequency of the capacitor. The change in self-resonant frequency observed between different package sizes, based on lead-length inductance in SMT packaging, was negligible and varied by ± 2–5 MHz.

When actual testing was performed in a laboratory environment on a large sample of capacitors, an interesting phenomenon was observed. The capacitors were self-resonant at the frequency analyzed, as expected. Based on a large sample size, the self-resonant frequency varied considerably. (There were too many plots to detail in this chapter or place within a table.) The self-resonant frequency varied because of the tolerance rating of the capacitor. Because of the manufacturing process, capacitors are provided with a tolerance rating of generally ±10%. More expensive capacitors are in the ± 2–5% range. Since the physical size of the capacitor is fixed, due to the manufacturing process used, the value of capacitance can change owing to the thickness and variation of the dielectric material and other parameters. With manufacturing tolerance for the capacitance part of the component, the actual self-resonant frequency will change based on the tolerance rating of the device. If a design requires an exact value of decoupling, the use of an expensive precision capacitor is required. The resonance equation easily illustrates this tolerance change.

TABLE 5.1 Approximate Self-resonant Frequencies of Capacitors (lead-length dependent)

Capacitor Value	Through-Hole* 0.25″ leads	Surface Mount** (0805)
1.0 μf	2.6 MHz	5 MHz
0.1 μf	8.2 MHz	16 MHz
0.01 μf	26 MHz	50 MHz
1000 pF	82 MHz	159 MHz
500 pF	116 MHz	225 MHz
100 pF	260 MHz	503 MHz
10 pF	821 MHz	1.6 GHz

*For through-hole, L = 3.75 nH (15 nH per/inch).

**For surface mount, L = 1 nH.

Leaded capacitors are nothing more than surface-mount devices with leads attached. A typical leaded capacitor has on the average approximately 2.5 nH of inductance for every 0.10 inch of lead length above the surface of the board. Surface-mount capacitors average 1 nH lead-length inductance.

An inductor does not change resonant response like a capacitor. Instead, the magnitude of impedance changes as the frequency changes. Parasitic capacitance around an inductor can, however, cause parallel resonance and alter response. The higher the frequency of the circuit, the greater the impedance. RF current traveling through an impedance causes an RF voltage. As a result, RF current is created in the device as related to Ohm's law, $V_{rf} = I_{rf} * Z_{rf}$. As examined above, one of the most important design concerns when using capacitors for decoupling lies in lead length inductance. SMT capacitors perform better at higher frequencies than radial or axial capacitors because of lower internal lead inductance. Table 5.2 shows the magnitude of impedance of a 15-nH inductor versus frequency. This inductance value is caused by the lead lengths of the capacitor and the method of placement of the capacitor on a typical PCB.

Figure 5.7 shows the self-resonant frequency of various capacitor values along with different logic families. It is observed that capacitors are capacitive until they approach self-resonance (null point) before going inductive. Above the point where capacitors go inductive, they proportionally cease to function for RF decoupling; however, they may still be the best source of charge for the device, even at frequencies where they are inductive. This is because the internal bond wire from the capacitor's plates to its mounting pad (or pin) must be taken into consideration. Inductance is what causes capacitors to become less useful at frequencies above self-resonance for decoupling purposes.

Certain logic families generate a greater spectrum of RF energy. This energy is generally higher in frequency than the self-resonant frequency range which a decoupling capacitor presents to the circuit. For example, a 0.1 μF capacitor will usually not decouple RF currents for an "ACT or F" logic device, whereas a 0.001 μF capacitor is a more appropriate choice due to the faster edge rate (0.8–2.0 ns minimum) typical of these higher-speed components.

TABLE 5.2 Magnitude of Impedance of a 15-nH Inductor versus Frequency

Frequency (MHz)	Z (ohms)
0.1	0.01
0.5	0.05
1.0	0.10
10.0	1.0
20.0	1.9
30.0	2.8
40.0	3.8
50.0	4.7
60.0	5.7
70.0	6.6
80.0	7.5
90.0	8.5
100.0	9.4

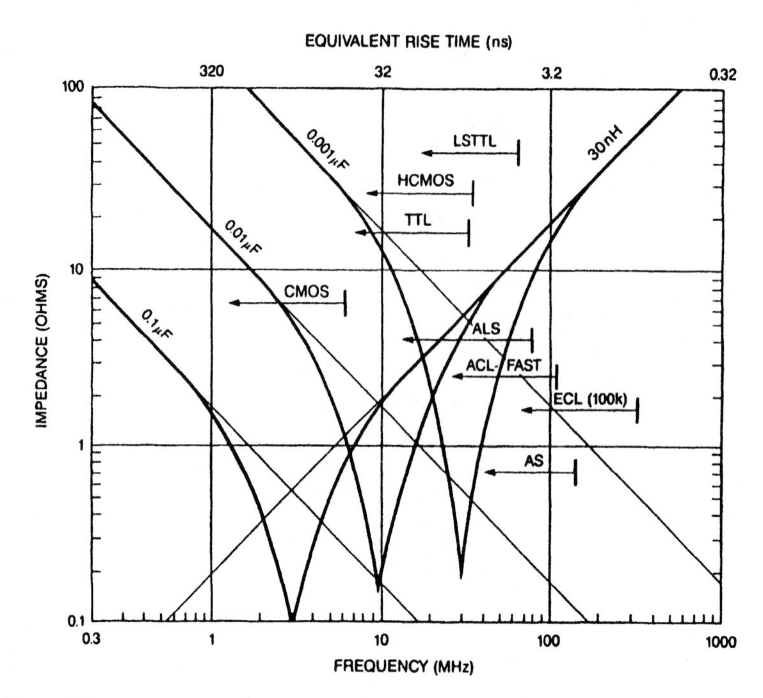

Figure 5.7 Self-resonant frequency of capacitors versus logic families. Capacitors provided with 30-nH series inductance (trace plus lead length). (Source: H. Ott, *Noise Reduction Techniques in Electronic Systems*. Copyright © 1988 John Wiley & Sons, reprinted with permission)

We now compare the difference between through-hole and surface-mount capacitors (SMT). Since SMT devices have much less lead-length inductance, the self-resonant frequency is higher than through-hole. Figure 5.8 illustrates a plot of the self-resonant frequency of various values of ceramic capacitors. All capacitors in the figure have the same lead-length inductance for comparison purposes.

Effective capacitive decoupling is achieved when capacitors are properly placed on the PCB. Random placement or excessive use of capacitors is a waste of material. Sometimes fewer capacitors strategically placed perform best for decoupling. In certain applications, two capacitors in parallel are required to provide greater spectral bandwidth of RF suppression. These parallel capacitors must differ by two orders of magnitude or value (e.g., 0.1 and 0.001 μF) or 100x for optimal performance. Use of parallel capacitors are discussed later in this chapter.

5.2.4 Benefits of Power and Ground Planes

A benefit of using multilayer PCBs is the placement of the power and ground planes adjacent to each other. The physical relationship of these two planes creates one large decoupling capacitor. This capacitor usually provides adequate decoupling for low-speed (slow edge rate) designs; however, additional layers add significant cost to the PCB. If components have signal edges (t_r or t_f) slower than 10 ns (e.g., standard TTL logic), use of high-

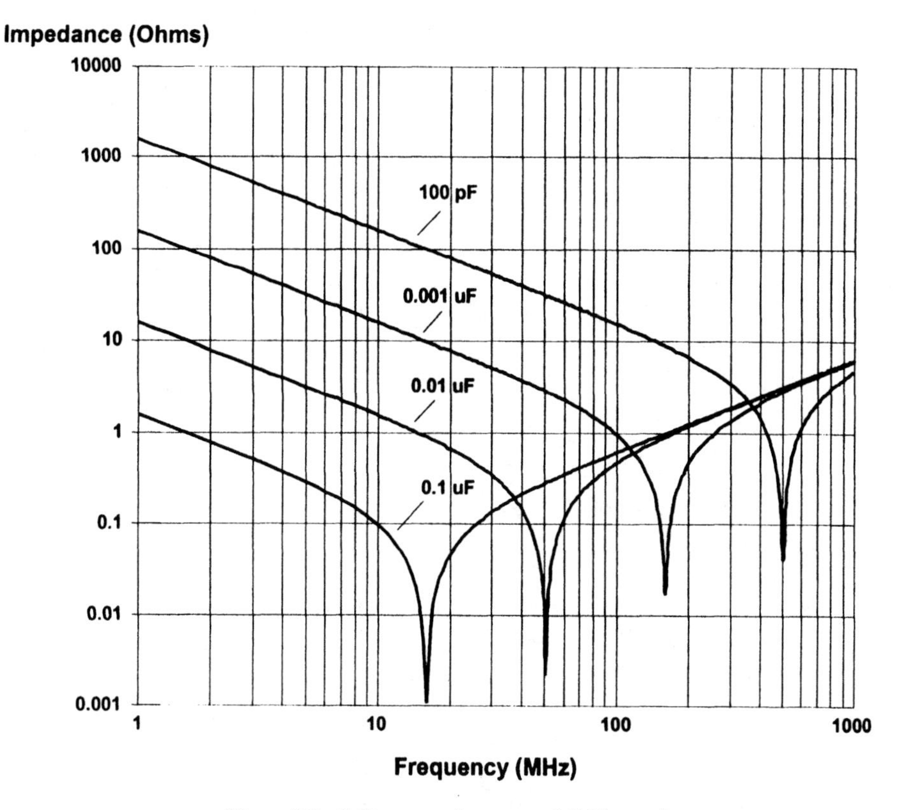

Figure 5.8 Self-resonant frequency of SMT capacitors. (ESL = 1nH)

performance, high self-resonant frequency decoupling capacitors is generally not required. Bulk capacitors are still needed, however, to maintain proper voltage levels. For performance reasons values such as 0.1 μF to 10 μF are appropriate for device power pins.

Another factor to consider when using power and ground planes as a primary decoupling capacitor is the self-resonant frequency of this built-in capacitor. If the self-resonant frequency of the power and ground planes is the same as the self-resonant frequency of the lumped total of the decoupling capacitors installed on the board, there will be a sharp resonance where these two frequencies meet. No longer will there be a wide spectral distribution of decoupling. If a clock harmonic is at the same frequency as this sharp resonance, the board will act as if little decoupling exists. When this situation occurs, the PCB may become an unintentional radiator with possible noncompliance with EMI requirements. Should this occur, additional decoupling capacitors (with a different self-resonant frequency) will be required to shift the resonance of the PCB's power and ground planes.

One simple method to change the self-resonant frequency of the power and ground planes is to change the distance spacing between these planes. Increasing or decreasing the height separation or relocation within the layer stackup will change the capacitance value of the assembly. Equations (5.7) and (5.8) provide this calculation. One disadvantage of using this technique is that the impedance of the signal routing layers may also change, which is a performance concern. Many multilayer PCBs generally have a self-resonant frequency between 200 and 400 MHz.

In the past, slower speed logic devices fell well below the spectrum of the self-resonant frequency of the PCB's power and ground planes. Logic devices used in newer, high-technology designs approach or exceed this critical resonant frequency. When both the impedance of the power planes and the decoupling capacitors approach the same resonant frequency, severe performance deterioration occurs. This degraded high-frequency impedance will result in serious EMI problems. Basically, the assembled PCB becomes an unintentional transmitter. The PCB is not really the transmitter; rather, the highly repetitive circuits or clocks are the cause of RF energy. Decoupling will not solve this type of problem (due to the resonance of the decoupling), requiring system-level containment measures to be employed.

5.3 CAPACITORS IN PARALLEL

It is common practice during a product design to make provisions for parallel decoupling of capacitors with the intent of providing greater spectral distribution of performance and minimizing ground bounce. Ground bounce is one cause of EMI created within a PCB. When parallel decoupling is provided, one must not forget that a third capacitor exists—the power and ground plane structure.

When DC power is consumed by components switching, a momentary surge occurs in the power distribution network. Decoupling provides a localized point source charge since a finite inductance exists within the power supply network. By keeping the voltage level at a stable reference point, false logic switching is prevented. Decoupling capacitors also minimize radiated emissions by providing a very small loop area for creating high spectral content switching currents instead of having a larger loop area created between the component and a remote power source.

Research on the effectiveness of multiple decoupling capacitors shows that parallel decoupling may not be significantly effective and that at high frequencies, only a 6-dB improvement may occur over the use of a single large-value capacitor.[1] Although 6-dB appears to be a small number for suppression of RF current, it may be all that is required to bring a noncompliant product into compliance with international EMI specifications. According to Paul,

> Above the self-resonant frequency of the larger value capacitor where its impedance increases with frequency (inductive), the impedance of the smaller capacitor is decreasing (capacitive). At some point, the impedance of the smaller value capacitor will be smaller than that of the larger value capacitor and will dominate thereby giving a smaller net impedance than that of the larger value capacitor alone.

[1]Paul, Clayton. "Effectiveness of Multiple Decoupling Capacitors," *IEEE Transactions on Electromagnetic Compatibility,* May 1992, vol. EMC-34, pp. 130–133.

This 6-dB improvement is basically the result of lower lead and device-body inductance provided by the capacitors in parallel. There are now two sets of parallel leads from the internal plates of the capacitors. These two sets provide greater trace width than would be available if only one set of leads were provided. With a wider trace width, there is less lead-length inductance. This reduced lead-length inductance is a significant reason why parallel decoupling capacitors work as well as they do.

Figure 5.9 shows a plot of two bypass capacitors, 0.01 μF and 100 pF, both individually and in parallel. The 0.01 μF capacitor has a self-resonant frequency at 14.85 MHz. The 100 pF capacitor has its self-resonant frequency at 148.5 MHz. At 110 MHz, there is a large increase in impedance due to the parallel combination. The 0.01 μF capacitor is inductive, while the 100 pF capacitor is still capacitive. We have both *L* and *C* in resonance—hence, an antiresonant frequency point, which is exactly what we do not want in a PCB if compliance to EMI requirements is mandatory.

> Between the self-resonant frequency of the larger value capacitor, 0.01 μF, and the self-resonant frequency of the smaller value capacitor, 100 pF, the impedance of the larger value capacitor is essentially inductive, whereas the impedance of the smaller value capacitor is capacitive. In this frequency range there exists a parallel resonant *LC* circuit and we should therefore expect to find an infinite impedance of the parallel combination. Around this resonant point, the impedance of the parallel combination is actually larger than the impedance of either isolated capacitor! [4, p.132]

In Fig. 5.9, observe that at 500 MHz, the impedances of the individual capacitors are virtually identical. The parallel impedance is only 6-dB lower. This 6-dB improvement is only valid over a limited frequency range from about 120 to 160 MHz.

To further examine what occurs when two capacitors are used in parallel, we look at a Bode plot of the impedance presented by two capacitors in parallel (see Fig. 5.10).

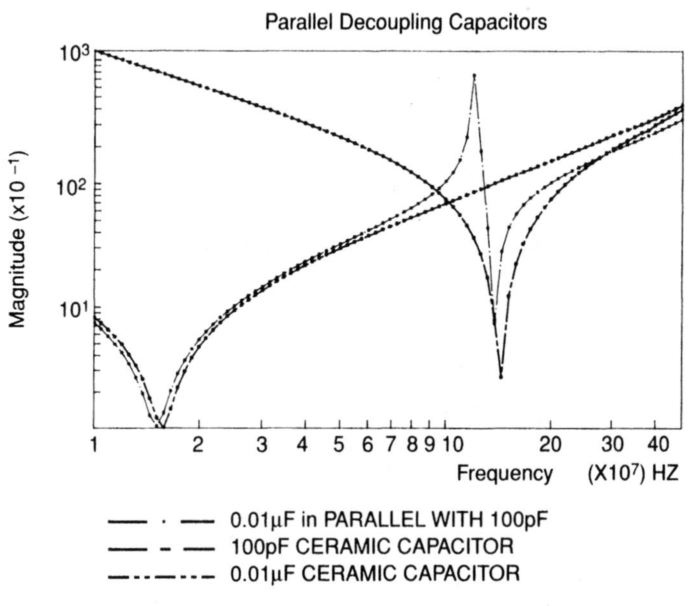

Figure 5.9 Resonance of parallel capacitors.

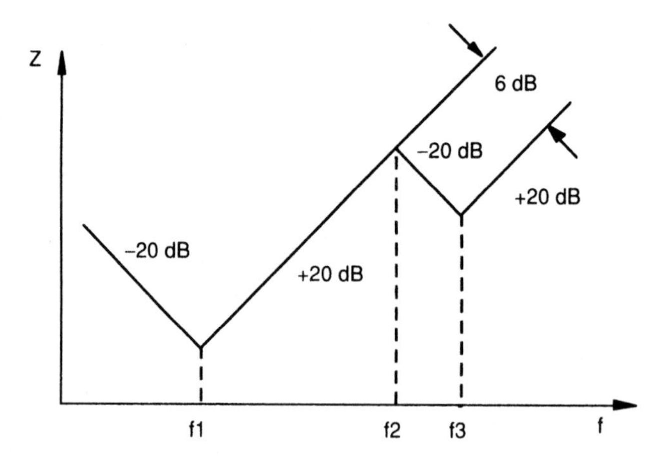

Figure 5.10 Bode plot of parallel capacitors.

For the bode plot of Fig. 5.10, the frequency responses of the magnitude at the various break frequencies are [4]

$$f_1 = \frac{1}{2\pi\sqrt{LC_1}} < f_2 = \frac{1}{2\pi\sqrt{LC_2}} < f_3 = \frac{1}{2\pi\sqrt{LC_3}} = 2f_2 \tag{5.6}$$

By shortening the lead-lengths of the larger value capacitor (0.01 μF), we can obtain the same results by a factor of 2. For this reason a single capacitor may be more optimal in a specific design than two, especially if minimal lead-length inductance exists.

To remove RF current generated by components switching all signal pins simultaneously (and it is desired to parallel decouple), it is common practice to place two capacitors in parallel (e.g., 0.1 μF and 0.001 μF) immediately adjacent to each power pin. If parallel decoupling is used within a PCB layout, one must be aware that the capacitance values should differ by two orders of magnitude, or 100x. The total capacitance of parallel capacitors is not important. Parallel reactance provided by the parallel capacitors (due to self-resonant frequency) is the important item. (See Tables 5.1 and 5.2.)

To optimize the effects of parallel bypassing and to allow use of only one capacitor, reduction in capacitor lead length inductance is required. A finite amount of lead length inductance will always exist when installing the capacitor on the PCB. Note that the lead length must also include the length of the via connecting the capacitor to the planes. The shorter the lead length from either single or parallel decoupling, the greater the performance. In addition, some manufacturers provide capacitors with significantly reduced "body" inductance internal to the capacitor.

5.4 POWER AND GROUND PLANE CAPACITANCE

The effects of the internal power and ground planes inside the PCB are not considered in Fig. 5.9. However, multiple bypassing effects are illustrated in Fig. 5.11. Power and ground planes have very little lead-length inductance equivalence and no ESR (Equiva-

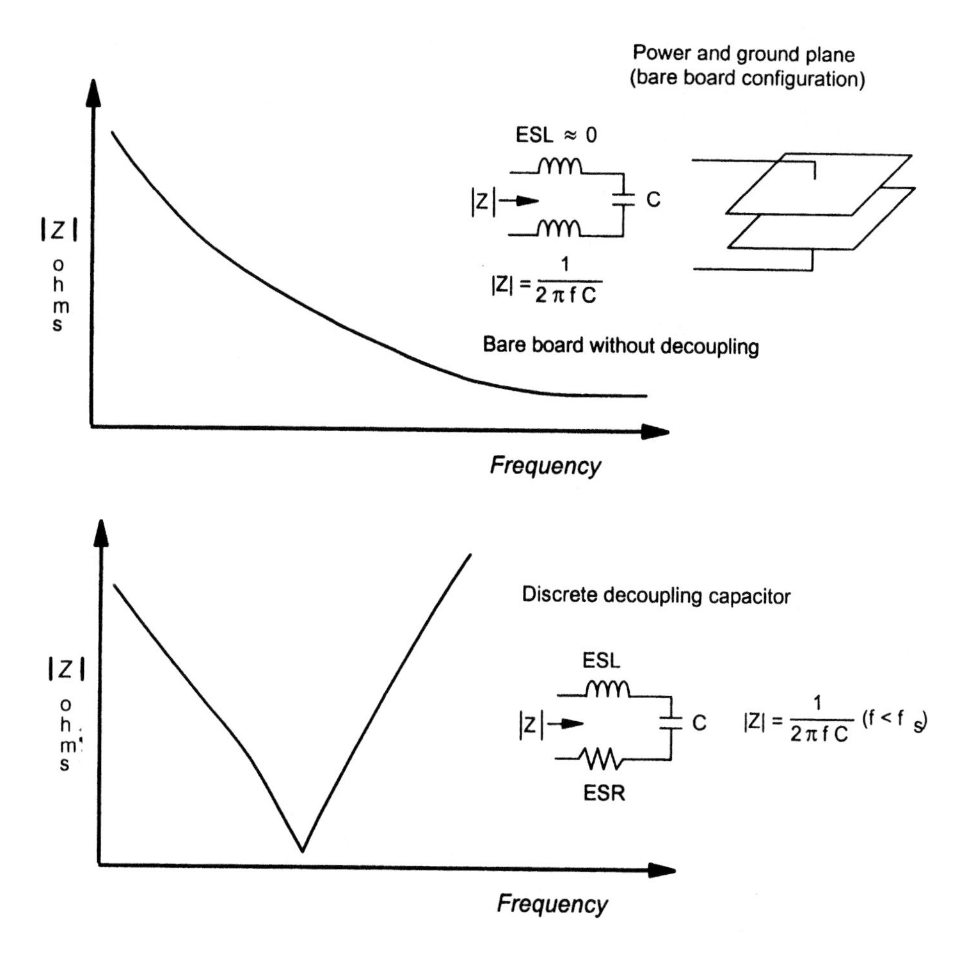

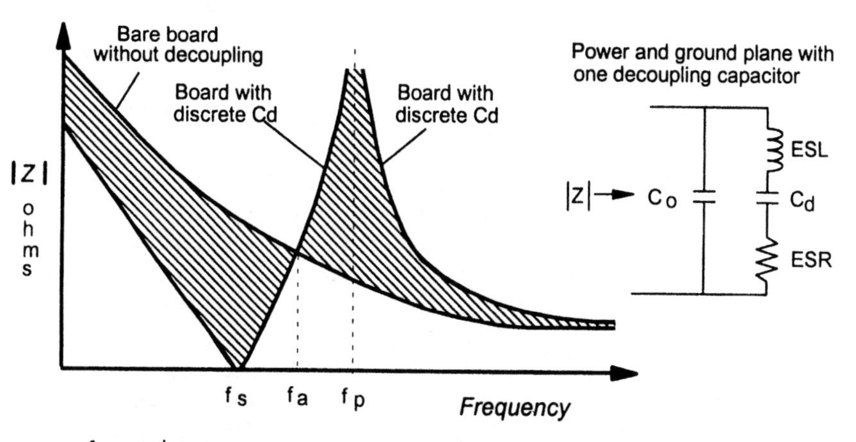

f_s = series resonance
f_p = parallel resonance
f_a = intersection of discrete capacitor and power plane capacitance

Figure 5.11 Decoupling effects of power and ground planes with discrete capacitors.

lent Series Resistance). Use of power planes as a decoupling capacitor reduces RF energy at frequencies generally in the higher frequency ranges.

On most multilayer boards, the maximum inductance of the planes between two components is significantly less than 1 nH. Conversely, lead-length inductance (e.g., the inductance associated with the traces connecting a component to its respective via plus the via themselves) is typically 2.5–10 nH or greater [8].

Capacitance will always be present between a voltage and ground plane pair. Depending on thickness of the core, the dielectric constant, and the placement of the planes within the board stackup, various values of internal capacitance can exist. Network analysis, mathematical calculations or modeling will reveal the actual capacitance of the power planes. This is in addition to determining the impedance of all circuit planes and the self-resonant frequency of the total assembly as potential RF radiators. This value of capacitance is easily estimated by Eqs. (5.7) and (5.8). This approximation may be used just to *estimate* the capacitance between planes since planes are finite, have multiple holes, vias, and the like. Actual capacitance is generally less than the calculated value.

$$C = \frac{\varepsilon_o \varepsilon_r A}{d} = \frac{\varepsilon A}{d} \tag{5.7}$$

where ε = permittivity of the medium between capacitor plates (F/m)
A = area of the parallel plates (m^2)
d = separation of the plates (m)
C = capacitance between the power and ground planes (pF)

Introducing relative permittivity ε_r of the dielectric material, and the value of the permittivity of free space, ε_o, we obtain the capacitance of the parallel-plate capacitor, namely, the power and ground plane combination.

$$C = 8.85 \frac{A\,\varepsilon_r}{d}\ (pF) \tag{5.8}$$

where ε_r = is the relative permittivity of the medium between the plates, typically ≈ 4.5 (varies for linear material, usually between 1 and 10)
and ε_o = permittivity of free space, $1/36\pi * 10^{-9}$ F/m = $8.85 * 10^{-12}$ F/m = 8.85 pF/m

Equations (5.7) and (5.8) show that the power and ground planes, when separated by 0.01 inch of FR-4 material, will have a capacitance of 100 pF/in^2.

Because discrete decoupling capacitors are common in multilayer PCBs, we must question the value of these capacitors when low-frequency, slow edge rate components are provided, generally in the frequency range below 25 MHz. Research into the effects of power and ground planes along with discrete capacitors reveals interesting results [6].

In Fig. 5.11, the impedance of the "bare board" closely approximates the ideal decoupling impedance that would result if only pure capacitance free of interconnect inductance and resistance could be added. This ideal impedance is given by $Z_c = 1/j\omega C_o$. The discrete capacitor becomes zero at the series resonant frequency, f_s, and infinite at the parallel resonance frequency, f_p, where n = number of discrete capacitors provided, C_d is the discrete capacitor, and C_o is the capacitance of the power and ground plane structure, conditioned by the source impedance of the power supply [6].

$$f_s = \frac{1}{2\pi\sqrt{LC}} \qquad f_p = f_s\sqrt{1 + \frac{nC_d}{C_o}} \tag{5.9}$$

For frequencies below the series resonance frequency, discrete decoupling capacitors behave as capacitors with an impedance of $Z = 1/j\omega C$. For frequencies near the series resonance frequency, the impedance of the loaded PCB is actually less than that of the ideal PCB. However, at frequencies above f_s, the decoupling capacitors begin to exhibit inductive behavior as a result of their associated interconnect inductance. Thus, the discrete decoupling capacitors function as inductors at frequencies above their series resonance frequency. The frequency at which the magnitude of the board impedance is the same with or without the decoupling capacitors (where the unloaded PCB intersects that of the loaded, nonideal PCB) is [6]

$$f_a = f_s\sqrt{1 + (nC_d / 2C_o)} \tag{5.10}$$

For frequencies above f_a, the additional number of "n" decoupling capacitors provides no additional benefit (as long as the switching frequencies of the components are within the decoupling range of the power and ground plane structure) since the bare board impedance remains far below that of the board that is loaded with discrete capacitors. At frequencies near the loaded board pole (parallel resonant) frequency, the magnitude of the loaded board impedance is extremely high, and the decoupling performance of the loaded board is far worse than that of the unloaded board (without additional decoupling capacitor). The analysis clearly indicates that minimizing the series inductance of the decoupling capacitor connection is crucial to achieving ideal capacitor behavior over the widest possible frequency range, which in the time domain corresponds to the ability to supply charge rapidly. Lowering the interconnect inductance increases the series and parallel-resonance frequency, thereby extending the range of ideal capacitor behavior [6].

Parallel resonances correspond to poles in the board impedance expression. Series resonances are null points. When multiple capacitors are provided, the poles and zeros will alternate so that there will be exactly one parallel resonance between each pair of series resonances. A parallel resonance will always exist between two series resonances.

Although good distributive capacitance exists when using a power and ground plane structure, adjacent close stacking of these two planes plays a critical part in the overall PCB assembly. If two sets of power and ground planes exist, for example, +5V/ground and +3.3V/ground, both with different dielectric spacing between the two planes, it is possible to have multiple decoupling capacitors built internal to the board. With proper selection of layer stackup, both high-frequency and low-frequency decoupling can be achieved without use of any discrete capacitors. To expand on this concept, a technology known as buried capacitance is finding use in extremely high-technology products that require high-frequency decoupling.

5.4.1 Buried Capacitance

Buried capacitance™ is a patented manufacturing process in which the power and ground planes are separated by a 0.001 inch (0.25 mm) dielectric.[2] With this small dielectric spacing, decoupling is effective up to 200–300 MHz. Above this frequency range, use

[2]Buried capacitance is a registered trademark of HADCO Corporation (which purchased Zycon Corporation, developers of this technology).

of discrete capacitors is required to decouple components that operate above the cutoff frequency of the buried capacitance. The important item to remember is that the closer the distance spacing is between the power and ground planes, the better the decoupling performance. It is to be remembered that, although buried capacitors may eliminate the employment and cost of discrete components, use of this technology may far exceed the cost of all discrete components that were removed.

To better understand the concept of buried capacitance, we should consider the power and ground planes as pure capacitance at low frequencies with very little inductance. These planes can be considered to be an equal-potential surface with no voltage gradient except for a small DC voltage drop. This capacitance is calculated simply as area divided by thickness times permittivity. For a 10-inch square board, with 2 mil FR-4 dielectric between the power and ground planes, we have 45 nF (0.045 μF).

At some frequency, a full wave will be observed between the power and ground planes along the edge length of the PCB. Assuming velocity of propagation to be 6 in/ns (15.24 cm/ns), we observe that the frequency will be 600 MHz for a 10 × 10 inch board. At this frequency, the planes are not at equal potential, for the voltage measured between two points can differ greatly as we move the test probe around the board. A reasonable transition frequency is one-tenth of 600 MHz or 60 MHz. Below this frequency, the planes can be considered as pure capacitance.

Knowing the velocity of propagation and capacitance per square area, we can calculate the plane inductance. For a 2-mil-thick dielectric, capacitance is 0.45 nF/inch^2, velocity of propagation = 6 inch/ns, and inductance is 0.062 nH/square. This inductance is a spreading inductance, similar in interpretation to spreading resistance, and is a very small number. This small number is the primary reason why power planes are mainly pure capacitance.

With inductance and capacitance, we calculate the impedance as $Z_o = \sqrt{L/C}$, which is 0.372 ohms-inch. A plane wave traveling down a long length of a 10-inch-wide board will see 0.372/10 = 0.0372 ohms impedance, again, a small number.

Decoupling capacitance is increased because the distance spacing between the planes *(d)* is in the denominator. Inductance is decreased because the velocity of propagation must remain constant and the total impedance is also decreased. The power and ground planes are the means of distributing power. Reducing the dielectric thickness is effective at increasing high-frequency decoupling capacitance and transporting high-frequency power through a lower impedance distribution system.

5.4.2 Calculating Power and Ground Plane Capacitance

Capacitance between a power and ground plane is described by

$$C_{pp} = k \frac{\varepsilon_r A}{d} \tag{5.11}$$

where C_{pp} = capacitance of parallel plates (pF)
ε_r = relative dielectric constant of the board material (vacuum = 1, FR4 material = 4.1 to 4.7)
A = common area between the parallel plates (square inches or cm)
d = distance spacing between the plates (inches or cm)
k = conversion constant (0.2249 for inches, 0.884 for cm)

One caveat in implementing this technology is that the inductance caused by the antipads (holes for through-vias) in the power and ground planes can minimize the theoretical effectiveness of this technique.

Because of the efficiency of the power planes as a decoupling capacitor, the use of high self-resonant frequency decoupling capacitors may not be required for standard TTL or slow-speed logic. This optimum efficiency exists, however, only when the power planes are closely spaced—less than 0.01 inch with 0.005 inch preferred for high-speed applications. If additional decoupling capacitors are not properly chosen, the power planes will go inductive below the lower cut-in range of the higher self-resonant frequency decoupling capacitor. With this gap in resonance, a pole is generated, causing undesirable effects on RF suppression. At this point, RF suppression techniques on the PCB become ineffective, and containment measures must be used at a much greater expense.

5.5 LEAD-LENGTH INDUCTANCE

All capacitors have lead and device body inductance. Vias also add to this inductance value. Lead inductance must be minimized at all times. When a signal trace plus lead-length inductance is combined, a higher impedance mismatch will be present between the component's ground pin and the system ground plane. With trace impedance mismatch, a voltage gradient is created between these two sources, creating RF currents. RF fields cause RF emissions on PCBs; hence, decoupling capacitors must be designed for minimum inductive lead length, including via and pin escapes (or pad connections from the component pin to the point where the device pin connects to a via).

In a capacitor, the dielectric material determines the magnitude of the zero for the self-resonant frequency of operation. All dielectric material is temperature sensitive. The capacitance value of the capacitor will change in relation to the ambient temperature provided to its case package. At certain temperatures, the capacitance may change substantially and may result in improper performance, or no performance at all when used as a bypass or decoupling element. The more temperature stable the dielectric material, the better performance of the capacitor.

In addition to the sensitivity of the dielectric material to temperature, the equivalent series inductance (ESL) and the equivalent series resistance (ESR) must be low at the desired frequency of operation. ESL acts like a parasitic inductor, whereas ESR acts like a parasitic resistor, both in series with the capacitor. ESL is not a major factor in today's small SMT capacitors. Radial and axial lead devices will always have large ESL values. Together, ESL and ESR degrade a capacitor's effectiveness as a bypass element. When selecting a capacitor, one should choose a capacitor family that publishes actual ESL and ESR values in their data sheet. Random selection of a standard capacitor may result in improper performance if ESL and ESR are too high. Most vendors of capacitors do not publish ESL and ESR values, so it is best to be aware of this selection parameter when choosing capacitors used in high-speed, high-technology PCBs.

Because surface-mount capacitors have essentially little ESL and ESR, their use is preferred over radial or axial types. Typically, ESL is <1.0 nH, and ESR should be 0.5 ohms or less. For decoupling capacitors, capacitance tolerance is not as important as the temperature stability, dielectric constant, ESL, ESR, and self-resonant frequency [1].

5.6 PLACEMENT

5.6.1 Power Planes

Multilayer PCBs generally contain one or more pairs of voltage and ground planes. Power planes function as a low-inductance capacitor that constrains RF currents generated from components and traces. Multiple chassis ground stitch connections to all ground planes minimizes voltage gradients between board and chassis and between/among board layers. These gradients also are a major source of common-mode RF fields. This is in addition to sourcing RF currents to chassis ground. In many cases, multiple ground stitch connections are not always possible, especially in card cage designs. In such situations, care must be taken to analyze and determine where RF loops will occur and how to optimize grounding of the power planes.

Power planes that are positioned next to ground planes provide for flux cancellation in addition to decoupling RF currents created from power fluctuations of components and noise injected into the power and ground planes. Components switching logic states cause a current surge during the transition. This current surge places a strain on the power distribution network. An image plane is a solid copper plane at voltage or ground potential located adjacent to a signal routing plane. RF currents generated by traces on the signal plane will mirror image themselves in this adjacent metal plane. This metal plane must not be isolated from the power distribution network [9]. To remove common-mode RF currents created within a PCB, all routing (signal) layers must be physically adjacent to the image plane. (Refer to Chapter 4 for a detailed discussion of image planes.)

5.6.2 Decoupling Capacitors

Before determining where to locate decoupling capacitors, the physical structure of a PCB must be understood. Figure 5.12 shows the electrical equivalent circuit of a PCB. In this figure, observe the loops that exist between power and ground caused by traces, IC wire bonds, lead frames of components, socket pins, component interconnect leads, and decoupling capacitor. The key to effective decoupling is to minimize R_2, L_2, R'_2, L'_2, R_3, L_3, R'_3, L'_3, R_4, L_4, R'_4, and L'_4. Placement of power and ground pins in the center of the component helps reduce R_4, L_4, R'_4, and L'_4. Basically, the impedance of the PCB structure must be minimized. The easiest way to minimize the resistive and inductive components of the PCB is to provide a solid plane. To minimize the inductance from the component, use of SMT, ball grid arrays, and flip chips is preferred. With less lead bond lengths from die to PCB pad, overall impedance is reduced.

Figure 5.12 [1] makes clear that EMI is a function of loop geometry and frequency, hence, the smallest closed-loop area is desired. We acquire this small area by placing a local decoupling capacitor, C_{pcb}, for current storage adjacent to the power pins of the IC. It is mandatory that the decoupling loop impedance be much lower than the rest of the power distribution system. This low impedance will cause the high-frequency RF energy developed by both traces and components to remain almost entirely within this small closed loop. As a result, low EMI emissions are observed.

If the impedance of the loop is smaller than the rest of the system, some fraction of the high-frequency RF component will transfer or couple to the larger loop formed by the power distribution system. With this situation, RF currents are developed and, hence, higher EMI emissions. This situation is best illustrated in Fig. 5.13.

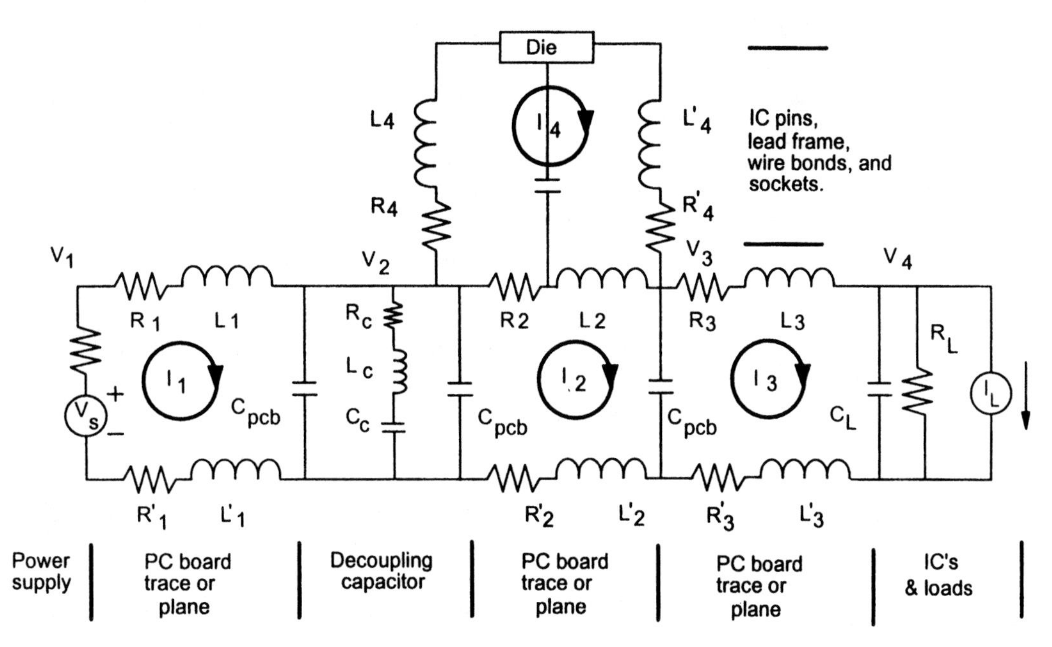

Figure 5.12 Equivalent circuit of a PCB.

To summarize,

> The important parameter when using decoupling capacitors is to minimize lead-length inductance and to locate the capacitors as close as possible to the component.

Decoupling capacitors must be provided for every device with edges faster than 2 ns and should be provided, placement wise, for "every component." Making provisions for decoupling capacitors is a necessity because future EMI testing may indicate a requirement for these devices. During testing, it may be possible to determine that there may be excess capacitors in the assembly. Having to add capacitors to an assembled board is difficult, if not impossible. Today, CMOS, ECL, and other fast logic families require additional decoupling capacitors besides the power and ground plane structure.

If a decoupling capacitor must be provided to a through-hole device after assembly, retrofit can be performed. Several manufacturers provide a decoupling capacitor assembly using a flat, level construction that resides between the component and PCB. This flat

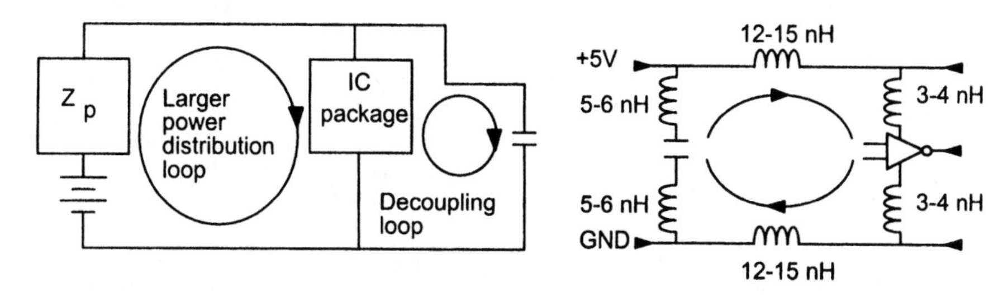

Figure 5.13 Power distribution model for loop control.

pack shares the same power and ground pins of the components. Because these capacitors are flat in construction, lead-length inductance is much less compared to capacitors with radial or axial leads. Since the capacitor and component share the same power and ground pins, R_2, L_2, R'_2 and L'_2 are also reduced. Some lead-length inductance will remain, which cannot be removed. The most widely used board level retrofit capacitors are known as Micro-Q™.[3] Other manufacturers provide similar products. An example of this type of capacitor is detailed in Fig. 5.14. Other configurations exist in pin grid array (PGA) form factor. For PGA retrofit decoupling capacitors, unique assemblies are available based on the particular pinout configuration of the device requiring this part.

A retrofit capacitor has a self-resonant frequency generally in the range of 10–50 MHz, depending on the capacitance of the device. Since DIP style leads are provided, higher frequency use cannot occur owing to excessive lead-length inductance. Although sometimes termed a "retrofit" device, the improved decoupling performance of these capacitors, compared to axial leaded capacitors on two-layer boards, makes them suitable for initial design implementation.

Poor planning during PCB layout and component selection may require use of Micro-Q devices. As yet, there is no equivalent retrofit for SMT packaged components.

When selecting a capacitor, we should consider not only the self-resonant frequency but the dielectric material as well. The most commonly used material is Z5U (barium titanite ceramic). This material has a high dielectric constant. This constant allows small capacitors to have large capacitance values with self-resonant frequencies from 1 MHz to 20 MHz, depending on design and construction. Above self-resonance, performance of Z5U decreases as the loss factor of the dielectric becomes dominant, which limits its usefulness to approximately 50 MHz.

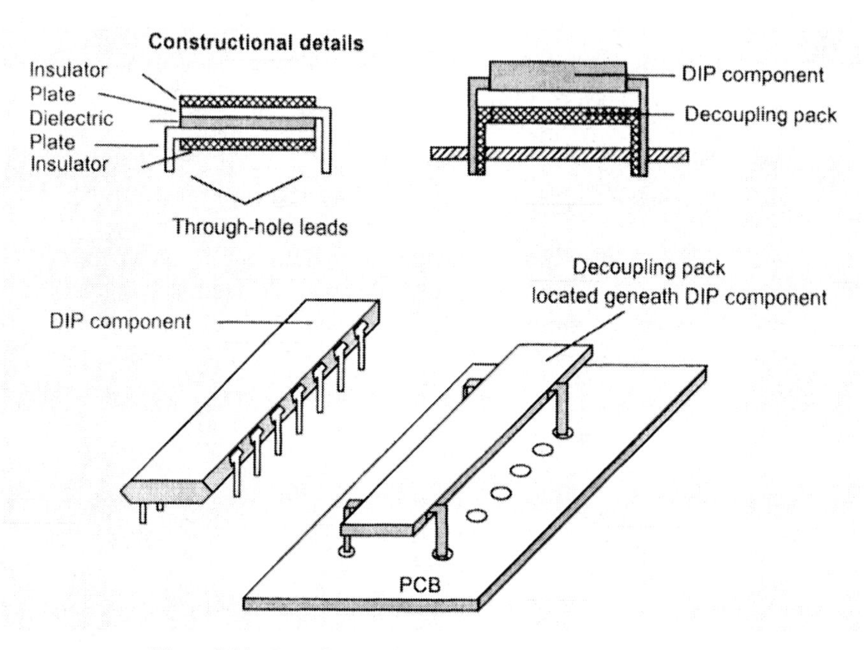

Figure 5.14 Retrofit decoupling capacitor—DIP mounting style.

[3]TM Micro-Q is a trademark of Circuit Components Inc. (formerly Rogers Corporation).

Another dielectric material commonly used is NPO (strontium titanite). This material has better high-frequency performance owing to a low dielectric constant. NPO is also a more temperature-stable dielectric. The capacitance value (and self-resonant frequency) is less likely to change when the capacitor is subjected to changes in ambient temperature or operating conditions.

Placement of 1 nF (1000 pF) capacitors (capacitors with a very high self-resonant frequency) on a 1-inch center grid throughout the PCB may provide additional protection from reflections and RF currents generated by both signal traces and the power planes, especially if a high-density PCB stackup is used [7]. It's not the exact location that counts in the placement of these additional decoupling capacitors. A lumped model analysis of the PCB will show that the capacitors will still function as needed, regardless of where the device is actually placed for overall decoupling performance. Depending on the resonant structure of the board, values of the capacitors placed in the grid may be as small as 30 to 40 pF [7].

VLSI and high-speed components (e.g., F, ACT, BCT, CMOS, ECL logic families) may require decoupling capacitors in parallel. As slew rates of components become steeper, a greater spectral distribution of RF currents is created. Parallel capacitors generally provide optimal bypassing of power plane noise, in addition to removing high-frequency RF energy. Multiple paired sets of capacitors are placed between the power and ground pins of VLSI components located around all four sides. These high-frequency decoupling capacitors are typically rated 0.1 μF in parallel with 0.001 μF for 50-MHz systems. Higher clock frequencies use a parallel combination of 0.01 μF and 100 pF components.

While the focus in this chapter is on multilayer boards, single- and double-sided boards also require decoupling. Figure 5.15 illustrates correct and incorrect ways of locating decoupling capacitors for a single- or double-sided assembly. When placing decoupling capacitors, ground loop control must be considered at all times. When using multilayer boards with internal power and ground planes, placement of the decoupling capacitor may be anywhere in the vicinity of the component's power pins [6], although this implementation may actually cause the PCB to become more RF active. This requirement is based on whether the component has its mounting pins via straight down to the power/ground plane, or whether a routed trace connects to the discrete capacitor. Location of the capacitor is not critical during placement for the previous statement because of the lumped distributed capacitance of the power planes and because the components themselves must via down to the power and ground plane—the same as the decoupling capacitor [6].

Another function of a decoupling capacitor is to provide localized energy storage, thereby reducing power supply radiating loops. When current flows in a closed-loop circuit, the RF energy produced is proportional to IAF, where I is the current in the loop, A is the area of the loop, and F is the frequency of the current. Because current and frequency are predetermined by the type of logic family selected, it becomes necessary to minimize the area of the logic current loop to reduce radiation. Minimal loop area can be accomplished by taking care in placement of decoupling capacitors. A good example of a large loop area is shown in Fig. 5.15.

In Fig. 5.15, V_{gnd} is LdI/dt induced noise in the ground trace flowing in the decoupling capacitor. This V_{gnd} now drives the ground structure of the board and contributes to the overall common-mode voltage across the entire board. One should minimize the ground path that is common with the board's ground structure and the decoupling capacitor.

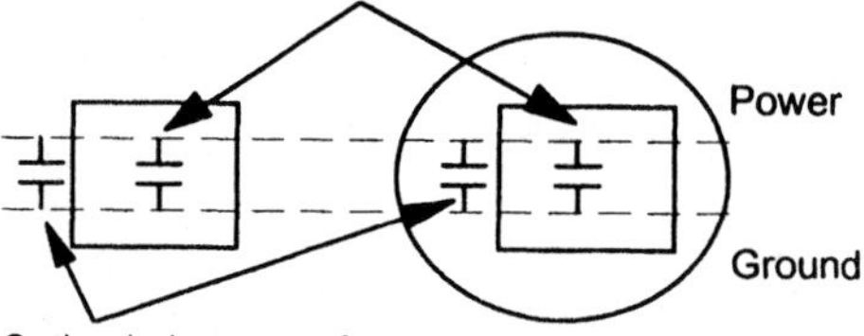

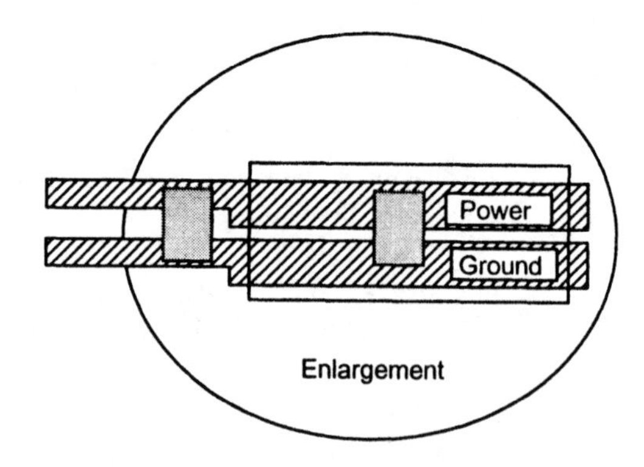

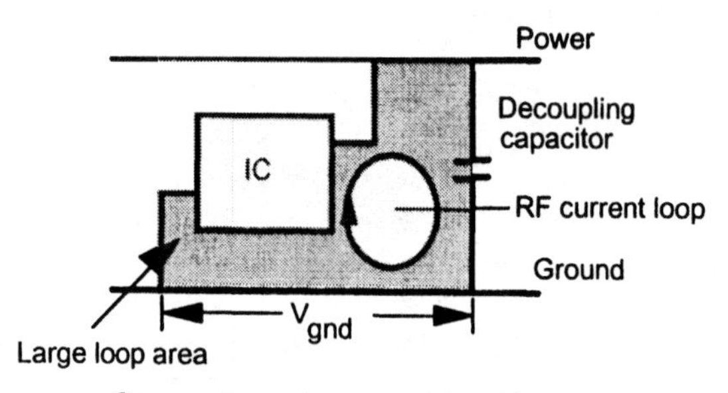

Figure 5.15 Placement of decoupling capacitors—two-layer board.

5.7 SELECTION OF A DECOUPLING CAPACITOR

Clock circuit components must be given emphasis to be RF decoupled. This is due to the switching energy generated by the component injected into the power and ground distribution system. This energy will be transferred to other circuits or subsections as common-mode or differential-mode RF. Bulk capacitors such as tantalum and high-frequency ceramic monolithic are both required, each for a different application. Furthermore, monolithic capacitors must have a self-resonant frequency higher than the clock harmonics requiring suppression. Typically, one selects a capacitor with a self-resonant frequency in the range of 10–30 MHz for circuits, with edge rates of 2 ns or less. Many PCBs are self-resonant in the 200–400 MHz range. Proper selection of decoupling capacitors, along with the self-resonant frequency of the PCB structure (acting as one large capacitor), will provide enhanced EMI suppression. Tables 5.1 and 5.2 are useful for axial or radial lead capacitors. Surface-mount devices have a much higher self-resonant frequency by approximately two orders of magnitude (or 100x) as the result of less lead-length inductance. Aluminum electrolytic capacitors are ineffective for high-frequency decoupling and are best suited for power supply subsystems or power line filtering.

It is common to select a decoupling capacitor for a particular application, usually the first harmonic of a clock or processor. Sometimes, a capacitor is selected for the third or fifth harmonic since this is where the majority of RF current is produced. There also needs to be plenty of larger discrete capacitors, bulk and decoupling. The use of common decoupling capacitor values of 0.1 μF in parallel with 0.001 μF can be too inductive and too slow to supply charge current at frequencies above 200–300 MHz.

When performing component placement on a PCB, one should make provisions for adequate high-frequency RF decoupling. One should also verify that all bypass and decoupling capacitor chosen are selected based on intended application. This is especially true for clock generation circuits. The self-resonant frequency must take into account all significant clock harmonics requiring suppression, generally considered to be the fifth harmonic of the original clock frequency. Finally, capacitive reactance (self-resonant reactance in ohms) of decoupling capacitors is calculated per Eq. (5.12).

$$X_c = \frac{1}{2\pi f C} \tag{5.12}$$

where X_c = capacitance reactance (ohms)
f = resonant frequency (Hertz)
C = capacitance value

5.7.1 Calculating Capacitor Values (Wave-Shaping)

Capacitors can also be used to wave-shape differential-mode RF currents on individual traces. These parts are generally used in I/O circuits and connectors and are rarely used in clock networks. The capacitor, C, alters the signal edge of the output clock line (slew rate) by rounding the time period the signal edge takes to transition from logic state 0 to logic state 1. This is illustrated in Fig. 5.16.

In examining Fig. 5.16, we should observe the change in the slew rate (clock edge) of the desired signal. Although the transition points remain unchanged, the time period t_r and t_f is different. This elongation or slowing down of the signal edge is a result of the capacitor charging and discharging. The change in transition time is described by the equations and illustration presented in Fig. 5.17. Note that a Thevenin equivalent circuit is shown without the load. The source voltage, V_b, and series impedance are internal to the driver or clock generation circuit. The capacitive effect on the trace, seen in the figure, is a result of this capacitor being located in the circuit. To determine the time rate of change of the capacitor detailed in Fig. 5.16, the equations in Fig. 5.17 are used.

When a Fourier analysis is performed on this signal edge (conversion from time to frequency domain), a significant reduction of RF energy is observed along with a decrease in spectral RF distribution. Hence, there is improved EMI compliance. Caution is required during the design stage to ensure that slower edges will not adversely affect functional operational performance.

The capacitive value for a decoupling capacitor can be calculated in two ways. Although the capacitance is calculated for optimal filtering at a particular resonant frequency, use and implementation depend on installation, lead length, trace length, and other parasitic parameters that may change the resonant frequency of the capacitor. The installed value of capacitive reactance is the item of interest. Calculating the value of capacitance will be in the ballpark and is generally accurate enough for actual implementation.

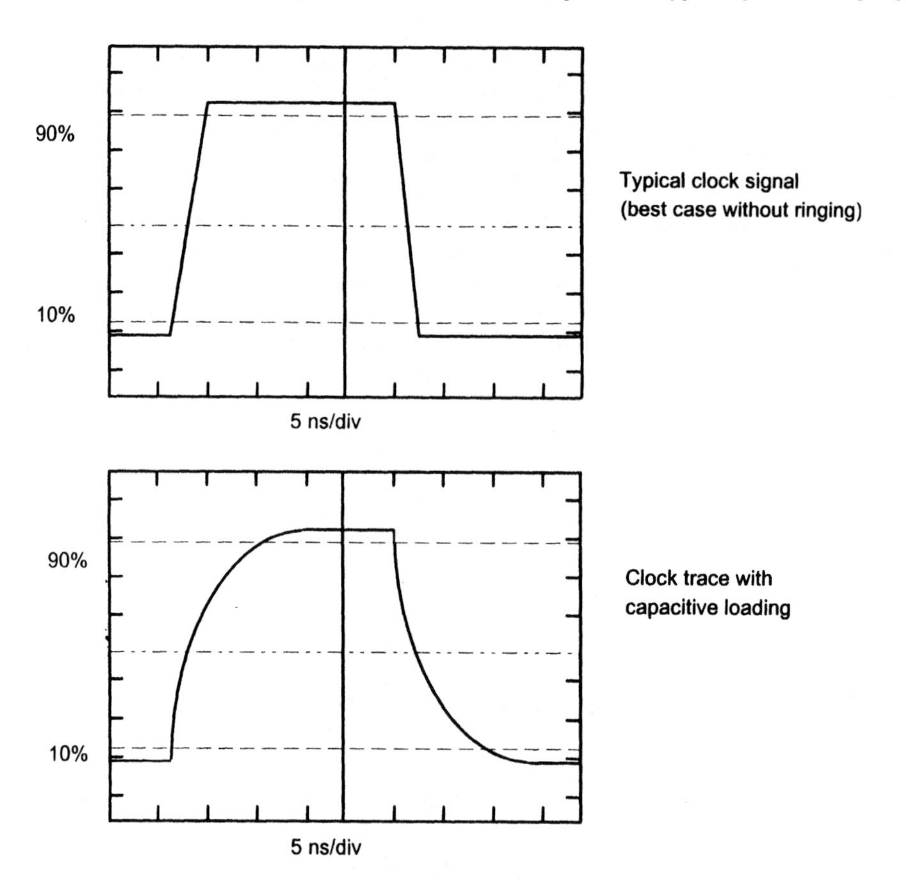

Figure 5.16 Capacitive effects on clock signals.

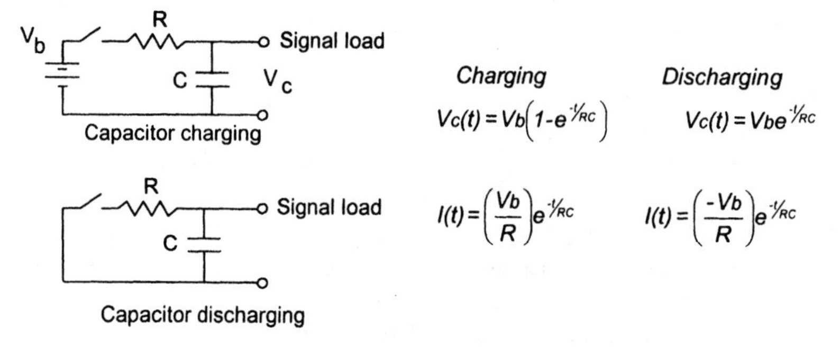

$$V_c(t) = V_b\left(1 - e^{-t/RC}\right) \qquad V_c(t) = V_b e^{-t/RC}$$

$$I(t) = \left(\frac{V_b}{R}\right) e^{-t/RC} \qquad I(t) = \left(\frac{-V_b}{R}\right) e^{-t/RC}$$

Figure 5.17 Capacitor equations.

Before calculating a decoupling capacitor value, the Thevenin impedance of the network should be determined. This impedance value should be equal to these two resistors values placed in parallel. Using a Thevenin equivalent circuit, we assume $Z_s = 150\ \Omega$ and $Z_L = 2.0\ \text{k}\Omega$.

$$Z_t = \frac{Z_s * Z_L}{Z_s + Z_L} = \frac{150 * 2000}{2150} = 140\Omega \tag{5.13}$$

Z_S Z_L

Method 1. Equation (5.14) is used to determine the capacitance value knowing the edge rate of the clock signal.

$$t_r = kR_tC_{\max} = 3.3 * R_t * C_{\max}$$
$$C_{\max} = \frac{0.3\ t_r}{R_t} \tag{5.14}$$

where t_r = edge rate of the signal (the faster of either the rising or falling edge)
R_t = total resistance within the network
$C_{\max}$ = maximum capacitance value to be used
k = one time constant

Note: C in nanofarads if t_r is in nanoseconds
C in picofarads if t_r is in picoseconds

The capacitor must be chosen so that $t_r = 3.3 * R * C$ equals an acceptable rise or fall time for proper functionality of the signal; otherwise baseline shift may occur. Baseline shift refers to the steady-state voltage level that is identified as logic low or logic high for a particular logic family. The number 3.3 is based on the value for the time constant of a capacitor charging based on the equation $\tau = RC$. Approximately three (3) time constants equals one (1) rise time. Since we are interested in only one time constant for calculating capacitance value, this value of k is $1/3t_r$, which becomes 3.3 when incorporated within the equation.

For example, if the edge rate is 5 ns and the impedance of the circuit is 140 Ω, we can calculate the maximum value of C as

$$C_{\max} = \frac{0.3 * 5}{140} = 0.01\ nF \quad \text{or} \quad 10\ pF \tag{5.15}$$

A 60-MHz clock with a period of 8.33 ns on and 8.33 ns off, $R = 33\ \Omega$ (typical for an unterminated TTL part) has an acceptable $t_r = t_f = 2$ ns (25% of the on or off value). Therefore,

$$\left(C = \frac{0.3 * t_r}{R_t}\right) \qquad C = \frac{0.3(2 * 10^{-9})}{33} = 18\ pF \tag{5.16}$$

Method 2

- Determine highest frequency to be filtered, $f_{\max}$.
- For differential pair traces, determine the maximum tolerable value of each capacitor to minimize signal distortion. Use Eq. (5.17).

$$C_{min} = \frac{100}{f_{max} * R_t}$$

$$\frac{1}{2\pi f_{max} * \frac{C}{2}} \geqslant 3 * R_t \qquad (5.17)$$

where C is in nanofarads and f in MHz.

To filter a 20-MHz signal with $R_L = 140\ \Omega$, the capacitance value would be

$$C_{min} = \frac{100}{20 * 140} = 0.036\ nF \quad \text{or} \quad 36\ pF \qquad (5.18)$$

with negligible source impedance, Z_c.

When using bypassing capacitors, the following should be implemented:

- If degradation of the edge rate is acceptable (generally three times the value of C), increase the capacitance value to the next highest standard value.
- Select a capacitor with proper voltage rating and dielectric constant for intended use.
- Select a capacitor with a tight tolerance level. A tolerance level of +80/–0% is acceptable for power supply filtering but is inappropriate as a decoupling capacitor for high-speed signals.
- Install the capacitor with minimal lead-length and trace inductance.
- Verify that the functionally of the circuit so that it still works with the capacitor installed. Too large a value capacitor can cause excessive signal degradation.

5.8 SELECTION OF BULK CAPACITORS

Bulk capacitors provide DC voltage and current to components when the devices are switching all data, address, and control signals simultaneously under maximum capacitive load. Switching components tend to cause current fluctuations within the power distribution network. These fluctuations can cause improper performance of components owing to voltage sags. Bulk capacitors provide energy storage for circuits to maintain optimal voltage and surge current requirements.

Bulk capacitors (usually tantalum dielectric) are often used in addition to higher self-resonant frequency decoupling capacitors to provide DC power for components and power plane RF modulation. One bulk capacitor should be placed for every two LSI and VLSI components in addition to the decoupling capacitors at the following locations:

- Power entry connector from the power supply to the PCB.
- Power terminals on I/O connectors for daughter or adapter cards, peripheral devices, and secondary circuits.
- Adjacent to power-consuming circuits and components.

- The furthest location from the input power connectors.
- High-density component placement remote from the DC input power connector.
- Adjacent to clock generation circuits and ripple sensitive devices.

When using bulk capacitors, the voltage rating should be calculated that the nominal voltage equals 50% of the capacitor's actual voltage rating requirement to prevent the capacitor from self-destruction if a voltage surge occurs. For example, with power at 5V, a capacitor with a minimum 10V rating should be used.

Table 5.3 shows the typical number of capacitors required for some popular logic families. This table is based on the maximum allowable power drop, which is equal to 25% of the noise immunity level of the circuit being decoupled. Note that for standard CMOS logic, this table is conservative since the trace wiring to the components cannot provide the required peak current without excessive voltage drop. The actual value of the capacitor used will be determined based on functional application.

Memory arrays require additional bulk capacitors owing to the extra current required for proper operation during a refresh cycle. The same is true for VLSI components with high pin counts. High-density pin grid array (PGA) modules also must have additional bulk capacitors provided, especially when all signal, address, and control pins switch simultaneously under maximum capacitive load.

Using Eq. (5.5) to calculate the peak surge current consumed by many capacitors, we observe that more is not necessarily better. An excessive number of capacitors could draw a large amount of current, which places a strain on the power supply.

Selection of a capacitor based on past experience with slower speed digital logic will generally not provide proper bypassing and decoupling when used with high-technology, high-speed designs. Consideration of resonance, placement on the PCB, lead-length inductance, existence of power planes, and the like must all be included when selecting a capacitor or capacitor combination.

For bulk capacitors, the following procedures are provided to determine optimal selection [2].

TABLE 5.3 Number of Decoupling Capacitors for Selected Logic Families

	Peak Transient Current Requirement (mA)		
Logic Family	Gate Overcurrent (mA)	1 Gate Drive (mA)	Number of Decoupling Capacitors for a Fanout of 5 Gates + 10 cm Trace Length
CMOS	1	0.3	1.0
TTL	16	1.7	2.6
LS-TTL	8	2.5	2.0
HCMOS	15	5.5	1.2
STTL	30	5	1.8
FAST	15	5.5	1.8
ECL	1	1.2	1.0

Source: Controlling Radiated Emissions by Design. Reprinted by permission, Van Nostrand Reinhold.

Method 1

1. Determine maximum current (ΔI) consumption anticipated on the board. Assume all gates switch simultaneously. Include the effect of power surges by logic crossover (cross-conduction currents).
2. Calculate maximum amount of power supply noise permitted (ΔV). Factor in a safety margin.
3. Determine maximum common-path impedance acceptable to the circuit.

$$Z_{cm} = \Delta V / \Delta I \tag{5.19}$$

4. If solid planes are used, allocate the impedance, Z_{cm}, to the connection between power and ground.
5. Calculate the impedance of the interconnect cable, L_{cable}, from the power supply to the board. Add this value to Z_{cm} to determine the frequency below which the power supply wiring is adequate $(Z_{\text{total}} = Z_{cm} + L_{\text{cable}})$.

$$f = \frac{Z_{\text{total}}}{2\,\pi\, L_{\text{cable}}} \tag{5.20}$$

6. If the switching frequency is below the calculated f of Eq. (5.20), the power supply wiring is fine. Above f, bulk capacitors, C_{bulk}, are required. Calculate the value of the bulk capacitor for an impedance Z_{total} at frequency f.

$$C_{\text{bulk}} = \frac{1}{2\,\pi\, f\, Z_{\text{total}}} \tag{5.21}$$

Method 2. A PCB has 200 CMOS gates (G), each switching 5 pF (C) loads within a 2-ns time period. Power supply inductance is 80 nH.

$$\Delta I = GC\frac{\Delta V}{\Delta t} = 200(5\,pF)\frac{5\text{V}}{2\text{ ns}} = 2.5\,A \text{ (worst case peak surge)}$$

$$\Delta V = 0.200\,V \text{ (from noise margin budget)}$$

$$Z_{\text{total}} = \frac{\Delta V}{\Delta I} = \frac{0.20}{2.5} = 0.08\,\Omega \tag{5.22}$$

$$L_{\text{cable}} = 80\text{ nH}$$

$$f_{ps} = \frac{Z_{\text{total}}}{2\,\pi\, L_{\text{cable}}} = \frac{0.08\,\Omega}{2\,\pi\, 80\text{ nH}} = 159\text{ kHz}$$

$$C = \frac{1}{2\,\pi\, f_{ps}\, Z_{\text{total}}} = 12.5\,\mu\text{F}$$

Capacitors commonly found on PCBs for bulk purposes are generally in the range of 10–100 μF.

Capacitance required for decoupling power plane RF currents due to the switching energy of components can be determined by knowing the resonant frequency of the logic circuits to be decoupled. The hardest part in calculating this resonant value is knowing the inductance of the capacitor's leads (ESL). If ESL is not known, an impedance meter or network analyzer may be used to measure the ESL value. The drawback of using an impedance meter is that low-frequency instruments may not catch higher frequency responses. ESL can be also approximated by knowing only the capacitance value and the self-resonant frequency parasitics.

5.9 DESIGNING A CAPACITOR INTERNAL TO A COMPONENT'S PACKAGE

Technology has progressed to the point where the majority of radiated emissions need not be caused by poor PCB layout, trace routing, impedance mismatches, or power supply corruption. Radiated emissions are the result of using digital components. What do we mean by the statement that digital components are the primary source of RF energy? The answer is simple. RF energy is produced by the Fourier spectra created by the switching transistors internal to the silicon wafer which is physically glued down inside a protective enclosure. This enclosure is commonly identified as the package, which consists of either plastic or ceramic material.

Recent advances in integrated circuit (IC) components such as microprocessors, digital signal processors, and applications-specific integrated circuits (ASICs) have become significant sources of electromagnetic noise. In recent years, clock rates have increased from 25 and 33 MHz to 200 through 500 MHz and beyond. With these clock rates, we have a resulting corresponding internal dynamic power dissipation increase due to switching currents that may exceed 10 watts within a VLSI device.

The silicon die demands current from a power distribution network and must drive a transmission line at certain levels of voltage and current. In addition, technology has progressed to the point where millions of transistors are provided within a single die on a wafer. Manufacturing technology has also approached the 0.18-micron line width, which allows for faster edge rate devices and die shrink. Die shrink is where the number of individual components on a silicon wafer increase, thus improving the yield and total number of devices from a single process batch. The cost of the product decreases when an increase in the number of functional units occurs. Because of smaller line widths, the propagation delay between the individual gates within the component package becomes shorter, along with corresponding faster edge rates. The faster the edge rate, the greater the ability of the device to create radiated emissions. With faster internal edges, the switching effects can cause greater losses across the inductance internal to the package.

With faster edge rates, DC current is demanded from the power distribution network at a quicker rate. This faster switching speed bounces the power distribution network, creating an imbalance in the differential-mode current between power and ground. With an imbalance in the differential-mode, common-mode currents are produced. Common-mode currents are observed during EMI tests radiating from cable assemblies, interconnects, or PCB components.

To address component-level problems, EMC engineers and component manufacturers must advance state-of-the-art principles in implementing suppression techniques for ICs, especially decoupling. Design and cost margins also play an important part in determining how an EMC solution will be implemented.

As presented earlier, decoupling capacitors provide an instantaneous point source of charge to a component for the time period that the device switches. A decoupling capacitor

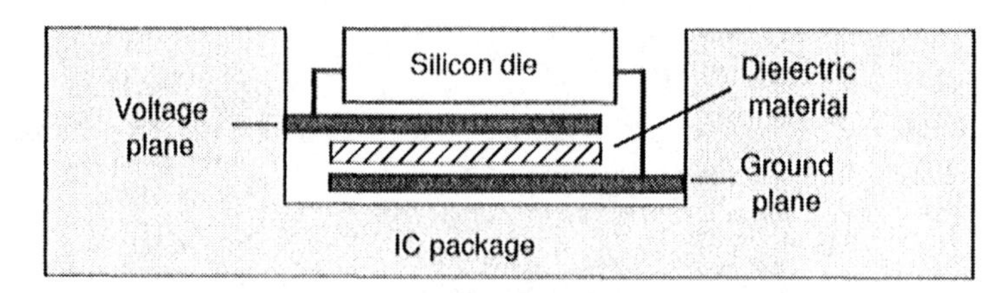

Figure 5.18 Decoupling capacitor internal to a silicon package.

must not only provide a voltage charge to the device at the speed that the device is switching at, but it must also recharge quickly. The self-resonant value of the capacitor depends on various parameters, which include not only the capacitance value but also ESL and ESR.

A component vendor can use various techniques to implement a decoupling capacitor internal to the component package. One approach is to implement a built-in decoupling capacitor before affixing the silicon die into the package, as illustrated in Fig. 5.18.

Two layers of metal film, separated by a thin layer of a dielectric material, will form a high-quality parallel plate capacitor. Since the applied voltage is extremely low, the dielectric layer can be very thin. This thin dielectric results in adequate capacitance for a very small area. The effective lead length approaches zero. The resonant frequency of the parallel plate configuration will thus be very high. The cost to manufacturers to implement this technique will be minimal compared to the overall cost of the IC. In addition to improved performance, the overall PCB cost may be reduced because use of discrete decoupling capacitors may not be required.

Another technique that will provide decoupling within a component package is by brute force. High-density, high-technology components often feature SMT capacitors located directly inside the device package. The use of discrete components is common in multichip modules. Depending on the inrush peak current surge of the silicon die, an appropriate capacitor is selected based on the current charge the device requires, in addition to providing decoupling of differential-mode currents at the self-resonant frequency of the component. Even with this internal capacitor, additional externally located discrete capacitors may be required. An example of how a discrete capacitor is provided in these modules is shown in Fig. 5.19.

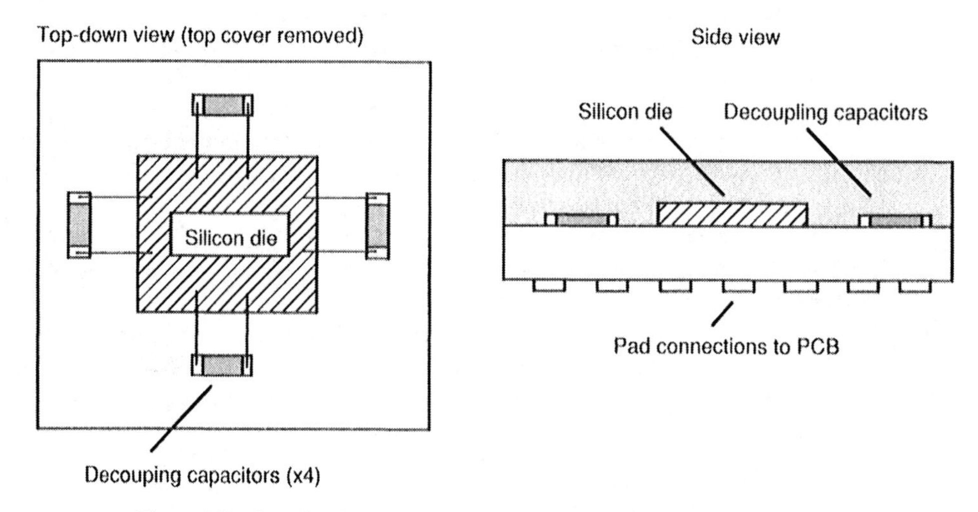

Figure 5.19 Locating decoupling capacitors internal to multichip module packaging.

For a synchronous design, CMOS power dissipation is a capacitive discharge effect. A device that consumes 2400 mW at 3.6V, running at 100 MHz, has an effective load capacitance of approximately 2000 pF, generally observed right after a clock event. If we allow a 10% drop in voltage, this means we require 20 nF of bulk capacitance for optimal performance. A gate capacitance of 6 to 7 $\mu F/m^2$ provides an area of approximately 3 square millimeters. This dimension is huge for a high density component. In addition, the capacitive value is not very large.

CMOS gates provide distributed capacitance by their input capacitance, both by coupling to the supply rails of the devices driving them and by the series capacitance of their own input transistors. This internal capacitance does not come close to the required value for functional operation. Silicon dies do not permit the extra silicon available to be used for bulk capacitance (floor space), since deep submicron designs consume routing space and are required to support the oxide layers of the assembly.

5.10 VIAS AND THEIR EFFECTS IN SOLID POWER PLANES

Use of vias in solid power planes will decrease the total capacitance based on the number of vias and the amount of real estate that has been etched out from the planes. A capacitor works by virtue of energy storage that is contained within a metallic structure. With less metal (copper plane), the current density distribution is decreased. As a result, less area exists to support the number of electrons that create the current density distribution. Figure 5.20 illustrates the value of capacitance between parallel power planes in two configurations: solid power planes and power planes with 30% of the area removed by vias and clearance pads.

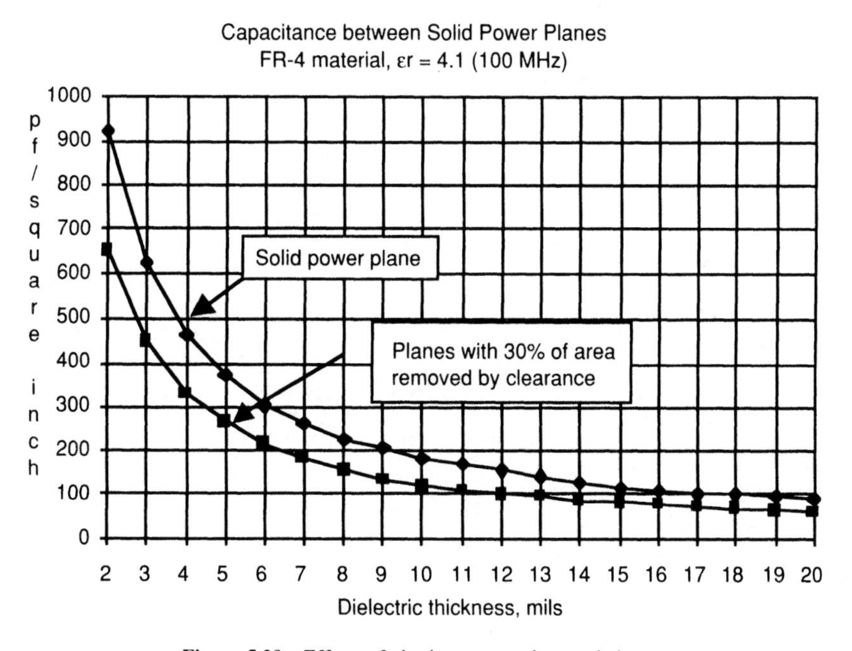

Figure 5.20 Effects of vias in power and ground planes.

REFERENCES

[1] Montrose, M. 1996. *Printed Circuit Board Design Techniques for EMC Compliance.* Piscataway, NJ: IEEE Press.

[2] Johnson, H. W., and M. Graham. 1993. *High Speed Digital Design.* Englewood Cliffs, NJ: Prentice Hall.

[3] Ott, H. 1988. *Noise Reduction Techniques in Electronic Systems.* 2nd ed. New York: John Wiley & Sons.

[4] Paul, C. R. 1992. "Effectiveness of Multiple Decoupling Capacitors," *IEEE Transactions on Electromagnetic Compatibility,* May, vol. EMC-34, pp. 130–133.

[5] Mardiguian, M. 1992. *Controlling Radiated Emissions by Design.* New York: Van Nostrand Reinhold.

[6] Drewniak, J. L., T. H. Hubing, T. P. Van Doren, and D. M. Hockanson. 1995. "Power Bus Decoupling on Multilayer Printed Circuit Boards." *IEEE Transactions on EMC* 37(2): 155–166.

[7] Montrose, M. 1991. "Overview on Design Techniques for PCB Layout Used in High Technology Products." *Proceedings of the IEEE International Symposium on Electromagnetic Compatibility,* 61–66.

[8] Van Doren, T., J. Drewniak, and T. Hubing. 1992. "Printed Circuit Board Response to the Addition of Decoupling Capacitors," Tech. Rep. TR92-4-007, UMR EMC Lab. (September 30).

[9] Montrose, M. 1996. "Analysis on the Effectiveness of Image Planes Within a Printed Circuit Board." *Proceedings of the IEEE International Symposium on Electromagnetic Compatibility,* 326–331.

6

Transmission Lines

6.1 OVERVIEW ON TRANSMISSION LINES

With today's high-technology products and faster logic devices, PCB transmission line effects become a limiting factor for proper circuit operation. A trace routed adjacent to a reference plane forms a simple transmission line. Consider the case of a multilayer PCB. When a trace is routed on an outer PCB layer, we have the microstrip topology, though it may be asymmetrical in construction. When a trace is routed on an internal PCB layer, the result is called stripline topology. Details on the effects of microstrip and stripline are provided in both Chapter 4 and this chapter.

A transmission line is a system of conductors, such as wires, waveguides, coaxial cables, or PCB traces suitable for conducting electric power or signals efficiently between two or more terminals.

To meet the challenges of high-speed digital processing, today's multilayer PCB must

- Reduce propagation delay between devices.
- Manage transmission line reflections and crosstalk (signal integrity).
- Reduce signal losses.
- Allow for higher density interconnections.

What are the electrical propagation modes that exist within a transmission line structure? A transmission line allows a signal to propagate from one device to another at or near the speed of light within a medium, as modified (slowed down) by the capacitance of the traces and by the active devices in the circuit. This signal contains some form of energy. Is this energy transmitted by electrons, line voltages and currents, or by something else? In a transmission line, electrons do not travel in the conventional sense. An *electromagnetic field* is the component that is present within and around a transmission line. The energy is carried along the transmission line by an electromagnetic field.

We usually place units of measurements for intelligence that exist within transmission lines. These units are voltage and current. *Voltage* is a unit of measurement whose spatial derivative describes the electrostatic force exerted on the electrons. *Current* is a unit of measurement that describes how many electrons flow in a transmission line structure during a specific time period. Neither unit describes the electromagnetic field or the electromagnetic wave present in the structure.

Typical electromagnetic fields consist of the following partial list.

- AM/FM radio waves
- Television waves
- Light waves
- Cellular telephone/pager waves
- Microwave and radar transmissions
- EMI/RFI created as a byproduct (unwanted energy) of digital components

All of these waves travel near the speed of light in a medium. EMI/RFI is included in this list to show that electromagnetic energy is a waveform that may cause harmful interference to other electronic equipment susceptible to electromagnetic disruption.

If a transmission line is not properly terminated, circuit functionality and EMI concerns can exist. These concerns include voltage droop, ringing, overshoot, and undershoot. All concerns will severely compromise switching operations and system signal integrity. Transmission line effects must be considered when the round-trip propagation delay exceeds the switching-current transition time. Faster logic devices and their corresponding increase in edge rates are becoming more common in the sub-nanosecond range. A very long trace in a PCB can become an antenna for radiating RF currents or cause functionality problems if proper circuit design techniques are not used early in the design cycle.

When dealing with transmission line effects, the impedance of the trace becomes an important factor in designing a product for optimal performance. A signal that travels down a PCB trace will be absorbed at the far end if, and only if, the trace is terminated in its characteristic impedance. If a proper termination is *not* provided, most of the transmitted signal will be reflected back in the opposite direction. If an *improper* termination exists, multiple reflections will occur, resulting in a longer signal-settling time because of multiple overshoots and undershoots. This condition is known as *ringing,* and is discussed later in this chapter.

When a high-speed electrical signal travels through a transmission line, a propagating electromagnetic wave will move down the line (e.g., a wire, coaxial cable, or PCB trace). A PCB trace looks very different to the signal source at high signal speeds than it does at DC or at low signal speeds. The characteristic impedance of the transmission line is identified by the letter Z_o. For a lossless line, the characteristic impedance is equal to the square root of L/C, where L is the inductance per unit length divided by C, the capacitance per unit length. Impedance is also the ratio of the line voltage to the line current, in analogy to Ohm's law. When we examine Eq. (6.1), we see subscripts for the line voltage and the line current. The ratio of line voltage to line current is constant with respect to the line distance x only for a matched termination. The (x) subscript indicates that variations in V and I will exist along the line, except for special cases.

$$Z_o = \sqrt{\frac{L_o}{C_o}} = \frac{V_{(x)}}{I_{(x)}} \tag{6.1}$$

We now examine *characteristic impedance*. As a logic signal transitions from a low to a high state, or vice versa, and propagates down a PCB trace, the impedance it encounters (the voltage to current ratio) is equal to a specific characteristic impedance. Once the signal has propagated up and down, trace reflections, if any, have died or become a nonissue related to signal integrity when the quiescent state is achieved. The characteristic impedance of the trace now has no effect on the signal. The signal becomes DC, and the line behaves as a typical wire.

To illustrate transmission line effects, let's assume a PCB trace has a propagation constant of 150 ps per inch, one way. Round-trip delay is this 300 ps per inch. If a clock driver has an edge rate of 2 ns, the transmission line characteristics of a short PCB trace is not a concern. This is because the signal will reflect back to the source long before the next edge-triggered event occurs. If properly terminated, the transmitted signal will have all possible reflections absorbed and dissipated within the network. Thus, a clean clock signal is available for the load device, which is the requirement for optimal functionality. If the clock trace is 10 inches in length, a serious problem could occur within the transmission line (the round-trip length of the trace is now 20 inches) since the reflected signal will return after the next edge-triggered event, causing possible functionality concerns if the trace is improperly terminated.

When a clock or strobe line drives multiple integrated circuits using a single trace, additional distributed capacitance and inductive elements are encountered because of additional components on the net. Each IC provides several pF of input shunt capacitance. This loading increases the capacitance value of the trace, thus increasing propagation delay of the signal. The increase in propagation delay occurs because distributed capacitance is proportional to the square root of the capacitance per unit length (Eq. 6.1). With a 2-ns and faster edge rate signal, transmission line effects become important for lead or trace lengths of no more than a few inches.

Figure 6.1 shows conceptually what a typical transmission line looks like within a PCB structure. The resistance, R, is omitted for simplicity.

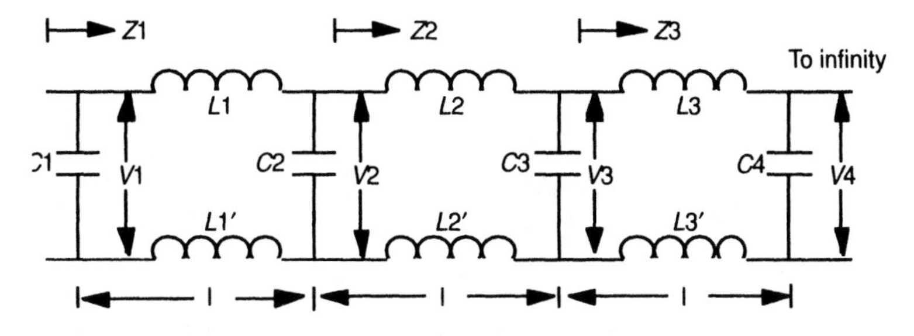

Figure 6.1 Transmission line equivalent circuit within a PCB.

6.2 TRANSMISSION LINE BASICS

How can a transmission line cause problems? Problems occur when a signal on a PCB trace encounters an impedance discontinuity or a change in geometry. We can consider transmission line effects as Ohm's law for high-speed, edge rate signals. When the output driver changes logic state, the voltage to current ratio in the structure will equal the characteristic impedance, Z_o, of the trace. As long as the impedance within the transmission line does not change, the signal smoothly propagates without changing the shape of the signal. If the end of the transmission line is open (infinite impedance), the current must go to zero. At the end of the trace, the voltage to current ratio inverts, and to satisfy Ohm's law, a reflected wave that is equal, but opposite in polarity (phase relationship), is created to cancel the current that exists within the trace. The reflected wave returns to the source driver. If the driver impedance does not match the transmission line impedance, another reflected wave will be created and be re-reflected when the signal reaches the source driver. This process keeps happening until all the energy in the signal is absorbed within the network. A descriptive example of this reflection is shown in Fig. 6.2.

The voltage that is reflected at the end of the transmission line will be greater than the initial voltage if the termination impedance is greater than the line impedance. The voltage level will be less than the initial voltage if the termination impedance is less than the line impedance. The amplitude of the reflected voltage at the end of the transmission line is calculated by Eq. (6.2).

$$V_r = V_i\left(\frac{R_t - Z_o}{R_t + Z_o}\right) = \rho\, V_i \tag{6.2}$$

where V_r is the reflected voltage at the far end, V_i is the initial voltage, R_t is the termination impedance, Z_o is the characteristic line impedance of the trace, and ρ is the reflection

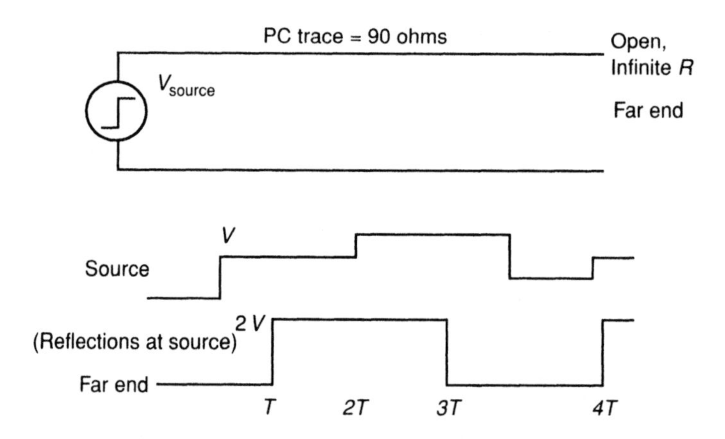

The top trace is the signal from the source driver.
The bottom trace illustrates the reflected waveform observed at the source if a mismatch exists within the transmission line structure.

Figure 6.2 Reflections in a transmission line.

coefficient. This equation identifies how much voltage gets reflected by the impedance mismatch. Notice that if $R = Z_o$, the reflection coefficient $\rho = 0$. There is no reflection and the voltage level does not change. If $R_t = \infty$, $\rho = +1$. This means that 100% of the voltage is reflected. This voltage will be double in value since the actual measured voltage is the sum of the initial voltage plus the reflected voltage. If $R_t = 0$, a short circuit exists, $\rho = -1$, and the voltage goes to zero. *The greater the mismatch, the greater the reflected voltage.* If both sides of the transmission line have mismatches, ringing will be created.

A circuit can be treated as a collection of lumped elements with capacitive and inductive components. This condition occurs when a signal path segment is small compared to the wavelength of a signal's highest frequency spectral component propagating down the trace. As the signal frequency increases, the circuit must be treated as a distributed transmission line. For this situation, controlled impedance, matched termination, and radiated emission effects must be considered.

6.3 TRANSMISSION LINE EFFECTS

For a high-speed transmission line, a fundamental concept exists called the *electrically long trace*. This means that as the length of the trace becomes greater than $\lambda/20$ (wavelength/20) of the signal (frequency domain), or the propagation delay becomes greater than rise time/4 (time domain), functionality concerns exist. The edge rate refers to a signal that changes logic in the period of dV/dt. We use the symbol t_r to identify edge rate. When using $\lambda/20$ and $t_r/4$ in the following discussions, we get roughly, but not exactly, the same line length. Depending on application, use of $\lambda/20$ *or* $t_r/4$ may provide a more accurate answer.

If the one-way "propagation time" distance from transmitter to receiver exceeds $\lambda/20$ (frequency domain) or the propagation delay is equal to or shorter than the propagation time of the trace with round-trip reflection, the PCB trace should be treated as a transmission line. These parameters are conservative. These traces may not fall within the electrically long trace requirement if the trace approximately equals these dimensions.

For a 1-ns edge rate signal, impedance matching may be required when the transmission line exceeds 9 cm (3.5 in.). Assuming a velocity of propagation, V_p that is 60% the speed of light ($c = 3 * 10^8$ m/s *or* $V_p = 1.8 * 10^8$ m/s), a line length greater than 9 cm must be treated as a transmission line which must include some type of termination. To determine this maximum line length, Eq. (6.3) is provided using the time domain analysis presented earlier.

$$\begin{aligned} l &= (t_r/2) * V_p \ (\textit{one way propagation travel}) \\ l &= (1 * 10^{-9}\ \text{sec}/2) * 1.8 * 10^8\ \text{m/sec} \\ l &= 0.09\ \text{m (9 cm or 3.5 in.)} \end{aligned} \tag{6.3}$$

where l = trace length (of the transmission line)
V_p = velocity of propagation (60% the speed of light)
t_r = edge transition rate of the signal

With $l = 9$ cm (3.5 in.), any trace longer than 9 cm for a 1-ns edge rate transition is considered to be electrically long. This value is for a signal that propagates in free space and not for a microstrip or stripline structure. A PCB will cause the propagated signal to travel at a much slower speed due to the dielectric constant of the core or prepreg material.

The length of a signal path, compared to the shortest wavelength of the signal trace, determines whether the circuit should be treated as a lumped or distributed configuration. If the one-way path length of the trace is less than $\lambda/20$, a low-frequency lumped circuit can be assumed. If the path length is greater than $\lambda/20$, we can assume that a high-frequency distributed circuit is present. With a distributed circuit, the characteristic impedance of the trace must be controlled and a matched termination is required for good signal integrity.

A frequently asked question is, When does impedance matching require $Z_L = Z_o$ or $Z_s = Z_o$? For Fig. 6.3, is it necessary to match both the source driver *and* load of a transmission line. If the signal flow is bidirectional, as found on bus structures, both ends must be terminated. For a single-ended circuit (e.g., a clock signal from an oscillator to a load), only one end of the trace requires termination. A decision must be made as to whether to terminate at the source or load end of the circuit. Termination methods and their applications are discussed in Chapter 8. Which end to terminate depends on several factors (again, discussed in Chapter 8).

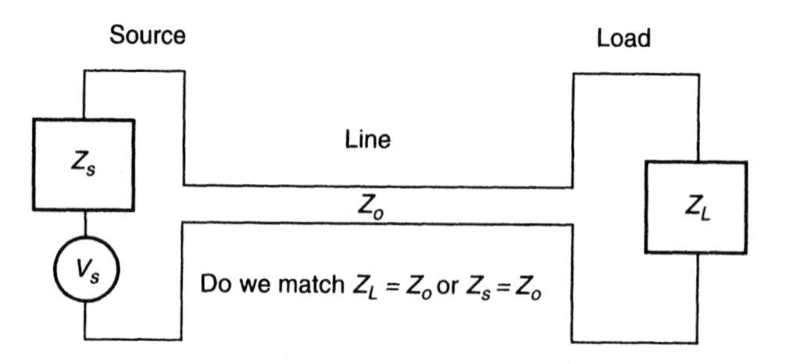

Figure 6.3 Impedance matching requirements of a circuit.

If the load value of Z_L is fixed, the transmission line is usually designed to match the load impedance, $Z_o = Z_L$. This transmission line connection occurs when preexisting equipment and the transmission line must be matched. If the load impedance is not known (which is often the case) or not predetermined, the optimal value for Z_o must be chosen so that the load is matched to the line ($Z_o = Z_L$). If Z_L is not known, termination pads to experimentally determine the correct component values for impedance matching should be provided on the PCB during layout.

To examine trace impedance in more detail, Fig. 6.4 illustrates a simple configuration. For a voltage pulse of amplitude V, driving a transmission line, Z_o, we have a drive current of $I = V/Z_o$. Assuming $V = 5\text{V}$, and $Z_o = 5\ \Omega$ (an unrealistic value), we observe

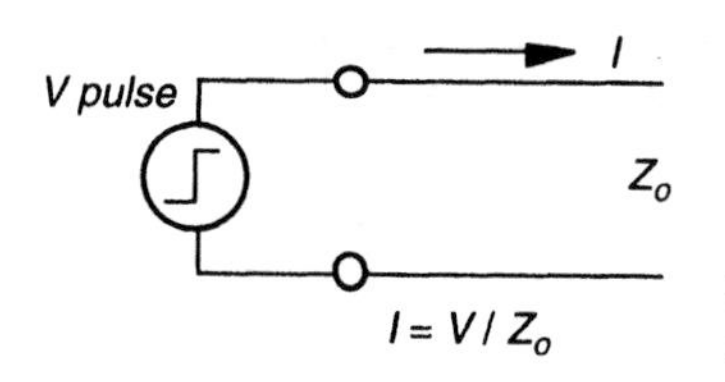

Figure 6.4 Simplified network for drive currents.

that the drive current would be 1A. If Z_o is now 50 Ω, a drive current of 100 mA is required. It is *not* advisable (for both functional purposes; power supply loading and EMI) to use source drivers with 100-mA drive capabilities for an application that generally requires only a few milliamps of drive current. For this reason, most components are designed to drive a minimal trace impedance of 30–65 Ω.

The propagation speed of a signal within a medium is finite. The propagation delay per unit length, δ, is equal to the square root of the inductance per unit length, L_o, times the capacitance per unit length, C_o. A typical PCB trace (with a dielectric constant of 4.6) has a propagation delay of 1.72 ns/ft (0.36 ns/cm or 0.143 ns/in.).

$$\delta = \sqrt{L_o C_o} \tag{6.4}$$

6.4 CREATING TRANSMISSION LINES IN A MULTILAYER PCB

Different logic families have different characteristic source impedance. Emitter-coupled logic (ECL) has a source and load impedance of 50 Ω. Transistor-transistor logic (TTL) has a source impedance range of 70 to 100 Ω. If a transmission line is to be created within a PCB, the engineer must seek to match the source impedance of the logic family being used.

Most high-speed traces must be impedance controlled. Calculations to determine proper trace width and separation to the nearest reference plane must occur. Board manufacturers and CAD programs can easily perform these calculations. If necessary, board fabricators can be consulted for assistance in designing the PCB, or a computer application program can be used to determine the most effective approach relative to trace width and distance spacing between planes for optimal performance. These approximate formulas may not be fully accurate because of manufacturing tolerances during the fabrication process. These formulas were simplified from exact models! Stock material may have a different thickness value and a different dielectric constant. The finished etched trace width may be different from a desired design requirement, or any number of manufacturing issues may exist. The board vendors know what the real variables are during construction and assembly. These vendors should be asked to provide the real or actual dielectric constant value, as well as the finished etched trace width for both base and crest dimensions, as shown in Fig. 6.5.

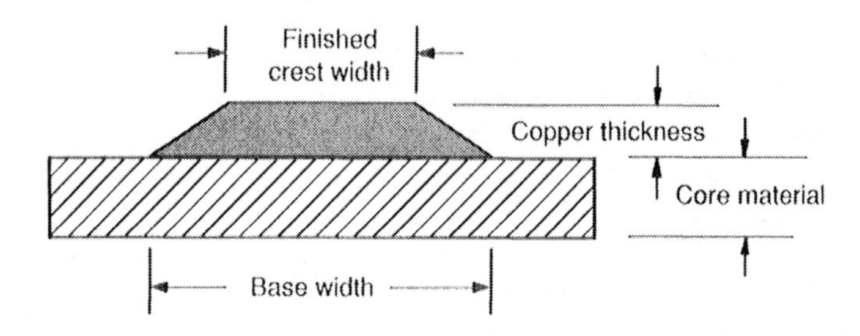

Figure 6.5 Finished trace width dimensions after etching.

6.5 RELATIVE PERMITTIVITY (DIELECTRIC CONSTANT)

Before giving a detailed description of how transmission lines are created within the PCB, we must examine the importance of the electrical parameter ε_r, also identified as *relative permittivity* or *dielectric constant.*

The relative dielectric constant ε_r is a measure of the amount of energy stored in the dielectric insulators per unit electric field, and hence a measure of the capacitance between a pair of conductors (trace-air, trace-trace, wire-wire, trace-wire, etc.) in the vicinity of the dielectric insulator compared to the capacitance of the same conductor pair in a vacuum. The relative dielectric constant of vacuum is 1.0. All materials have a dielectric constant greater than one. The larger the number, the more energy stored per unit insulator volume. The higher the capacitance, the slower the wave travels down the transmission line. The relationship between the capacitance and propagation speed was presented in Eq. (6.4).

Electromagnetic waves propagate at a speed that is dependent on the electrical properties of the surrounding medium. Propagation delay is typically measured in units of picoseconds/inch. Propagation delay is the inverse of velocity of propagation (the speed at which data is transmitted through conductors in a PCB), as presented in Chapter 2. The dielectric constant varies with several material parameters. Factors that influence the relative permittivity of a given material include the electrical frequency, temperature, extent of water absorption (also forming a dissipative loss), and the electrical characterization technique. In addition, if the PCB material is a composite of two or more laminates, the value of ε_r may vary significantly as the relative amount of resin and glass of the composite is varied [8].

In air, or vacuum, the velocity of propagation is the speed of light. In a dielectric material, the velocity of propagation is slower (approximately 0.6 times the speed of light for common PCB laminates). Both velocity of propagation and the effective dielectric constant are given by Eq. (6.5).

$$V_p = \frac{C}{\sqrt{\varepsilon_r}} \quad \text{(velocity of propagation)}$$

$$\varepsilon'_r = \left(\frac{C}{V_p}\right)^2 \quad \text{(dielectric constant)} \qquad (6.5)$$

where C = $3 * 10^8$ meters per second, or about 30 cm/ns (12 in./ns)
ε'_r = effective dielectric constant
V_p = velocity of propagation

The effective relative permittivity ε'_r, is the relative permittivity that is experienced by an electrical signal transmitted along a conductive path. Effective relativity permittivity can be determined by using a Time Domain Reflectometer (TDR) or by measuring the propagation delay for a known length line and calculating the value.

The propagation delay and dielectric constant of common PCB base materials are presented in Table 6.1. Coaxial cables often use a dielectric insulator to reduce the effective dielectric insulator inside the cable to improve performance. This dielectric insulator lowers the propagation delay while simultaneously lowering the dielectric losses.

TABLE 6.1 Propagation Delay in Various Transmission Media

Medium	Propagation Delay (ps/in)	Relative Dielectric Constant
Air	85	1.0
FR-4 (PCB), microstrip	141–167	2.8–4.5
FR-4 (PCB), stripline	180	4.5
Alumina (PCB), stripline	240–270	8–10
Coax (65% velocity)	129	2.3
Coax (75% velocity)	113	1.8

FR-4, currently the most common material used in the fabrication of a PCB, has a dielectric constant that varies with the frequency of the signal within the material. Most engineers generally assume that ε_r is in the range of 4.5 to 4.7. These values, referenced by designers, have been published in various technical reference manuals for many years and are based on measurements taken with a 1-MHz reference signal. Measurements were not made on FR-4 material under actual operating conditions, especially with today's high-speed designs. What worked over 20 years ago is insufficient for twenty-first-century products. Knowledge of the correct value of ε_r for FR-4 must now be introduced. A more accurate value of ε_r is determined by measuring the actual propagation delay of a signal in a trace using a TDR. The values in Table 6.2 are based on a typical, high-speed edge rate signal.

Figure 6.6 shows the "real" value of ε_r for FR-4 material based on research by the Institute for Interconnecting and Packaging Electronics Circuits Organization (IPC). This chart has been published in document IPC-2141, *Controlled Impedance Circuit Boards and High Speed Logic Design.*

TABLE 6.2 Dielectric Constants and Wave Velocities of PCB Materials

Material	ε_r (at 30 MHz)	Velocity (inches/ns)	Velocity (ps/inch)
Air	1.0	11.76	85.0
PTFE/glass (Teflon)™	2.2	7.95	125.8
RO 2800	2.9	6.95	143.9
CE/custom ply (Cyanide ester)	3.0	6.86	145.8
BT/custom ply (Beta-Triazine)	3.3	6.50	153.8
CE/glass	3.7	6.12	163.4
Silicon dioxide	3.9	5.97	167.5
BT/glass	4.0	5.88	170.1
Polyimide/glass	4.1	5.82	171.8
FR-4 glass	4.5	5.87	170.4
Glass cloth	6.0	4.70	212.8
Alumina	9.0	3.90	256.4

Note: Values measured at TDR frequencies using velocity techniques. Values were not measured at 1 MHz, which provides faster velocity values. Units for velocity are different due to scaling and are presented in this format for ease of presentation.

Source: IPC-2141, *Controlled Impedance Circuit Boards and High Speed Logic Design,* Institute for Interconnecting and Packaging Electronics Design. © 1996. Reprinted with permission.

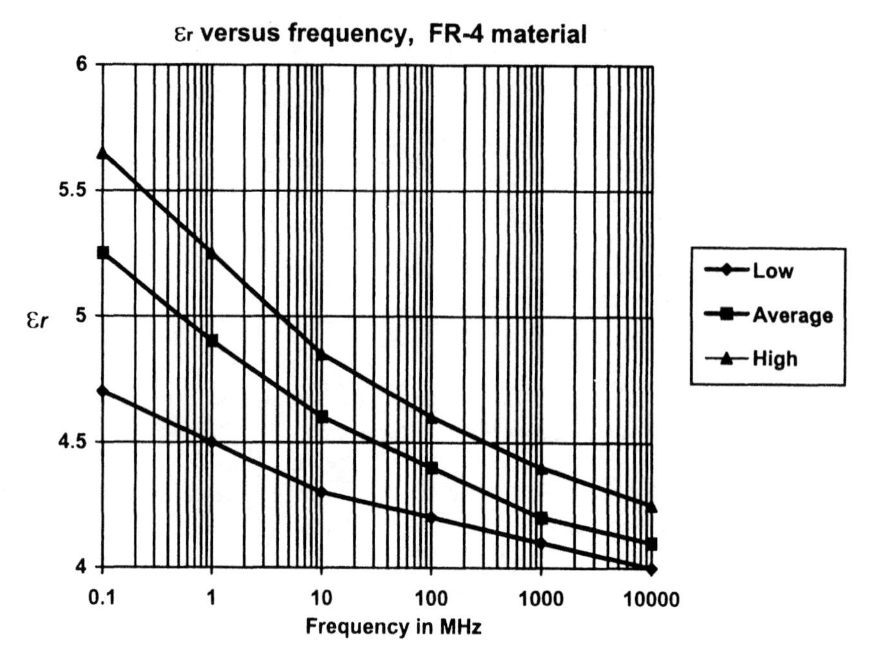

Figure 6.6 Actual dielectric constant values for FR-4 materials. (*Source:* IPC-2141, *Controlled Impedance Circuit Boards and High Speed Logic Design,* Institute for Interconnecting and Packaging Electronics Circuits. © 1996. Reprinted with permission.)

Figure 6.6 shows the frequency range from 100 kHz to 10 GHz for FR-4 laminate with a glass-to-resin ratio of approximately 40:60 by weight. The value of ε_r for this laminate ratio varies from about 4.7 to 4.0 over this frequency range. This change in the magnitude of ε_r is due principally to the frequency response of the resin and is reduced if the proportion of the glass ratio in the composite is increased. In addition, the frequency response will also be changed if an alternative resin system is selected. Material suppliers typically quote values of dielectric properties determined at 1 MHz, not at actual system frequencies that now easily exceed 100 MHz [9].

If a TDR is used for measuring the velocity of propagation, it is appropriate to use a frequency corresponding to the actual operating conditions of the PCB for comparing dielectric parameters. The TDR is a wideband measurement technique using time domain analysis. The location of the TDR on the trace being measured may affect measurement values. IPC-2141 provides an excellent discussion of how to use a TDR for propagational delay measurements [9].

The dielectric constant of various materials used to manufacture a PCB is provided in Table 6.2. These values are based on measurements using a TDR, and are not based on published, limited-basis reference information.

For microstrip topology, the dielectric constant is usually lower than the number provided by the manufacturer of the material. The reason is that part of the energy flow is in air or soldermask, and part of the energy flows within the dielectric medium. As a result, the signal will propagate faster down the trace than for the stripline configuration.

When a stripline conductor is surrounded by a single dielectric that extends to the reference planes, the value of ε'_r may be equated to that of ε_r for the dielectric measured under appropriate operating conditions. If more than one dielectric exists between the conductor and reference plane, the value of ε'_r is determined from a weighted sum of values of ε_r for all contributing dielectrics. Use of an electromagnetic field solver is required for a more accurate ε'_r value [16, 17, 18]. For purposes of evaluating the electrical characteristics of PCB, a composite such as a reinforced laminate, with a specific ratio of compounds, is usually regarded as a homogeneous dielectric with an associated relative permittivity.

For microstrip with a compound dielectric medium consisting of board material and air, Kaupp [14] derived an empirical relationship that gives the effective relative permittivity as a function of board material. Use of electromagnetic field solver is required for a more accurate answer [16].

$$\varepsilon'_r = 0.475\varepsilon_r + 0.67 \quad \text{for } 2 < \varepsilon_r < 6 \tag{6.6}$$

In this expression, ε_r relates to values determined at 25 MHz.

Trace geometries also affect the electromagnetic field within a PCB structure. These geometries determine if the electromagnetic field is radiated into free space or if it will stay internal to the assembly. If the electric field stays local to, or in the board, the effective dielectric constant becomes greater and signals propagate more slowly. The dielectric constant value will change internal to the board based on where the electric field shares its electrons. For microstrip (see Section 6.6.1) the electric field shares its electrons with free space, whereas stripline configurations capture free electrons. Microstrip permits faster transitions of electromagnetic waves. These electric fields are associated with capacitive coupling (Chapter 2) owing to the field structure within the PCB. The greater the capacitive coupling between a trace and its reference plane, the slower the propagation of the electromagnetic wave.

6.5.1 How Losses Occur Within a Dielectric

When we speak about losses within a dielectric structure, the term *lossy dielectric* implies an energy loss or joule heating in the dielectric material. How does this energy loss or joule heating occur? An examination of the atomic mechanism of dielectric behavior reveals the causes of heating.

A dielectric material is any substance that resists the penetration of an electric field into its interior. Consider one plate of a dielectric within which we attempt to establish an electric field by piling negative charges along one side and positive charges along the other. The dielectric will resist the electric field by producing opposite charges at its surfaces wherever the charges appear. The more charge provided, the more the dielectric will counter with opposing charges. The effect of this action is to cancel out some of the electric fields that would otherwise have been produced inside the dielectric. The dielectric constant of a material is defined by the magnitude of this reduction. This magnitude is the ratio of the field which would have been produced without these opposing charges to the field which is actually produced. For example, if only one-third of the expected field occurs within the dielectric, its dielectric constant is 3. The dielectric constant of a vacuum is 1.

A dielectric behaves similar to a capacitor. As we pile charges onto one end and take them out of the other, the dielectric inside the capacitor is busy piling up canceling charges right next to the ones we put in. This process is performed at both ends of the capacitor and prevents a voltage potential from being established across the capacitor. To the outside observer it appears that the capacitor is "gobbling up" the charges (storing them).

From an atomic viewpoint, how does a dielectric do this? All substances consist of atoms, which in turn contain a positively charged nucleus and an equally (but negatively) charged circle of electrons. In a *conductor*, electrons are free to swim about (almost like a fluid) within the crystal lattice. In a *dielectric*, however, the electrons are bound to the nuclei. In the presence of an electric field, the electrons may shift slightly to one side (this is an extreme simplification of quantum mechanics). This tiny amount of shifting, though very small in comparison to the size of the atom, takes place for every atom within the dielectric. This shifting creates the effect of a bulk displacement of all positive charges toward the negative pole of the applied field and vice versa for the negative charge. The total amount of charge in any common substance is astounding but we are unaware of it because the positive and negative charges are always in near perfect balance. Because of this balance, even an exceedingly slight shift can cause a significant total effect to be observed.

This shifting effect is similar to a team sport in which the ball is reversed on the field. Every player takes two or three steps to the side. The net effect is the same as if a single player were moved from one side of the field to the other. The same thing happens in a dielectric. The net effect of shifting electrons is the same as if some charges were transferred from one side of the material to the other (positive charges appears on one surface, negative on the other). This shifting is called *polarization*.

How does "loss" come into all this, and what does it have to do with the resonant frequency of polarization? It all has to do with time delay. When a field is applied, the aforementioned shifting cannot occur instantaneously; there must be some time delay. This time delay occurs at the resonant frequency of polarization. If we apply a rapidly reversing field to a dielectric, a phase lag will take place between the applied field and the resulting polarization field. As the frequency of the applied field is increased, the absolute lag is constant; therefore, the phase lag increases. When the phase lag reaches $\pi/2$ (90 degrees) we are at resonance.

There is a frequency-dependent phase lag between the applied electric field and polarization. This polarization is the mechanism by which a dielectric resists the application of the electric field. This cancellation gives the appearance that the dielectric is "gobbling up" charge (e.g., drawing excess current). Thus, a frequency-dependent phase lag will occur between the applied electric field and the resulting current flowing to establish those fields.

For a molecular explanation of what is going on, we can imagine a dielectric sample to which we apply one half cycle of a very high-frequency electric field (single short electric pulse). As soon as the pulse arrives, the electrons begin shifting over toward the positive side of the field. Since the pulse is so short, the electric field disappears by the time the electrons have started moving. We now have many electrons that have acquired kinetic energy. This energy has to go somewhere. As the electrons move around to return to their normal equilibrium state, some of the energy stored in the polarization field is transferred into the crystal lattice as vibrational energy (heat). This all happens because the pulse was so short that the electrons couldn't track it fast enough (phase lag). If the electric field were continuously oscillating instead of a single pulse, this effect would occur continuously. The shifting of the electrons would be out of phase with the applied electric field. It basically comes down to energy being pumped into the time-varying electric field between the atoms. This energy must eventually be dissipated into the crystal lattice, causing heating.

6.6 ROUTING TOPOLOGIES

Several techniques are available for creating a transmission line structure in a multilayer PCB. Two basic topologies are used, each of which has two configurations: microstrip (single and embedded) and stripline (single and dual).

Note: None of the equations provided in the next section for microstrip and stripline is applicable to PCBs constructed of two or more dielectric materials, excluding air, for example, or fabricated with more than one type of laminate. All equations are extracted from IPC-D-317A, *Design Guidelines for Electronic Packaging Utilizing High-Speed Techniques* [8].[1]

6.6.1 Microstrip Topology

Microstrip topology is a popular method used to provide trace-controlled impedance on a PCB for digital circuits. Microstrip lines are exposed to both air and a dielectric material referenced to a planar structure. The approximate formula for surface microstrip impedance is provided in Eq. (6.7) for the configuration of Fig 6.7. The intrinsic line capacitance is shown in Eq. (6.8).

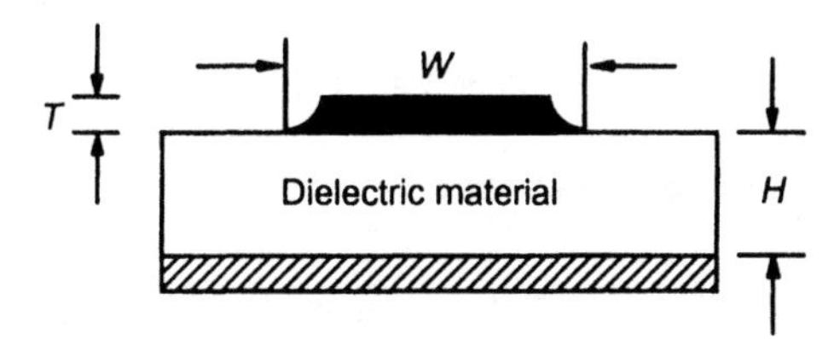

Figure 6.7 Surface microstrip topology.

[1]Within the IPC standards, typographical and mathematical errors exist in the section related to impedance. Before applying equations detailed within IPC-D-317, study and identify all errors before literal use. Equations presented herein have been verified for accuracy.

$$Z_o = \left(\frac{87}{\sqrt{\varepsilon_r + 1.41}}\right) ln\left(\frac{5.98\,H}{0.8\,W + T}\right)\Omega \quad \text{Valid for } 15 < w < 25 \text{ mils}$$
$$Z_o = \left(\frac{79}{\sqrt{\varepsilon_r + 1.41}}\right) ln\left(\frac{5.98\,H}{0.8\,W + T}\right)\Omega \quad \text{Valid for } 5 < w < 15 \text{ mils} \tag{6.7}$$

$$C_o = \frac{0.67(\varepsilon_r + 1.41)}{ln\left(\frac{5.98H}{0.8W + T}\right)} \quad \text{pf/inch} \tag{6.8}$$

where Z_o = characteristic impedance (ohms)
W = width of the trace (inches)
T = thickness of the trace (inches)
H = distance between signal trace and reference plane (inches)
C_o = intrinsic capacitance of the trace (pF/inch)
ε_r = dielectric constant of the planar material

Equation (6.7) is typically accurate to ± 5% when the ratio of W to H is 0.6 or less. When the ratio of W to H is between 0.6 and 2.0, accuracy typically drops ± 20%.

When measuring (or calculating trace impedance), the width of the line should technically be measured at the middle of the trace thickness. Depending on the manufacturing process, the finished line width after etching may be different from that specified (Fig. 6.5). The width of the copper on the top of the trace may be etched away, thus making the trace width smaller than desired. Using the average between top and bottom of the trace thickness, we find that a more typical, accurate impedance number is possible. With respect to the measurement of a trace's width, with an *ln* (natural logarithm) expression, how much significance should we give to accuracy of trace impedance for the majority of designs? Most manufacturing tolerances are well within 10% of desired impedance.

The propagation delay of a signal routed microstrip is described by Eq. (6.9) which has a variable of only ε_r. This equation states that the speed of a signal within a trace is related only to the effective permittivity of the dielectric material. Kaupp derived this equation for the propagation delay function under the square root radical [14].

$$t_{pd} = 1.017\sqrt{0.475\,\varepsilon_r + 0.67} \quad \text{(ns/ft)}$$
or
$$t_{pd} = 85\sqrt{0.475\,\varepsilon_r + 0.67} \quad \text{(ps/in.)} \tag{6.9}$$

6.6.2 Embedded Microstrip Topology

The embedded microstrip is a modified version of standard microstrip. The difference lies in providing a dielectric material on the top surface of the copper trace. This material may include another routing layer (core or prepreg material). If the embedded trace is surrounded by a material, such as soldermask, conformal coating, potting, or other material containing the same dielectric constant, with a thickness of 0.008 to 0.010 inch (8 to 10 mils) placed on top of the trace, air or the environment will have little effect on impedance calculations. Another way to view embedded microstrip is to compare it to a single, asymmetric stripline with one plane infinitely far away.

Coated microstrip uses the same conductor geometry as the uncoated except that the effective relative permittivity will be higher. Coated microstrip refers to placing a substrate on the outer microstrip layer. This substrate can be soldermask, conformal coating, or another material, including another microstrip layer. The dielectric on top of the trace may be asymmetrical to the host material. The difference between coated and uncoated microstrip is that the conductors on the top layer are fully enclosed by a dielectric substrate. The equations for embedded microstrip are the same as those for uncoated microstrip with a modified permittivity, ε'_r. If the dielectric thickness above the conductor is more than a few thousandths of an inch, ε'_r will need to be determined either by experimentation or by use of an electromagnetic field solver. For "very thin" coatings, such as soldermask or conformal coating, the effect is negligible. Masks and coatings may drop the impedance of the trace by several ohms.

The approximate formula for embedded microstrip impedance is provided by Eq. (6.10). For embedded microstrip, particularly those with asymmetrical dielectric heights, knowledge of the base and crown widths after etching will improve accuracy. These formulas are reasonable as long as the thickness of the upper dielectric material $[B - (T + H)]$ is greater than 0.004 inch (0.001 mm). If the coating is thin, or if the relative dielectric coefficient of the coating is different (e.g., conformal coating), the impedance will typically be between those calculated between microstrip and embedded microstrip.

Embedded microstrip is described in Eq. (6.10) for the configuration shown in Fig. 6.8. The intrinsic capacitance of the trace is defined in Eq. (6.11).

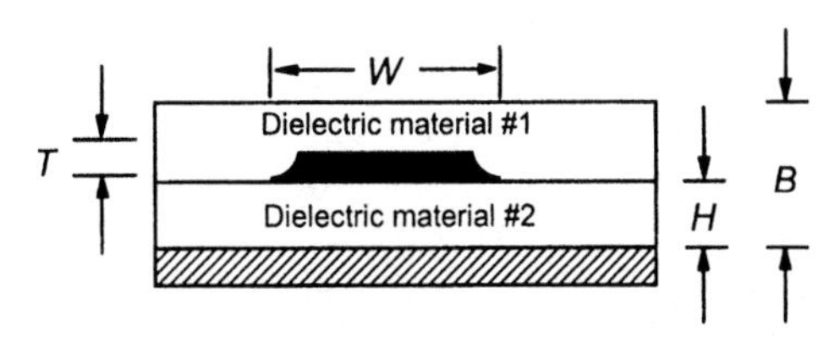

Figure 6.8 Embedded microstrip. NOTE: Thickness of the dielectric material may asymmetrical.

$$Z_o = \left(\frac{87}{\sqrt{\varepsilon'_r + 1.41}}\right) ln\left(\frac{5.98\,H}{0.8\,W + T}\right)\ \Omega \tag{6.10}$$

$$\text{where:}\quad \varepsilon'_r = \varepsilon_r\left\{1 - e^{\left(\frac{-1.55B}{H}\right)}\right\}$$

$$C_o = \frac{(1.41\,\varepsilon'_r)}{ln\left(\frac{5.98\,H}{0.8\,W + T}\right)}\ \text{pF/inch} \tag{6.11}$$

where Z_o = characteristic impedance (ohms)
C_o = intrinsic capacitance of the trace (pF/inch)
W = width of the trace (inches)
T = thickness of the trace (inches)
H = distance between signal trace and reference plane (inches)
B = overall distance of both dielectrics (inches)
ε_r = dielectric constant of the planar material

The propagation delay of a signal-routed embedded microstrip is given in Eq. (6.12). For a typical embedded microstrip, with FR-4 material and a dielectric constant that is 4.1, propagation delay is 0.35 ns/cm or 1.65 ns/ft (0.137 ns/in.). This propagation delay is the same as single stripline, discussed next, except with a modified ε'_r.

$$t_{pd} = 1.017\sqrt{\varepsilon'_r} \quad \text{(ns/ft)}$$

or

$$t_{pd} = 85\sqrt{\varepsilon'_r} \quad \text{(ps/in.)}$$

where

$$\varepsilon'_r = \varepsilon_r\left(1 - e^{\left(\frac{-1.55B}{H}\right)}\right) \tag{6.12}$$

$$0.1 < W/H < 3.0$$

$$1 < \varepsilon_r < 15$$

6.6.3 Single Stripline Topology

Stripline refers to a trace that is located between two planar conductive structures with a dielectric material completely surrounding the trace (Fig. 6.9). As a result, stripline traces, routed internal to the board, are not exposed to the external environment.

Routing stripline traces compared to microstrip has several advantages, namely, it captures fields and minimizes crosstalk, and it also provides an RF current reference return plane for magnetic field flux cancellation. Any radiated emissions that may occur from a routed trace will be captured by the reference plane and be prevented from radiating to the outside environment, provided the correct routing rules (e.g., the 3-W rule) are followed; see Chapter 7.7. Radiated emissions will still exist from components located on the outside layers of the board and their bond lead wires, not from the traces themselves buried within the PCB.

When measuring (or calculating) trace impedance, the microstrip section should be consulted for a discussion of why we should measure trace impedance of the line at the middle of the trace thickness after etching.

The approximate formula for single stripline impedance is provided in Eq. (6.13) for the illustration in Fig. 6.9. Intrinsic capacitance is presented in Eq. (6.14). Note that Eq. (6.14) is based on variables chosen for an optimal value. In actual board construction, the impedance may vary by as much as ± 5%.

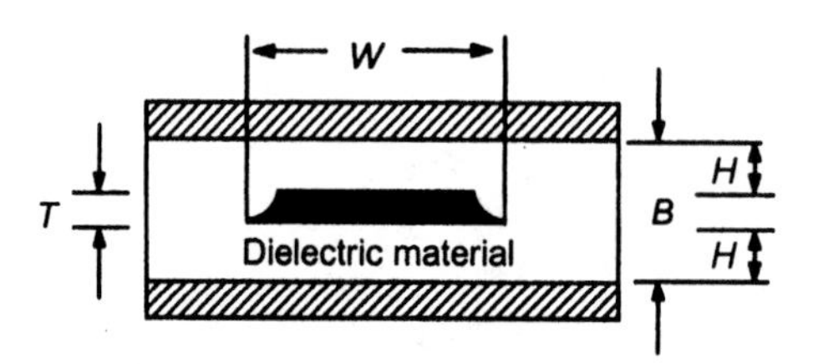

Figure 6.9 Single stripline topology.

$$Z_o = \left(\frac{60}{\sqrt{\varepsilon_r}}\right) ln\left(\frac{1.9B}{(0.8W + T)}\right)\Omega = \frac{60}{\sqrt{\varepsilon_r}}\ ln\left(\frac{1.9(2H + T)}{(0.8W + T)}\right) \tag{6.13}$$

$$C_o = \frac{1.41\varepsilon_r}{ln\left(\frac{3.81h}{0.8W + T}\right)} \quad \text{pF/inch} \tag{6.14}$$

where Z_o = characteristic impedance (ohms)
W = width of the trace (inches)
T = thickness of the trace (inches)
B = distance between both reference planes (inches)
h = distance between signal plane and reference plane (inches)
C_o = intrinsic capacitance of the trace (pF/inch)
ε_r = dielectric constant of the planar material
$W/(H-T) < 0.35$
$T/H < 0.25$

The propagation delay of signal stripline is described by Eq. (6.15), which has only ε_r as a variable.

$$t_{pd} = 1.017\sqrt{\varepsilon_r} \quad \text{(ns/ft)}$$

or

$$t_{pd} = 85\sqrt{\varepsilon_r} \quad \text{(ps/in.)} \tag{6.15}$$

6.6.4 Dual Stripline Topology

A variation on the single stripline is the dual stripline, which increases coupling between the circuit plane and the nearest reference plane. When the circuit is placed approximately in the middle one-third of the interplane region, the error caused by assuming the circuit to be centered will be quite small.

The approximate formula for dual stripline impedance provided in Eq. (6.16) is for the illustration of Fig. 6.10. This equation is a modified version of that used for a single stripline. Note that the same approximation reason as dual stripline is used to compute Z_o.

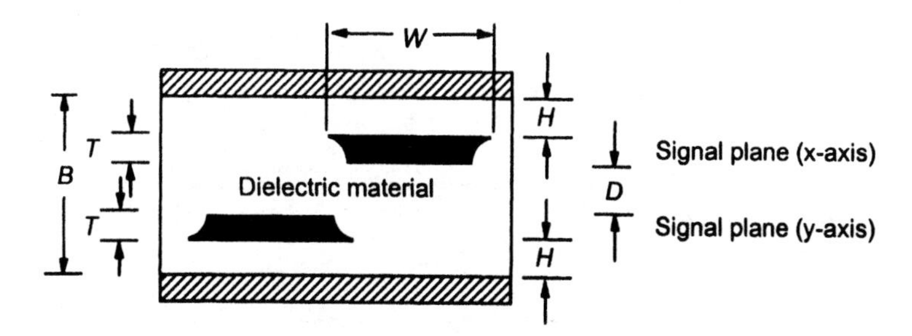

Figure 6.10 Dual stripline topology.

$$Z_o = \left(\frac{80}{\sqrt{\varepsilon_r}}\right) ln\left[\frac{1.9\,(2H + T)}{(0.8\,W + T)}\right]\left[1 - \frac{H}{4(H + D + T)}\right] \tag{6.16}$$

$$C_o = \frac{2.82\,\varepsilon_r}{ln\left[\frac{2(H - T)}{0.268W + 0.335T}\right]} \tag{6.17}$$

where Z_o = characteristic impedance (ohms)
W = width of the trace (inches)
T = thickness of the trace (inches)
D = distance between signal plane (inches)
H = dielectric thickness between signal plane and reference plane
C_o = intrinsic capacitance of the trace (pF/inch)
ε_r = dielectric constant of the planar material
$W/(H - T) < 0.35$
$T/H < 0.25$

Equation (6.16) can be applied to asymmetrical (single) stripline configuration when the trace is not centered equally between the two reference planes. In this situation, H is the distance from the center of the line to the nearest reference plane. The letter D would become the distance from the center of the line being evaluated to the other reference plane.

The propagation delay for the dual stripline configuration is the same as that for the single stripline, since both configurations are embedded in a homogeneous dielectric material.

$$t_{pd} = 1.017\sqrt{\varepsilon_r} \quad \text{(ns/ft)}$$

$$\text{or} \qquad (6.18)$$

$$t_{pd} = 85\sqrt{\varepsilon_r} \quad \text{(ps/in.)}$$

Note: When using the dual stripline, both routing layers must be routed orthogonal to each other. This means that one routing layer is provided for x-axis trace routing, while the other layer is used for y-axis traces. Routing these layers at 90 degree angles prevents crosstalk from occurring between the two routing planes with wide busses or with high-frequency traces causing data corruption to the alternate routing layer.

The actual operating impedance of a line can be significantly influenced (e.g., ≈ 30%) by multiple high-density crossovers of orthogonally routed traces, increasing the loading on the net and reducing the impedance of the transmission line. This impedance change occurs because these routed traces include a loaded impedance to the image plane, along with capacitance to the signal trace under observation. This is best illustrated by Fig. 6.11.

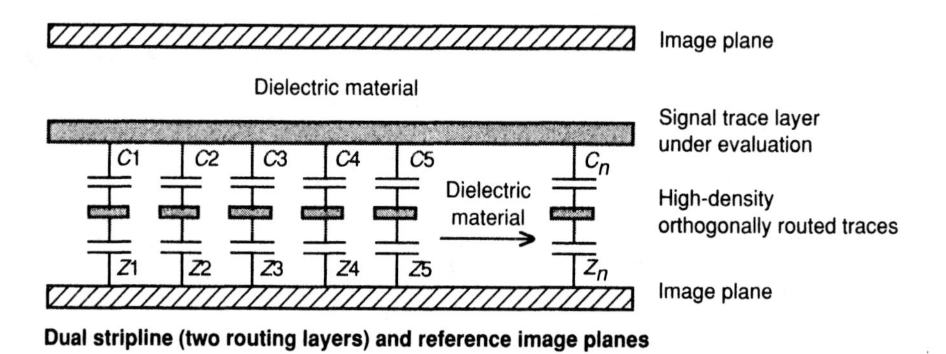

Figure 6.11 Impedance influences on dual stripline routing planes.

6.6.5 Differential Microstrip and Stripline

Differential traces have conductors routed adjacent to each other throughout the entire trace route. The impedance for differentially routed traces is not the same as a single-ended routed trace unless the position of the images obeys the *10-W* rule. The *10-W* rule refers to the distance separation between the two traces measured at 10 times the width (of an individual trace) from the centerline of one trace to the centerline of the other. For this configuration, sometimes only line-to-ground (or reference plane) impedance is considered as if the traces were routed single-ended. The concern should also be with the line-to-line impedance between the two traces operating in differential mode.

For Fig. 6.12, differential traces are shown. If the configuration is microstrip, the upper reference plane is not provided. For stripline, both reference planes are provided, with equal center spacing between the parallel traces and the two planes.

When calculating differential Z_o (Z_{diff}), trace width W should be adjusted to alter Z_{diff}. The user should not adjust D, which should be the minimal spacing specified by the PCB vendor [15].

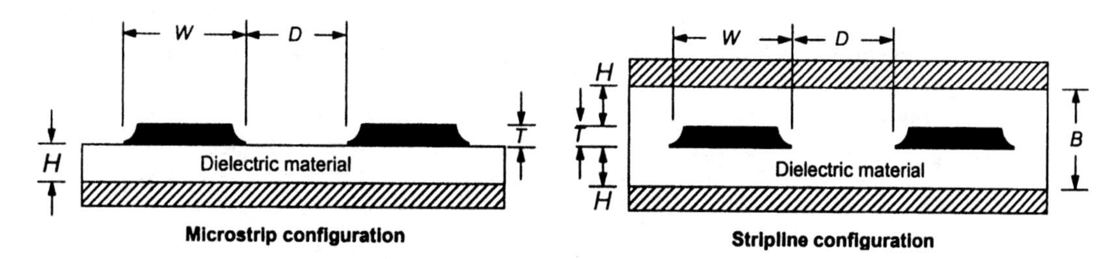

Figure 6.12 Differential trace routing topology.

$$Z_{diff} \approx 2 * Z_o\left(1 - 0.48e^{-0.96\frac{D}{h}}\right) \text{ ohms} \quad \text{(microstrip)}$$
$$Z_{diff} \approx 2 * Z_o\left(1 - 0.347e^{-2.9\frac{D}{B}}\right) \text{ohms} \quad \text{(stripline)} \tag{6.19}$$

where

$$Z_o = \frac{87}{\sqrt{\varepsilon_r + 1.41}}\, ln\left(\frac{5.98H}{(0.8W + T)}\right) \text{ohms} \quad \text{(microstrip)}$$
$$\frac{60}{\sqrt{\varepsilon_r}}\, ln\left(\frac{1.9B}{(0.8W + T)}\right) = \frac{60}{\sqrt{\varepsilon_r}}\, ln\left(\frac{1.9(2H + T)}{(0.8W + T)}\right) \quad \text{(stripline)} \tag{6.20}$$

where B = plane separation
W = width of the trace
T = thickness of the trace
D = trace edge-to-edge spacing
h = distance spacing to nearest reference plane

Note: Use consistent dimensions for the above (inches or centimeters).

6.7 ROUTING CONCERNS

In regards to multiple loads daisychained on a net when a signal travels along a transmission line, the transition voltage will change at different propagation times. The difference in the reception time at the loads located at various propagational positions along the net is referred to as clock skew (see Chapter 3). Since the component closest to the driver will receive the signal before a load at the end of a long trace, synchronous clocking of multiple devices becomes difficult, especially if the edge rate is extremely fast and the trace length is electrically broken up into different propagational lengths.

If clock skew is an important consideration for multiple loads on a bus structure, microstrip is preferred. This is because signal propagation for microstrip is faster than stripline (1.65 ns/ft vs. 2.06 ns/ft with a dielectric constant of 4.1). Microstrip is faster than stripline by approximately 25%. This is because stripline has twice the capacitance per unit length owing to the routing layers sandwiched between two planar structures (compared to microstrip with one adjacent plane). The effective ε_r is lower, thus higher $\upsilon = c/\sqrt{\varepsilon_r}$. Propagation time, δ, per unit length is proportional to the square root of the product of inductance and capacitance per unit length; see Eq. (6.4).

When radiated emissions is a concern, routing a trace using stripline is preferred. Unfortunately, single (centered) stripline assemblies are more difficult to handle from a manufacturing viewpoint. The PCB would become extremely thick if a single stripline configuration and a large layer stackup were provided. With increased layers, dual stripline is an optimal choice. Emissions performance is enhanced since two routing layers exist between two planar "image" (shield layers). Crosstalk is also minimized at the same time because the routing layers are always placed orthogonally to each other, (one layer in the x-axis, the other layer in the y-axis) or separated by interplaced image planes.

With different logic families and impedance concerns within a PCB, the characteristic impedance of a trace may also have to be different. An example of changing trace impedance in an impedance controlled assembly follows.

EXAMPLE

A motherboard is designed for a controlled 50-Ω impedance for all logic functions. Video circuitry must be routed at 75 Ω. How does one create two different impedance traces on the same PCB structure? This operation can be performed by changing physical dimensions within the PCB assembly. To change the impedance of a trace within a PCB, the following techniques are available from easiest (lower costs in board construction) to a more costly implementation method.

- Changing trace width referenced to a plane.
- Changing distance spacing between the routing layer and the reference plane.
- Removing a portion of the reference plane underneath the signal trace and allowing the trace to be referenced to another plane within the structure at a distance further away than the original reference plane (requires absence of copper over what would normally be a solid plane and a trace-void zone in adjacent layers).
- Changing thickness of the PCB layers (core material).
- Using a different dielectric constant (core or prepreg) between planar structures.

TABLE 6.3 Impedance of Different Conductor Pairs

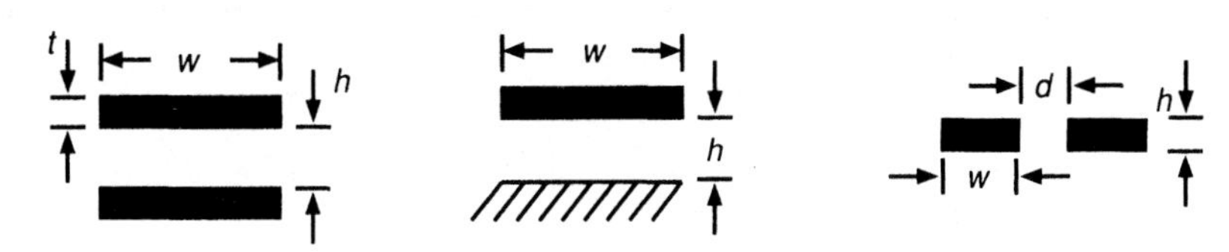

w/h or d/w	Parallel Traces Z_{o1}	Trace over Ground plane Z_{o2}	Traces Side by Side Z_{o3}
0.5	377	377	NA
0.6	281	281	NA
0.7	241	241	NA
0.8	211	211	NA
0.9	187	187	NA
1.0	169	169	0
1.1	153	153	25
1.2	140	140	34
1.5	112	112	53
1.7	99	99	62
2.0	84	84	73
2.5	67	67	87
3.0	56	56	98
3.5	48	48	107
4.0	42	42	114
5.0	34	34	127
6.0	28	28	137
7.0	24	24	146
8.0	21	21	153
9.0	19	[illegible]	160
10.0	17	[illegible]	166
12.0	14	[illegible]	176
15.0	11.2	[illegible]	188
20.0	8.4	[illegible]	204
25.0	6.7	[illegible]	217
30.0	5.6	[illegible]	227
40.0	4.2	[illegible]	243
50.0	3.4	[illegible]	255
100	1.7	[illegible]	293

N/A = not applicable

w/h = 3.5
h = 1.5
w = 5.25
h = 2.0
w = 7

microstrip
$Z_0 = 50$
w/h = 3.0
h = 2
w = 6.0
h = 1.5
w = 4.5

$$Z_{o1} = \left(377/\sqrt{\varepsilon_r}\right)(h/w), \text{ for } W > 3h \text{ and } h > 3t$$

$$Z_{o2} = \left(377/\sqrt{\varepsilon_r}\right)(h/w), \text{ for } W > 3h$$

$$Z_{o3} = \left(120/\sqrt{\varepsilon_r}\right) ln_e\,(d/w + \sqrt{(d/w^2 - 1}, \text{ for } W >> 1$$

$d >>$ nearby ground plane

diff $Z_0 = 50$
d/w = 1.5
w = 3
d = 4.5

Changing the width of the trace is the easiest method if a different impedance is required. If different logic families are provided in an assembly, poor performance or signal integrity can exist if impedance-controlled traces are routed on the same layer. Table 6.3 illustrates different impedance values for three common trace configurations [13].

6.8 CAPACITIVE LOADING

Capacitive input loading affects trace impedance and will increase with gate loading (additional devices added to the routed net). The unloaded propagation delay for a transmission line is defined by $t_{pd} = \sqrt{L_o C_o}$. If a lumped load, C_d, is placed in the transmission line (includes all loads with their capacitance added together), the propagation delay of the signal trace will increase by a factor of

$$t'_{pd} = t_{pd}\sqrt{1 + \frac{C_d}{C_o}} \text{ ns/length} \tag{6.21}$$

where t_{pd} = unmodified propagation delay, nonloaded circuit
t'_{pd} = modified propagation delay when capacitance is added to the circuit
C_d = input gate capacitance from all loads
C_o = characteristic capacitance of the transmission line

For example, let's assume a load of five CMOS components are on a signal route, each with 10-pF input capacitance (total of C_d = 50 pF). With this capacitance value on a glass epoxy board, 25 mil traces, and a characteristic board impedance Z_o = 50 Ω (t_r = 1.65 ns/ft), there exists a value of C_o = 35 pF. The modified propagation delay is:

$$t'_{pd} = 1.65 \text{ ns/ft}\sqrt{1 + \frac{50}{35}} = 2.57 \text{ ns/ft} \tag{6.22}$$

This equation states that the signal arrives at its destination 2.57 ns/ft (0.54 ns/cm) later than expected. The characteristic impedance of this transmission line, altered by gate loading, Z'_o, is:

$$Z'_o = \frac{Z_o}{\sqrt{1 + \frac{C_d}{C_o}}} \tag{6.23}$$

where Z_o = original line impedance (ohms)
Z'_o = modified line impedance (ohms)
C_d = input gate capacitance—sum of all capacitive loads
C_o = characteristic capacitance of the transmission line

For the example above

$$Z'_o = \frac{50}{\sqrt{1 + \frac{50}{35}}} = 32\ \Omega$$

Typical values of C_d are 5 pF for each ECL input, 10 pF for each CMOS device, and 10–15 pF for TTL. Typical C_o values of a PCB trace are 2–2.5 pF/inch. These C_o values are subject to wide variations due to the physical geometry and the length of the trace. Sockets and vias also add to the distributed capacitance (sockets ≈ 2 pF and vias ≈ 0.3–0.8 pF each). Given that $t_{pd} = \sqrt{L_o * C_o}$ and $Z_o = \sqrt{L_o / C_o}$, C_o can be calculated as

$$C_o = 1000\left(\frac{t_{pd}}{Z_o}\right) \text{pF / length} \tag{6.24}$$

This loaded propagation delay value is one method that may be used to decide if a trace should be treated as a transmission line ($2 * t'_{pd}$ * trace length > t_r or t_f) where t_r is the rising edge of the signal and t_f is the falling edge.

C_d, the distributed capacitance per length of trace, depends on the capacitive load of all devices including vias and sockets, if provided. To mask transmission line effects, slower edge times are recommended. A heavily loaded trace slows the rise and fall times of the signal due to an increased time constant ($\tau = ZC$) associated with increased distributed capacitance and filtering of high-frequency components from the switching device. Notice that the impedance, Z, is used, and not R (pure resistance) for the time constant equation. This is because Z consists of real resistance and inductive reactance. Inductive reactance, ($j\omega L$), is much greater than R in the trace structure at RF frequencies, which must be taken into consideration. Heavily loaded traces seem advantageous until the loaded trace condition is considered in detail.

A high C_d increases the loaded propagation delay and lowers the loaded characteristic impedance. The higher loaded propagation delay value increases the likelihood that transmission line effects will not be masked during rise and fall transition states. A lower loaded characteristic impedance often exaggerates impedance mismatches between the driving device and the PCB trace. Thus, the apparent benefits of a heavily loaded trace are not realized unless the driving gate is designed to drive large capacitive loads [2].

Loading alters the characteristic impedance of the trace. As with the loaded propagation delay, a high ratio between distributed capacitance and intrinsic capacitance exaggerates the effects of loading on the characteristic impedance. Because $Z_o = \sqrt{L_o / (C_o + C_d)}$, the additional load, C_d, adds capacitance. The loading factor $\sqrt{1 + C_d/C_o}$ divides in Z_o, and the characteristic impedance is lowered when the trace is loaded. Reflections on a loaded trace, which cause ringing, overshoots, undershoots, and switching delays, are more extreme when the loaded characteristic impedance differs substantially from the driving device's output impedance and the receiving device's input impedance. The units of measurements used for both capacitance and inductance are *per inch* or *cm* units. If the capacitance used in the L_o equation is pF/inch, the resulting inductance will be in pH/inch.

With knowledge of added capacitance lowering the trace impedance, it becomes apparent that if a device is driving more than one line, the active impedance of each line must be determined separately. This determination must be based on the number of loads and the length of each line. Careful control of circuit impedance and reflections for trace routing and load distribution must be given serious consideration during the design and layout of the PCB.

If capacitive input loading is high, compensating a signal may not be practical. Compensation refers to modifying the transmitted signal to enhance the quality of the received signal pulse using a variety of design techniques. For example, use of a series resistor, or a different termination method to prevent reflections or ringing that may be present in the transmission line, is one method to compensate a distorted signal. Reflections in multiple lines from a single source must also be considered.

The low impedance often encountered in the PCB sometimes prevents proper Z_o (impedance) termination. If this condition exists, a series resistor as large as possible should be put in the trace (without corrupting signal integrity). Even a 10-Ω resistor is helpful; however, 33-Ω is commonly used.

REFERENCES

[1] Coombs, C. 1996. *Printed Circuits Handbook.* New York: McGraw-Hill.

[2] Montrose, M. 1996. *Printed Circuit Board Design Techniques for EMC Compliance.* Piscataway, NJ: IEEE Press.

[3] Johnson, H. W., and M. Graham. 1993. *High Speed Digital Design.* Englewood Cliffs, NJ: Prentice Hall.

[4] Motorola, Inc. 1989. *Transmission Line Effects in PCB Applications (#AN1051/D).*

[5] Motorola, Inc. 1989. *Low Skew Clock Drivers and Their System Design Considerations (#AN1091).*

[6] Motorola, Inc. 1996. *ECL Clock Distribution Techniques (#AN1405).*

[7] Motorola, Inc. 1988. *MECL System Design Handbook (#HB205).* Chapters 3 and 7.

[8] IPC-D-317A. 1995, January. *Design Guidelines for Electronic Packaging Utilizing High-Speed Techniques.* Institute for Interconnecting and Packaging Electronic Circuits (IPC).

[9] IPC-2141. 1996, April. *Controlled Impedance Circuit Boards and High Speed Logic Design.* Institute for Interconnecting and Packaging Electronic Circuits.

[10] IPC-TM-650. 1996, April. *Characteristic Impedance and Time Delay of Lines on Printed Boards by TDR.* Institute for Interconnecting and Packaging Electronic Circuits.

[11] Van Doren, T. 1995. *Circuit Board Layout to Reduce Electromagnetic Emission and Susceptibility.* Seminar Notes.

[12] Paul, C. R. 1992. *Introduction to Electromagnetic Compatibility.* New York: John Wiley & Sons.

[13] Violette, M. 1986, March-April. *EMI Control in the Design and Layout of Printed Circuit Boards.* EMC Technology Magazine.

[14] Kaupp, H. R. 1967, April. "Characteristics of Microstrip Transmission Lines." *IEEE Transactions,* EC-16, No. 2.

[15] National Semiconductor. 1996. *LVDS Owner's Manual.*

[16] Booton, R. 1992. *Computational Methods for Electromagnetics and Microwaves.* New York: John Wiley and Sons.

[17] Collin, R. E. 1992. *Foundation for Microwave Engineering.* 2nd ed. New York: McGraw-Hill.

[18] Sadiku, M. 1992. *Numerical Techniques in Electromagnetics.* Boca Raton, FL: CRC Press.

7

Signal Integrity and Crosstalk

7.1 NEED FOR SIGNAL INTEGRITY

Every year, clocks and system speeds become faster. Clocks in computer systems that now operate at frequencies above 300 MHz will rise into the GHz range in the near future. Processors currently in development will reach new levels of high-speed performance. Computer systems, local-area networks, wide-area networks, and cellular and optical-fiber systems are becoming more common every day.

As the demand for more and faster computer chips increases, the chip vendors are also struggling to increase the yield and decrease the cost of the individual die obtained from each raw wafer. One method for simultaneously achieving all of these goals is to make more die on each silicon wafer using a process known as "die shrink." In die shrink, the photolithographic manufacturing steps are adjusted in order to reduce the size of each transistor gate without having to redesign the individual transistor components or their metallic interconnects. When the distances between individual gates on a die decrease, the time it takes for electrons to propagate between the gates also decreases. In addition, as the transistors are made smaller, the transistors switch faster. As a result, the component becomes faster with the same functional performance, all at practically zero additional cost per semiconductor wafer.

Today's chip manufacturing processes commonly use transistors and interconnects with submicron dimensions. Components made in the 1980s used fabrication technologies with 2- to 5-micron line widths. With 5-micron technologies, edge transition rates were in the 20-ns range. With 0.12- to 0.25-micron process technologies, speeds in the low picoseconds are now possible and will become the standard for most semiconductor products in the near future.

Chip manufacturing equipment is very expensive. Major new semiconductor fabrication facilities typically cost from $1 billion to $2 billion. The time it takes to incorporate a manufacturing process for a new silicon wafer is typically six to eight weeks. As a result, chip manufacturers are constantly changing their process steps in order to accommodate a smaller component and to achieve a higher and more uniform throughput of semiconductor die. It is much less expensive for a chip manufacturer to die shrink a standard CMOS logic component by 50% and label it with a slower speed rating than it is to redesign or retool the manufacturing process. The shrunken chip will still operate as desired; however, the "new" chip edge rate transitions are much faster than the "old" ones.

TTL logic is becoming obsolete with high-technology products; hence, newer logic families with higher frequency components are being developed. The component will operate as desired, but the "real" edge rate transitions are much faster. Recall that EMI spectral profiles are dominated by the edge rate transition times and their related harmonic frequency components. The process of marking chip components with a slower speed rating explains why a second source of components for the same function may result in significant EMI concerns that are not encountered with the original part. If a product once complied with EMI regulatory requirements for emissions and immunity, and the product now fails, it could be that the vendor took the device through the die shrink process without the customer knowing it. A typical logic designer will usually accept a manufacturer's faster component without considering EMI or signal integrity issues.

When components operate at high frequencies, signal edge rates become faster to accommodate the smaller clock pulse intervals. As a result, RF spectral distribution increases. When this level of technology is reached, signal integrity CAD tools (circuit simulation programs with high-frequency models) must be used to determine the effects of fast edge rates, long trace lengths, and parasitic capacitance and inductance of circuit elements. This includes the behavioral characteristics of the source and load devices, the conductor impedances, the physical and electrical parameters of the PCB materials, and numerous other parameters that become a prime concern for the designer during the circuit simulation and layout cycle.

To define high-speed design characteristics, a simple driver-receiver circuit is used to illustrate signal integrity problem, shown in Fig. 7.1 [5]. As long as the interconnects are "short" and the clock rates are "low," the receivers act as loads for the driver. This loading can affect the receiver's output waveform. The effect of loading the traces is counted as lumped capacitance to ground in addition to the capacitance provided by the receiver loads.

When the trace length becomes electrically "long,"—that is, when the edge transition time of the signal is less than the time it takes for the signal to travel from source to load and return from load to source—signal integrity concerns increasingly arise. Within the time period that the signal's transition occurs between the high and low state, the impedance of the trace becomes the actual load for the driver. This load is in addition to the input impedance of the receivers. Transmission line effects such as reflections, overshoot, undershoot, and crosstalk will distort the transmitted signal. Classical lumped circuit theory no longer applies under these conditions, and a distributed circuit model must be used in the circuit simulation design program [13].

To better describe this signal integrity concern, we reexamine why edge rate, and not clock frequency, is of primary concern. For example, a relatively low-frequency 1-MHz clock with a rise time of 1 ns has transmission line effects only during the transition time as the periodicity of the waveform is long. The signal will eventually attain steady state. Because of the fast edge rate, if the transmission line length is long, the sig-

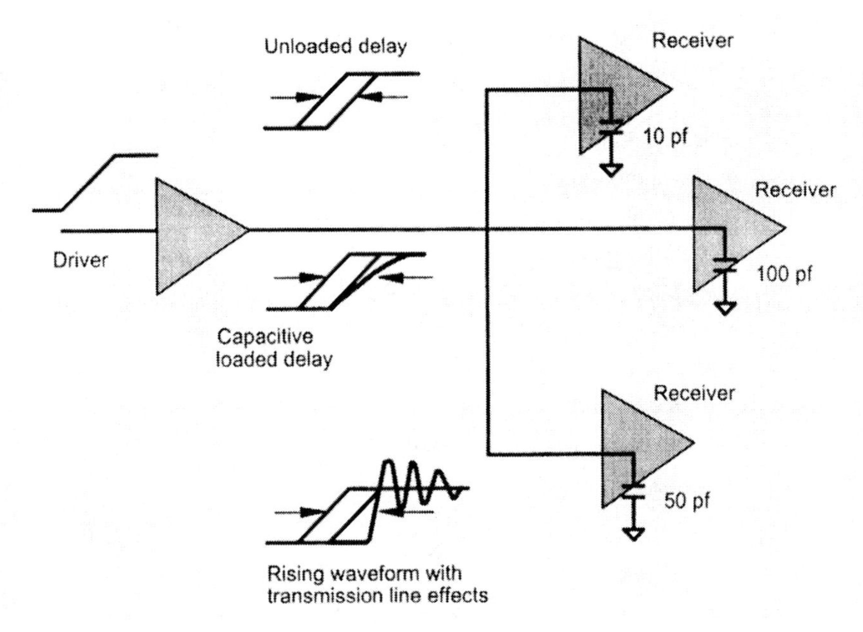

Figure 7.1 Transmission line effects in a trace.

nal may falsely trigger circuits, for the threshold voltage will vary at different portions of the timing cycle. Ringing and reflections may also exist.

A trapezoidal signal with a finite rise and fall time has a spectral distribution based on the Fourier component of the signal. The fundamental RF frequency is that of the clock itself. The magnitude of the spectral distribution decreases at the rate of 20 dB per decade up to the frequency that corresponds to the rise time. For example, a 5V, 100 MHz clock with a 50% duty cycle and a 1-ns edge rate has a fundamental RF frequency with an amplitude of approximately 3V. The 1-GHz component of this clock will have an amplitude of approximately 0.3V. Depending on the application of the circuit, undesired effects may occur along with switching noise and ground bounce. In mixed logic designs, this is a concern, especially if a 0.3V noise margin requirement exists. As a rule of thumb, the trace length should be between 20 and 25% of the signal rise and fall time.

For example, all logic and analog components have some sensitivity to changes in power supply voltages. This sensitivity may be reflected in changes in the output levels or in the switching thresholds at the input pins. The amount of performance degradation as a function of power supply voltage must be determined during the design cycle. A tradeoff between power supply accumulation and noise erosion needs to be performed, taking into consideration all other components and their unique noise margin requirements.

Lack of an optimal 0V reference structure is common in high-speed networks and designs. Inductance will always be present between the virtual and actual ground because of interconnect (trace) and lead inductance. If many clock drivers switch simultaneously, a voltage proportional to the rate of change of current with time is induced in the trace. This simultaneous switching may cause false switching of devices on victim traces. Vias, bond wires, and package connector pins also contribute to this inductive effect, which is also known as ground bounce or delta I noise. Component packages with *no* internal ground planes are the worst offender!

The following are potential sources of noise that may cause signal functionality concerns or make a logic signal unusable.

- Reflections
- Ground bounce
- Crosstalk
- Reference accuracy
- Thermal offsets
- Ground offsets
- Power supply variations
- Trace IR drop
- Ground IR drop
- Terminator noise

7.2 REFLECTIONS AND RINGING

Reflections are an unwanted byproduct in digital logic designs. Ringing within a transmission line contains both overshoot and undershoot before stabilizing to a quiescent level and is a manifestation of the same effect. *Overshoot* is the effect of excessive voltage above the power rail or below the ground reference. Excessive voltage levels below ground reference is not undershoot. Undershoot is a condition where the voltage level does not reach the desired amplitude for both maximum and minimum transition levels. Components must have a sufficient tolerance rating for voltage margin requirements. Both overshoot and undershoot can be controlled by proper terminations and proper PCB and IC package design. Overshoot and undershoot, if severe enough, can overstress devices and cause damage or even failure.

For an unterminated transmission line, ringing and reflected noise are one and the same. This can be observed with measurement equipment at the frequency associated as a quarter wave length of the transmission line, as is most apparent in an unterminated, point-to-point trace. The driven end of the line is commonly tied to AC ground with a low-impedance (5–20 ohms) load. This transmission line closely approximates a quarter wavelength resonator (stub shorted on one end, open on the other). Ringing is the resonance of that stub.

As signal edges become faster, consideration must be given to propagation and reflection delays of the routed trace. If the propagation time and reflection within the trace are longer than the edge transition time from source to load, an *electrically long trace* will exist. This electrically long trace can cause signal integrity problems depending on the type and nature of the signal. These problems include crosstalk, ringing, and reflections. EMI concerns are usually secondary to signal quality when referenced to electrically long lines. Although long traces can exhibit resonances, other suppression and containment measures implemented within the product may mask the EMI energy created. As a result, the device may cease to function properly if impedance mismatches exist in the system between source and load. Reflections are frequently both a signal quality and an EMI issue when the edge time of the signals constitutes a significant percentage of the propagation time between the device load intervals. Solutions to reflection problems may require extending the edge time (slowing the edge rate) or decreasing the distance between load device intervals.

Reflections from signals on a trace are one source of RF noise within a network. Reflections are observed when impedance discontinuities exist. These discontinuities consist of

- Changes in trace width
- Improperly matched termination networks
- Lack of terminations
- T-stubs or bifurcated traces[1]
- Vias between routing layers
- Varying loads and logic families
- Large power plane discontinuities
- Connector transitions
- Changes in impedance of the trace

When a signal propagates down a transmission line, a fraction of the source voltage will initially propagate down the trace. This source voltage is a function of frequency, edge rate, and amplitude. Ideally, all traces should be treated as a transmission line. Transmission lines are described by their characteristic impedance, Z_o, and propagation delay, t_{pd}. These parameters are dependent on the inductance and capacitance per unit length of the trace, the actual interconnect component, the physical dimensions of the interconnect, the RF return path, and the permittivity of the insulator between them. Propagation delay is also a function of the length of the trace and dielectric constant of the material. When the load impedance at the end of the interconnect equals that of the characteristic impedance of the trace, no signal is reflected.

A typical transmission line is shown in Fig. 7.2. Here we notice that

- Maximum energy transfer occurs when $Z_{out} = Z_o = Z_{load}$.
- Minimum reflections will occur when $Z_{out} = Z_o$ and $Z_o = Z_{load}$.

If the load is not matched to the transmission line, a voltage waveform will be reflected back toward the source. The value of this reflected voltage is

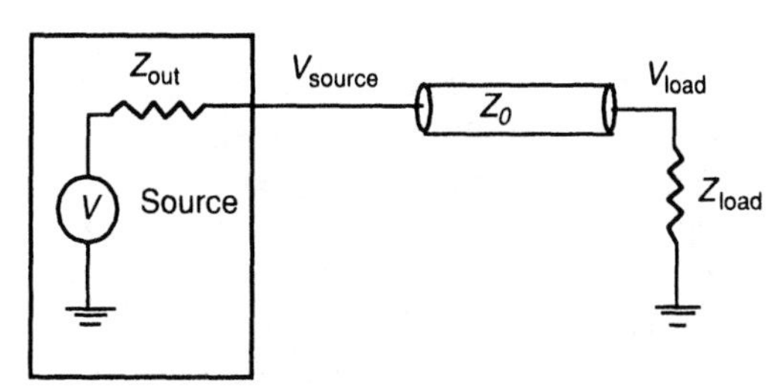

Figure 7.2 Typical transmission line system.

[1] A bifurcated trace is a single trace that is broken up into two traces routed to different locations.

$$V_r = V_o\left(\frac{Z_L - Z_o}{Z_L + Z_o}\right) \tag{7.1}$$

where V_r = reflected voltage
V_o = source voltage
Z_L = load resistance
Z_o = characteristic impedance of the transmission path

When Z_{out} is less than Z_o, a negative reflected wave will be created. If Z_L is greater than Z_o, a positive wave is observed. The wave will repeat itself at the source driver if the impedance is different from the line impedance, Z_o.

Equation (7.1) relates the reflected signal in terms of voltage. When a portion of the propagating signal reflects from the far end, this component of energy will travel back to the source. As it reflects back, the reflected signal may cross over the tail of the incoming signal. At this point, both signals will propagate simultaneously in opposite directions, neither interfering with each other.

We can derive an equation for the reflected wave. The reflection equation, Eq. (7.2), is for the fraction of the propagating signal that is reflected back toward the source.

$$\% \text{ reflection} = \left(\frac{Z_L - Z_o}{Z_L + Z_o}\right) \times 100 \tag{7.2}$$

This equation applies to any impedance mismatch, regardless of voltage levels. Use Z_o for the signal source of the mismatch and Z_L for the load. To improve the noise margin budget and requirements for logic devices, positive reflections are acceptable as long as they do not exceed V_{Hmax} of the receive component.

A forward-traveling wave is initiated at the source in the same manner as the incoming backward-traveling wave, which is the original pulse returned back to the source by the load. The corresponding points in the incoming wave are reduced by the percentage of the reflection on the line. The process of repeated reflections can continue as re-reflections at both the source and load. At any point in time, the total voltage (or current) becomes the *sum* of all the individual voltage (or current) sources present. It is for this reason that we may observe a 7V signal on the output of a source driver while the power supply is operating at 5V. The term *ringback* is the effect of the rising edge of a logic transition that meets or exceeds the logic level required for functionality, and then recrosses the threshold level before settling down. Ringback can be caused by a mismatch of logic drivers and receivers, poor termination techniques, and impedance mismatches of the network [14].

Sharp transitions in a trace may be observed through use of a Time Domain Reflectometer (TDR). Multiple reflections caused by impedance mismatches are observed by a sharp jump in the signal voltage level. These abrupt transitions usually have rise and fall times that can be comparable to the edge of the original pulse. The time delay from the original pulse to the occurrence of the first reflection can be used to determine the location of the mismatch. A TDR determines discontinuities within a transmission line structure. An oscilloscope observes reflections. Both types of discontinuities are shown in Fig. 7.3. Although several impedance discontinuities are shown, only one reflection is illustrated [16].

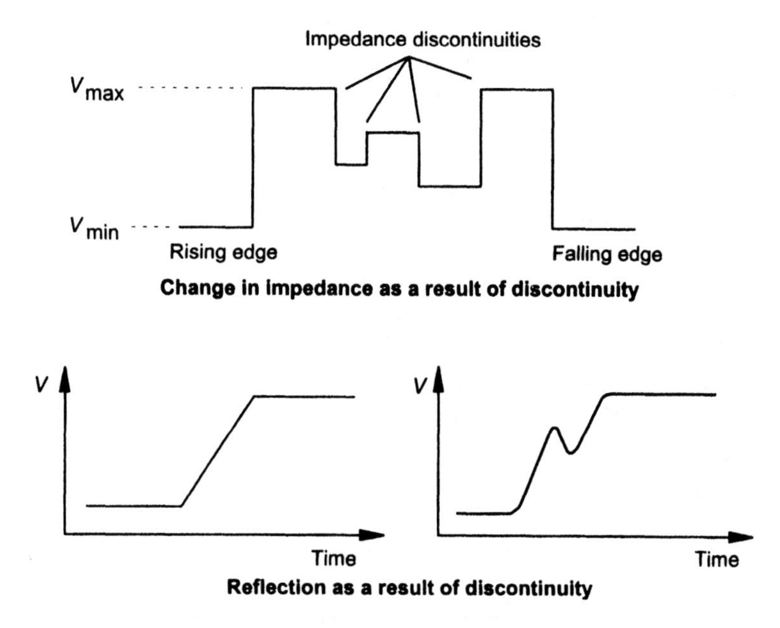

Figure 7.3 Discontinuities in a transmission line.

7.2.1 Identification of Signal Distortion

The shape of signal distortion can indicate the type of signal quality problem. When a signal deviates from its desired shape, the waveform will indicate the specific problem. The distortion on the leading or trailing edge of a pulse is often referred to as a *glitch*. Usually, little attention is paid to the detailed shape of the glitch. We examine two common waveforms that indicate signal quality problems in Fig. 7.4.

Ringing is caused by reflections with significant impedance mismatch (a result of resonant effects) in the trace and will corrupt signal quality and cause possible nonfunctionality of the circuit. Ringing within a trace also indicates that excessive inductance may be present in the network. The signal that is ringing will either add or subtract (voltage phasing of the signal). Depending on the net result of phasing, the signal may be degraded

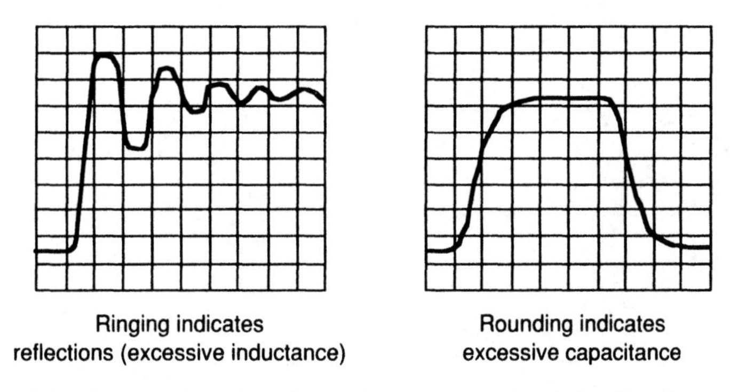

Figure 7.4 Ringing and rounding of a signal within a trace (identifying distortion).

to the point where it becomes an invalid or metastable logic state. Knowledge and proper use of transmission line theory allow signals to travel between the source and load without creating a functionality concern.

An *underdamped* circuit with excessive trace inductance can cause the edges of a signal pulse to ring. Ringing is a damped sinusoidal oscillation or resonance. For a circuit to ring or oscillate, there must be capacitance. *Capacitance* is a necessary part of the circuit load and will always be present. Excessive wiring inductance is also a cause of ringing. Ringing may be minimized by adding series resistance or providing proper termination.

Rounding indicates excessive capacitance. When the pulse edge is rounded, the circuit is *overdamped.* Shunt capacitance always exists in both the trace and the input to the load. The parallel combination of these two capacitors (trace and load) determines total shunt capacitance. Excessive series resistance in the signal source may also cause rounding due to the time constant $\tau = RC$. The impedance ratios between source, line, and load are an important requirement for circuit design parameters.

7.2.2 Conditions That Create Ringing

Figure 7.5 illustrates a typical circuit using lumped components. A source driver with internal series resistance, R_s, is shown along with inductance of the trace, L (which includes component lead wires), and distributed capacitance from trace to ground, X_c. This is in addition to the internal capacitance of the receiver. Within this circuit, additional resistance, inductance, and capacitance may also exist but is not included in this simple example.

Assume there is capacitive reactance of the trace plus load ($X_c = 1/\omega C$) which is much less than the load resistance (R_L) at high frequency. When a trace is physically short, the semiconductor package and decoupling capacitor lead-length inductance become the dominant cause of ringing. This is analyzed in terms of lumped circuitry where the damping of a simple RLC series circuit applies. The condition for ringing (underdamped) is

$$\text{Ringing} = R^2 X_c/4 > 1 \tag{7.3}$$

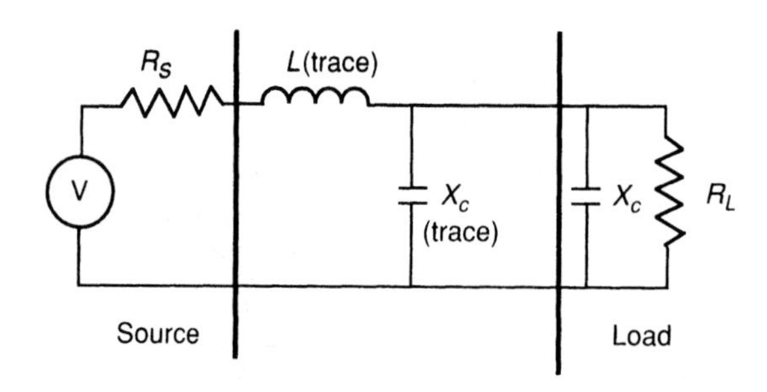

Figure 7.5 Equivalent circuit for ringing or rounding.

For rounding (overdamped)

$$\text{Rounding} = X_c > 4L/R^2 \tag{7.4}$$

The inductance that causes ringing is often very small. Sometime 0.5 nH is more than enough to cause signal functionality concerns. A 1-inch trace, located directly above a ground plane, could easily exceed this inductance value.

Figure 7.6 shows what occurs in a typical PCB layout as it relates to signal integrity: reflections and ringing. Notice the overshoot of up to 7V at the load and possible false triggering of a low logic state at 3.5V. At the source, the reflected signal also has an overshoot of −1.0V and ringing at 3V. False trigger can occur with 3V ringing for certain logic families. For a trapezoidal waveform, the ringing illustrated in Fig. 7.6 distorts the spectra, emphasizing the frequency of the ringing or the spectra of the complex waveform.

Figure 7.6, plot A, shows that if a properly terminated transmission line is present, a smooth signal pulse will be propagated down the trace from source to load. Ringing, which always exists in some manner, is usually minor compared to what can happen when a transmission line is not matched or has impedance discontinuities. Active components always exhibit some ringing generated by the output switching transistors. These transistors are generally nonideal or have nonperfect drive characteristics. These nonideal or nonperfect characteristics are due to the manufacturing process and design of the circuit. The behavioral models used for signal integrity analysis usually are considered as being ideal. In actual usage, behavioral models often may not illustrate real, important transmission line effects.

Figure 7.6, plot B, shows an active load circuit with an electrically long trace. Ringing (overshoot and undershoot) is present. Ringback, if severe enough, can falsely trigger the load into believing that a logic 0 state is present. This false triggering can cause im-

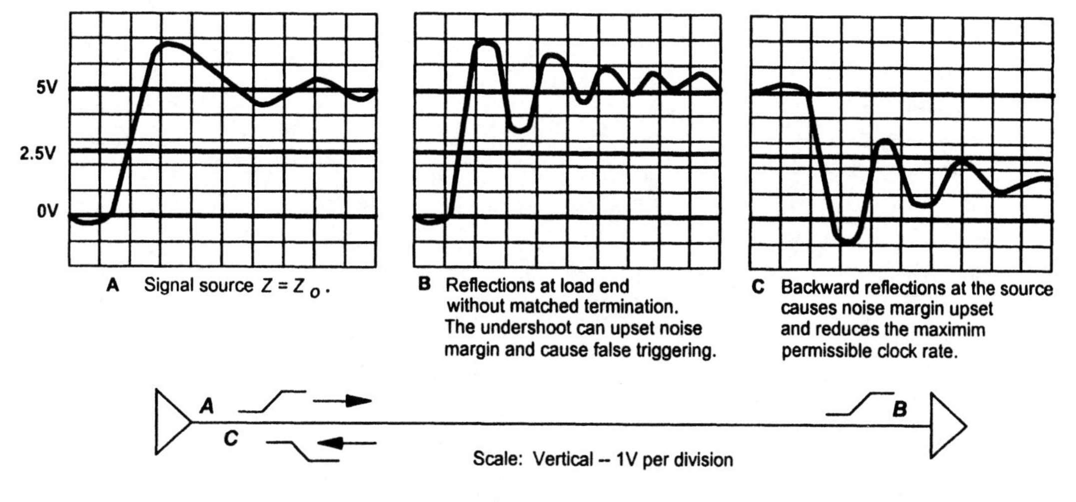

Figure 7.6 Ringing on traces.

proper operation of the circuit. If the length of the trace is very long with respect to the propagation delay of the signal (source-to-load and load-to-source) or if there are long intervals between devices, reflections will be created and bounced back and forth between these end points.

Figure 7.6, plot C, indicates what happens back at the source driver of an unterminated transmission line. Backward reflections can cause noise margin upset and will corrupt the quality of the desired signal if another clock transition occurs when the reflected pulse reaches the driver at the wrong point in time. These reflections are also created by an electrically long signal trace (or long loading intervals) as described for Fig. 7.6, plot B. When backward reflections occur, the edge rate desired for proper operation is reduced to a slower time period. This signal degradation may be sufficient to prevent other sections of the PCB from functioning at the intended speed of operation. Hence, performance is degraded, or the circuit may become nonfunctional.

The relationship between PCB line length and logic families is illustrated in Table 7.1. Details on how this table is created are presented later in this chapter.

TABLE 7.1 Logic Families and Important Characteristic Parameters

Logic Family (Sample List)	Rise/Fall Time (Approx.) T_r/T_f	Maximum Non-transmission Line Trace Length (Microstrip) $L_{max} = 9 * T_r$	Maximum Non-transmission Line Trace Length (Stripline) $L_{max} = 7 * T_r$
74L xxx	31–35 ns	279 cm (110″)	217 cm (85.4″)
74C xxx	25–60 ns	225 cm (88.5″)	175 cm (69″)
74HC xxx	13–15 ns	117 cm (46″)	91 cm (36″)
74 xxx	10–12 ns	90 cm (35.5″)	70 cm (27.5″)
(flip-flop)	15–22 ns	135 cm (53″)	105 cm (41″)
74LS xxx	9.5 ns	85.5 cm (34″)	66.5 cm (26″)
(flip-flop)	13–15 ns	117 cm (46″)	91 cm (36″)
74H xxx	4–6 ns	36 cm (14.2″)	28 cm (11″)
74S xxx	3–4 ns	27 cm (10.6″)	21 cm (8.3″)
74HCT xxx	5–15 ns	45 cm (18″)	35 cm (14″)
74ALS xxx	2–10 ns	18 cm (7″)	10 cm (4″)
74ACT xxx	2–5 ns	18 cm (7″)	10 cm (4″)
74F xxx	1.5–1.6 ns	10.5 cm (4″)	10.5 cm (4″)
ECL 10K	1.5 ns	10.5 cm (4″)	10.5 cm (4″)
ECL 100K	0.75 ns	6 cm (2.4″)	5.25 cm (2″)
BTL	1.0	9 cm (3.5″)	7 cm (2.8″)
LVDS	0.3*	2.7 cm (1.1″)	2.1 cm (0.8″)
GTL+	0.3*	2.7 cm (1.1″)	2.1 cm (0.8″)
GaAs	0.3*	2.7 cm (1.1″)	2.1 cm (0.8″)

Assume 1.7 ns/ft (0.14 ns/in. or 0.36 ns/cm) for microstrip. Propagation delay for FR-4, $\varepsilon_r = 4.6$.
Assume 2.2 ns/ft (0.18 ns/in. or 0.47 ns/cm) for stripline. Propagation delay for FR-4, $\varepsilon_r = 4.6$.
T_r and T_f depends greatly on load capacitance, supply voltage, and IC complexity.
Note: T_r and T_f will differ between device manufacturers because of the fabrication process used.
*These are the fastest edge rate values.

7.3 CALCULATING TRACE LENGTHS (ELECTRICALLY LONG TRACES)

When designing a transmission line, PCB designers need a method that allows quick determination if a trace routed on a PCB can be considered electrically long during component placement. A simple calculation is available that determines whether the approximate length of a routed trace is electrically long under typical conditions. When determining whether a trace is electrically long, we must think in the *time domain.* The equations provided below are best used when doing preliminary component placement on a PCB. For extremely fast edge rates, detailed calculations are required based on the actual dielectric constant value of the core and prepreg material. Chapter 6 provides equations if more accuracy is required.

Assuming a typical velocity of propagation that is 60% the speed of light, we can calculate the maximum permissible unterminated line length per Eq. (7.5). This equation is valid when the two-way propagation delay (source-load-source) equals the signal rise time.

$$l_{max} = \frac{t_r}{2\,t'_{pd}} \tag{7.5}$$

where t_r is edge rate (ns)
t'_{pd} is propagation delay (ns)
l_{max} maximum routed track length (cm)

Figure 7.7 illustrates this equation for quick reference with a dielectric constant of 4.6.

To simplify Eq. (7.5), we use the real value of propagation delay (actual dielectric constant based on frequency of interest; see Chapter 6) from FR-4 material using microstrip and stripline impedance equations (factoring in propagation delay and constant). Equations (7.6) and (7.7) are presented for determining the maximum electrical line length before termination is required. This length is for round-trip distance. *The one-way length, from source to load is one-half the value of* l_{max} *calculated below.* The following calculations are for a dielectric constant of 4.6.

$$\begin{aligned} l_{max} &= 9 * t_r && \text{(for microstrip topology—in cm.)} \\ l_{max} &= 3.5 * t_r && \text{(for microstrip topology—in in.)} \end{aligned} \tag{7.6}$$

$$\begin{aligned} l_{max} &= 7 * t_r && \text{(for stripline topology—in cm.)} \\ l_{max} &= 2.75 * t_r && \text{(for stripline topology—in in.)} \end{aligned} \tag{7.7}$$

For example, if a signal edge is 2 ns, the maximum round-trip, unterminated trace length when routed on microstrip is

$$l_{max} = 9 * t_r = 18 \text{ cm } (7'')$$

When this same clock trace is routed on stripline, the maximum unterminated trace length of this 2-ns signal edge becomes

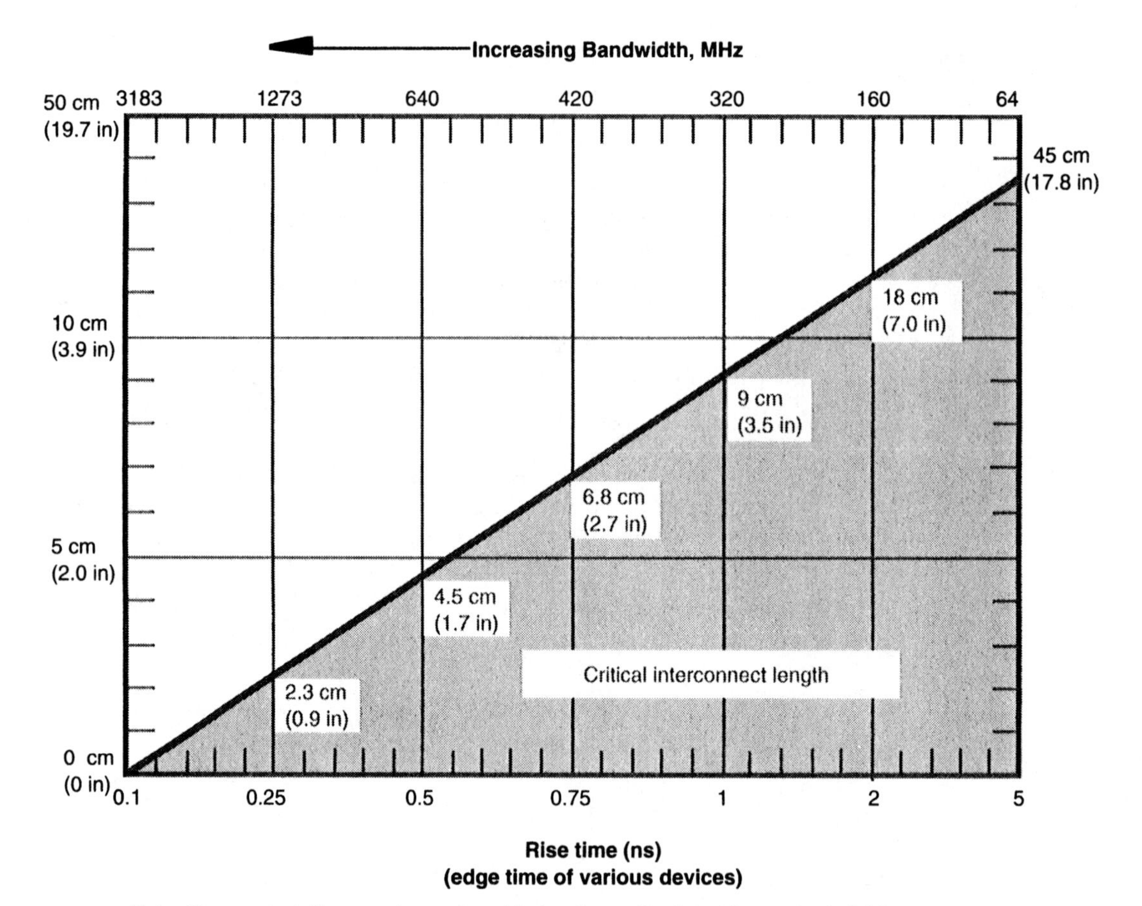

Figure 7.7 Maximum unterminated line length versus signal edge rate (FR-4 material).

$$l_{max} = 7 * t_r = 14 \text{ cm } (5.5'')$$

These equations are also useful when we are evaluating the propagational time between load intervals on a line with multiple loads.

To calculate the constant, k, either 7 or 9, for l_{max}, use the following example.

EXAMPLE

$$k = x\left(\frac{1}{[t_{pd}/a]}\right) \quad \text{(cm)}$$

where $a = 30.5$ for cm, 12 for inches; $x = 0.5$, converts transmission line to one way path

$t_{pd} = 1.017\sqrt{0.475\,\varepsilon_r + 0.67}$ (for microstrip); $t_{pd} = 1.017\sqrt{\varepsilon_r}$ (for stripline)

example for $\varepsilon_r = 4.6$, $k = 8.9$ for microstrip (in cm) or 3.5 (in inches)

$k = 6.9$ for stripline (in cm) or 2.7 (in inches)

If a trace or interval is longer than l_{max}, then termination should be implemented, for signal reflections (ringing) may occur in this electrically long trace. Even with good termination, a finite amount of RF currents can still be in the trace. For example, use of a series resistor (source location) will achieve the following:

- Minimize RF currents within the trace.
- Absorb reflections (ringing).
- Match trace impedance.
- Minimize overshoot and undershoot.
- Reduce RF energy generated by slowing down the edge rate of the clock signal.

When placing PCB components during layout that use clocks or periodic waveform signals, these components must be located so that the signal traces are routed for the best straight-line path possible with minimal trace length and number of vias in the route. Each via will add inductance and discontinuities to the trace (approximately 1–3 nH each). Inductance in a trace may cause signal integrity concerns, impedance mismatches, and potential RF emissions. Inductance in a trace allows this wire to act as an antenna. The faster the edge rate of the clock signal, the more important this design rule becomes. If a periodic signal or clock trace must traverse from one routing plane to another, this transition should occur at a component lead (pin escape) and not anywhere else. If possible, additional inductance presented to the trace can be reduced from using two less vias.

Equation (7.8) is used to determine if a trace or loading interval is electrically long and requires termination.

$$l_d < l_{max} \tag{7.8}$$

where l_{max} is the calculated maximum trace length and l_d is the length of the trace route as measured in the actual board layout. Keep in mind that l_d is the round-trip length of the trace.

Ideally, trace impedance should be kept at ± 10%. In some cases, ±20–30% may be acceptable only after careful consideration has been given to performance. The width of the trace, its height above a reference plane, dielectric constant of the board material, plus other microstrip and stripline constants determine the impedance of the trace (see Eqs. 6.7 through 6.20). It is always best to maintain constant impedance control at all times for any dynamic signal condition.

An example used to determine whether it is necessary to terminate a signal trace using characteristic impedance, propagation delay, and capacitive loading is now presented.

MICROSTRIP EXAMPLE

A 5-ns edge rate device is provided on a 5-inch surface microstrip trace. Six loads (components) are distributed throughout the route. Each device has an input capacitance of 6 pF. Is termination required for this route?

Geometry

Trace width, $W = 0.010$ in.
Height above a plane, $H = 0.012$ in.

Trace thickness, $T = 0.002$ in.
Dielectric constant, $\varepsilon_r = 4.6$

A. Calculate characteristic impedance and propagation delay detailed in Chapter 6 [Eqs. (6.7) and (6.9)].

$$Z_o = \left(\frac{79}{\sqrt{\varepsilon_r + 1.41}}\right) Ln \left(\frac{5.98\,H}{0.8\,W + T}\right)$$

$$Z_o = \left(\frac{79}{\sqrt{4.6 + 1.41}}\right) Ln \left(\frac{5.98 \times 12}{0.8 \times 10 + 2}\right) = 63.5\ \Omega$$

$$t_{pd} = 1.017 * \sqrt{0.475\,\varepsilon_r + 0.67} = 1.72\text{ ns / ft}\quad (0.143\text{ ns/in.})$$

B. Analyze capacitive loading.

Calculate C_d, distributed capacitance (total normalized input capacitance divided by length).

$$C_d = 6 * C_d/\text{trace length} = (6 * 6\text{ pF}) / 5\text{ in.} = 7.2\text{ pF/in.}$$

Calculate intrinsic capacitance of the trace—Eq. (6.24).

$$C_o = 1000\,(t_{pd}/Z_o) = 1000\,(1.72/63.5)\text{ ns/ft} = 27.0\text{ pF/ft} = 2.26\text{ pF /in.}$$

Calculate one-way propagation delay time from the source driver—Eq. (6.21).

$$t'_{pd} = t_{pd}\sqrt{1 + C_d/C_o}$$

$$t'_{pd} = 0.143\sqrt{1 + 7.2 / 2.26} = 0.29\text{ ns/in. } (3.5\text{ ns/ft})$$

C. Perform a transmission line analysis.

Ringing and reflections are masked during edge transitions if

$$(2*t'_{pd})*\text{trace length} \le t_r \text{ or } t_f$$

For this situation,

$$(2 * t'_{pd})*\text{ trace length} = (2*0.29\text{ ns/in.}) * 5\text{ in.} = 2.9\text{ ns}$$

Given that the edge rate of the component is $t_r = t_f = 5$ ns and propagation delay is 2.9 ns, termination is not required. Sometimes the guideline of ($3 \times t'_{pd} \times$ trace length) is used as a margin of safety. For this case, propagation delay would be 4.35 ns; hence, termination would still not be needed.

Assume now that the trace is routed stripline. Is termination required?

From above:

$$t_{pd} = 1.017 \times \sqrt{\varepsilon_r} = 2.18\text{ ns/ft} = 0.18\text{ ns/in.}$$

$$C_o = t_{pd} / Z_o = 2.18 / 63.5 = 34.3\text{ pf/ft } (2.86\text{ pF/in.})$$

$$C_o = 1000\,(t_{pd}/Z_o) = 1000\,(2.18/63.5) = 34.3\text{ pF/ft } (2.86\text{ pF/in.})$$

C_d is the same as above (7.2 pF/in.)

$$t'_{pd} = t_{pd}\sqrt{1 + C_d/C_o} = 4.05\text{ ns/ft } (0.34\text{ ns/in.})$$

$$2 \times t'_{pd} \times \text{trace length} = 2 \times 0.34\text{ ns/in.} \times 5\text{ in.} = 3.4\text{ ns}$$

Again, this trace would not require termination since 3.4 ns ≤ 5 ns. The propagation delay for stripline is 1.60 ns longer because t_{pd} (unloaded) is substantially greater than microstrip (0.65-ns margin). This factor helps prevent transmission line effects from being masked during edge rate changes.

STRIPLINE EXAMPLE

A 2-ns edge rate device on a 10-inch stripline trace is used. Five logic devices are distributed throughout the route. Each device has an input capacitance of 12 pF. Is termination required for this route?

Geometry

Trace width, $W = 0.006$ in.
Distance from a plane, $B = 0.020$ in.
Trace thickness, $T = 0.0014$ in.
Dielectric constant, $\varepsilon_r = 4.6$

A. Calculate characteristic impedance and propagation delay detailed in Chapter 6. [Use Eqs. (6.13) and (6.15).]

$$Z_o = \left(\frac{60}{\sqrt{\varepsilon_r}}\right) Ln\left(\frac{1.9\,B}{0.8W + T}\right)$$

$$Z_o = \left(\frac{60}{\sqrt{4.6}}\right) Ln\left(\frac{1.9 \times 20}{0.8(6) + 1.4}\right) = 50.7\ \Omega$$

$$t_{pd} = 1.017\sqrt{\varepsilon_r} = 2.18\ \text{ns/ft}\quad (0.182\ \text{ns/in.})$$

B. Analyze capacitive loading.

Calculate C_d, distributed capacitance (total input capacitance divided by length).

$$C_d = 6 * C_d \,/\, \text{trace length} = (6 * 12\ \text{pF}) \,/\, 10\ \text{in.} = 7.2\ \text{pF/in.}$$

Calculate intrinsic capacitance of the trace.

$$C_o = 1000\,(t_{pd} / Z_o) = 1000\,(0.182/50.7) = 3.58\ \text{pF/in.}\quad (43.0\ \text{pF/ft})$$

Calculate one-way propagation delay time from the source driver.

$$t'_{pd} = 0.182\sqrt{1 + 7.2 / 3.58} = 0.32\ \text{ns/in.}\ (3.79\ \text{ns/ft})$$

C. Perform transmission line analysis.

The important condition of interest is $(2 * t'_{pd}) *$ trace length $\leq t_r$ or t_f.

$$(2 * t'_{pd}) * \text{trace length} = (2 * 0.32\ \text{ns/in.}) * 10\ \text{in.} = 6.4\ \text{ns}$$

Since the edge rate of the component $t_r = t_f = 2$ ns, and propagation delay (6.4 ≥ 2), termination is required to absorb transmission line effects.

Assume the trace is routed microstrip. Is termination required?

From above:

$$t_{pd} = 1.017 \times \sqrt{0.475\,\varepsilon_r + 0.67} = 0.14\ \text{ns/in.}\ (1.72\ \text{ns/ft})$$

$$C_o = 1000\,(t_{pd}/Z_o) = 1000\,(0.14/50.7) = 2.76\ \text{pF/in.}\ (33\ \text{pF/ft})$$

C_d is the same as above (7.2 pF/in.)

$$t'_{pd} = t_{pd}\sqrt{1 + C_d / C_o} = 0.26 \text{ ns/in. } (3.19 \text{ ns/ft})$$

$$2 \times t'_{pd} \times \text{trace length} = 2 \times 0.26 \text{ ns/ft} \times 10 \text{ in.} = 5.20 \text{ ns}$$

Again, this trace would require termination since 5.20 ns ≥ 2 ns.

7.4 LOADING DUE TO DISCONTINUITIES

Depending on the routed configuration of a net along with component placement, the effects of a transmission line discontinuity must be examined. For a discontinuity to exist, a finite distance is required between source and load as well as the time of propagation across the distance interval with respect to edge times. This environment includes the difference between both a lumped and a distributed capacitive loaded transmission environment. This difference is dependent on the spacing at which the loads start to affect each other with a dependency of edge time of the signal [8]. The return path discontinuities also need to be considered. If a trace switches routing layers internal to the PCB, impedance control may be disrupted.

The load separation distance interval defines the point where reflections from one load on a transmission line starts to affect adjacent loads.

Within a PCB, logic input has an effective input capacitance associated with it. In a transmission line structure, a point discontinuity occurs whenever a capacitive load is provided. Each discontinuity will allow the propagated signal to pass down the line with a small portion of the signal reflected back toward the source. If the reflected signal is large, signal integrity concerns exist, including reflections and crosstalk. The width of the reflected pulse is a function of the edge transition of the incident pulse.

For a noticeable effect to be observed by point discontinuities, there must be a finite amount of distance separation. We identify this unit of separation as l_{sep} between adjacent loads. If the distance, l_{sep} (propagation time) is small with respect to edge times, the reflected pulses will not add together or combine into one large discontinuity, which may be measured with use of a TDR.

To calculate l_{sep} between two points, consider the reflected pulse width. This pulse width is directly equal to the propagation time between points A and B. Using knowledge of an electrically long trace which includes round-trip propagation delay (source-to-load and return path from load-to-source), use Eq. (7.9), based on Fig. 7.8 [8].

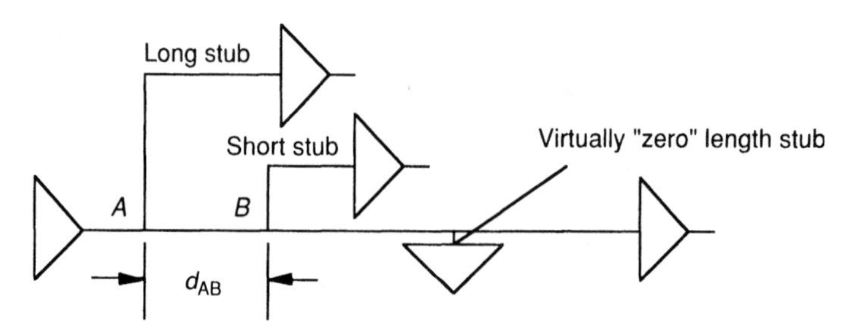

Figure 7.8 Point discontinuities and distance separation.

$$\gamma = 2 * d_{AB} * t_{pd} = 1.7 * t_r \quad (7.9)$$

where γ = width of the reflected pulse from A
d_{AB} = distance between points A and B
t_{pd} = unloaded propagation delay of the transmission line
t_r = 10%–90% edge transition rate (rise time)

Solving for d_{AB}, calculate minimum distance separation before a point discontinuity is observed.

$$d_{AB} = (1.7 * t_r)/(2 * t_{pd}) = 0.85\, t_r / t_{pd} \quad (7.10)$$

Assume an edge transition time of 0.8 ns (typical of a high-speed edge rate component) and FR-4 material. If a trace is routed stripline, with a dielectric constant 4.2, ($t_{pd} = 1.017\sqrt{4.2} = 2.08$ ns/ft or 0.17 ns/in.). The distance between two points acceptable before point discontinuity affects the signal is

$$d_{AB} = (1.7 * 0.8\text{ns}) / (2 * 0.17 \text{ ns/in.}) = 4.0 \text{ in. } (10.2 \text{ cm}) \quad (7.11)$$

The above example illustrates that if there are to be no point discontinuities, or a reflected pulse, the distance separation between points A and B must be less than 4.0 in., actual routed length.

What is meant by the term *reflected pulse?* A reflected pulse is a combination of two pulses as shown in Fig. 7.9. When there is a combination of multiple reflected signals, a single pulse effect will be observed within the circuit. If two pulses are allowed to overlap each other within the minimum distance separation, the maximum overlap of the two pulses will be less than the maximum amplitude of either pulse. This occurs when the pulse reaches a maximum amplitude and width. One-half of the pulse overlaps with another identical pulse at its halfway point. With this situation, distance separation can be reduced without affecting signal content.

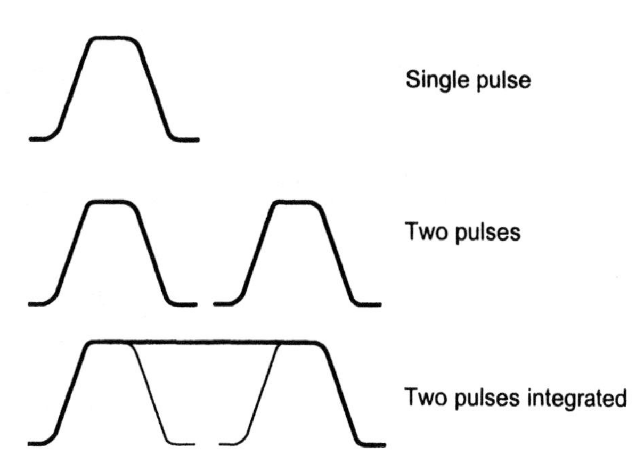

Figure 7.9 Reflected pulses from point discontinuities.

7.5 RF CURRENT DISTRIBUTION

A 0V reference plane allows RF current to return to its source from a load. This 0V plane completes the closed-loop circuit requirements discussed in Chapter 2. Current distribution along microstrip traces tends to spread out within the ground plane structure as illustrated in Fig. 7.10. This distribution will always exist in both the forward direction and the return path. This current distribution will share a common impedance between the trace and plane (or trace-to-trace), which results in mutual coupling due to the current spread. The peak current density lies directly beneath the trace and falls off sharply from each side of the trace into the ground plane structure.

When the distance spacing is far apart between trace and plane, the loop area between the forward and return path increases. This return path increase raises the inductance of the circuit where inductance is proportional to loop area. Equation (7.12) describes the current distribution that is optimum for minimizing total loop inductance for both the forward and return current path. The current that is described in Eq. (7.12) also minimizes the total amount of energy stored in the magnetic field surrounding the signal trace [3].

$$i(d) = \frac{I_o}{\pi H} \cdot \frac{1}{1 + \left(\frac{D}{H}\right)^2} \tag{7.12}$$

where $i(d)$ = signal current density (A/in.)
I_o = total current (A)
H = height of the trace above the ground plane (inches)
D = perpendicular distance from the centerline of the trace (inches)

The mutual coupling factor is highly dependent on frequency of operation and the skin depth effect of the ground plane impedance. As the skin depth increases, the resistive component of the ground plane impedance will also increase. This increase will be observed with proportionality at relatively high frequencies [2, 14].

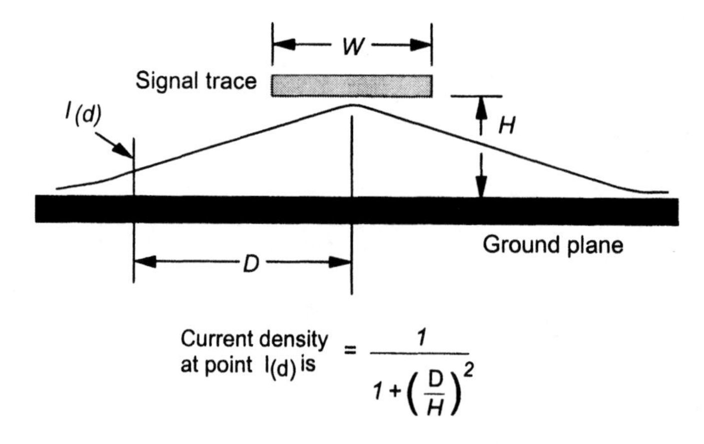

Figure 7.10 Current density distribution from trace to reference plane.

7.6 CROSSTALK

Crosstalk is one of several important aspects of a PCB design that must be considered during any design cycle. *Crosstalk* refers to the unintended electromagnetic coupling between traces, wires, trace-to-wire, cable assemblies, components, and any other electrical component subject to electromagnetic field disturbance. Crosstalk is caused by currents and voltages in a network and is similar to antenna coupling. When coupling occurs, near-field effects are observed.

Crosstalk between wires, cables, and traces affects intrasystem performance [2]. Intrasystem refers to both source and receptor being located within the same system or assembly. A product must be designed to be self-compatible. Hence, crosstalk may be identified as EMI internal to a system that must be minimized or eliminated. Crosstalk is an undesirable feature usually associated not only with clock or periodic signals, but also with data, address, control, and I/O traces.

Crosstalk is generally considered to be a functionality concern (signal quality) by causing a disturbance between traces. In reality, crosstalk can be a major contributor in the propagation of EMI. High-speed traces, analog circuits, and other high-threat signals may be corrupted by crosstalk induced from external sources. These EMI sensitive circuits may, however, also unintentionally couple their RF energy to the I/O section. This I/O coupling can result in radiated or conducted EMI that may be present within the enclosure or cause functionality problems between circuits and subsystems.

For crosstalk to occur, typically three or more conductors are required. These three conductors are identified in Fig. 7.11. Two lines carry the signal of interest, and the third line is a reference conductor which gives the circuits the ability to talk (communicate) to each other by capacitive or inductive coupling. If a two-wire system is provided, one wire pair is usually at a reference potential, while the other is differential, which prevents crosstalk from naturally occurring [2].

Figure 7.11 illustrates coupling between two circuits due to the result of a nonzero impedance in the mutual ground reference structure. This impedance is a prime reason why it is important to keep a low impedance between connecting points at 0V reference or ground.

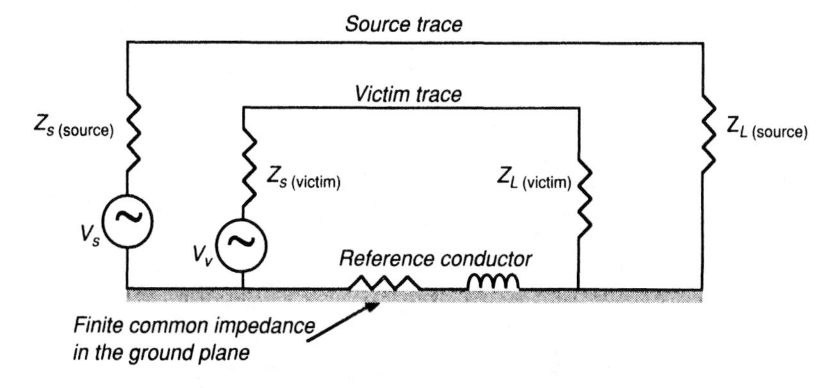

Figure 7.11 Three-conductor representation of a transmission line illustrating crosstalk.

Another visual representation of trace-to-trace coupling with capacitive and inductive components within a PCB trace is shown in Fig. 7.12. This is a detailed schematic of what occurs in a three-wire configuration. There are two parallel traces with mutual coupling mechanisms. The coupling on the source trace occurs through the common ground impedance, Z_g, the mutual capacitance between the traces, C_{sv}, and the mutual inductance M_{sv} between traces. Capacitive coupling between a trace and a reference plane is identified as C_{sg} (source-ground) and C_{vg} (victim-ground).

Crosstalk evaluation requires frequency domain analysis. What occurs in a logic circuit is both capacitive and inductive coupling of an electromagnetic field that interacts with other circuits. In Fig. 7.11, V_s is the source that generates an electromagnetic field that interacts between source and victim circuit, or trace. This interaction induces current and voltages at the input terminals of the termination point Z_s and Z_L. This termination is attached to the source and load ends of the circuit, respectively. The designer's responsibility is to determine if the crosstalk is near end (Z_s) or far end (Z_L). Near end refers to the

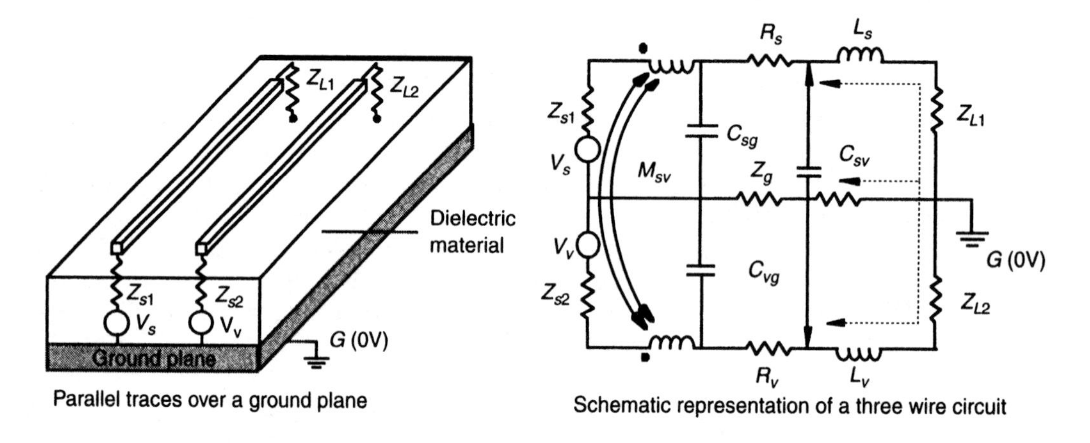

C_{sv} = Capacitance between source trace and victim trace
C_{vg} = Capacitance between victim trace and ground
C_{sg} = Capacitance between source trace and ground

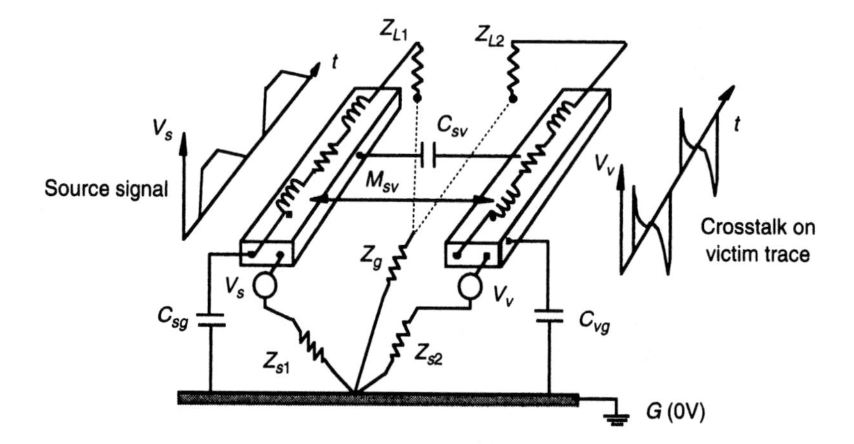

Figure 7.12 Trace-to-trace coupling within a PCB structure.

point of the circuit that is identified as source. Far end is the point identified as the load side of the circuit.

Time domain crosstalk analysis determines the time form of the receptor's terminal voltages, whereas frequency domain crosstalk analysis determines the magnitude and phase of the receptor voltages for a sinusoidal source voltage (electromagnetic field) [2]. This discussion examines only frequency domain analysis.

Crosstalk involves capacitive and inductive coupling. Capacitive coupling usually results from traces lying one on top of the other or above a reference plane. This coupling is a direct function of the distance spacing between the trace and an overlap area. The discussion on RF current distribution in the previous section illustrates the field density that occurs between a trace and reference plane. Coupled signals may exceed design limits with a very short trace route. This coupling may also be so severe that overlapping parallelism should be avoided at all times.

Inductive crosstalk involves traces that are physically located in close proximity to each other. With parallel routed traces, two forms of crosstalk will be observed, forward and backward. In a PCB, backward crosstalk is considered to be a greater concern than forward crosstalk. The high impedance in the circuit between source and victim trace will produce a high level of crosstalk. The preferred method for preventing crosstalk must be implemented during routing of the traces, or their physical location relative to a cable, I/O interconnect, and similar circuits subject to corruption. Inductive crosstalk can be controlled by increasing trace edge-to-edge separation between offending transmission lines or wires, or by minimizing the height separation distance of the trace above the reference plane.

Capacitive and inductive coupling are shown in Fig. 7.13. If a signal is sent from source-to-load, trace *C-D*, the signal will capacitively couple to the adjacent line, trace

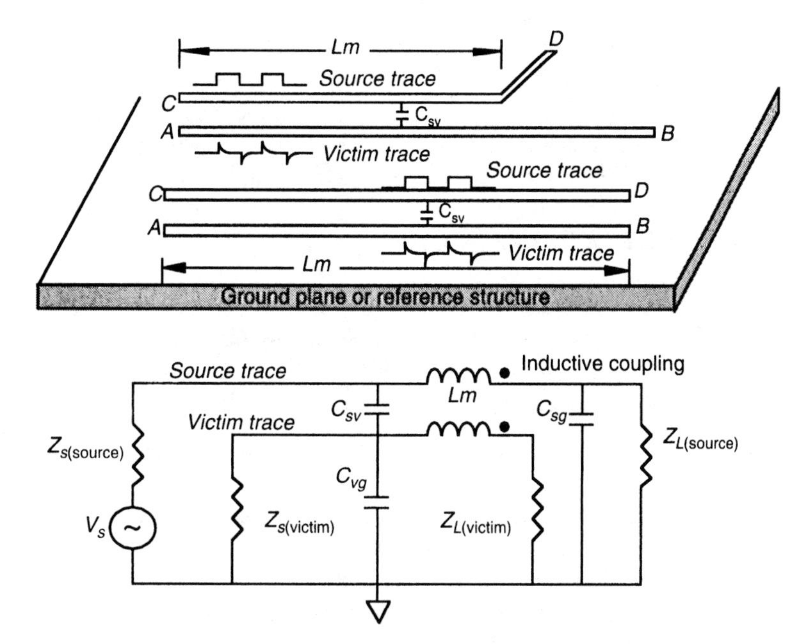

$$Z_v = \frac{Z_s(v) \times Z_L(v)}{Z_s(v) + Z_L(v)}$$

C_{sv} = Capacitance between source trace and victim trace
C_{vg} = Capacitance between victim trace and ground
C_{sg} = Capacitance between source trace and ground

Figure 7.13 Fundamental representation of crosstalk.

A-B, only if the two lines are parallel to each other and in close proximity. The larger the capacitance between the two traces (mutual capacitance), the tighter the coupling that occurs with crosstalk energy transferred between the two. The coupled voltage on the victim trace, *A-B*, causes a current to flow from the "coupling point" toward both ends of the trace. The current going back toward the source, *A*, is backward crosstalk, whereas the signal traveling to the load, *B*, is forward crosstalk. The two traces also have mutual inductance between them, causing inductive coupling, L_m, of the current in the direction of backward crosstalk. If the output impedance of the driver, *A*, is normally low compared to the transmission line impedance, most of the backward crosstalk is reflected back toward the driver, *A*. Because a capacitor conducts RF energy (current) efficiently at high frequencies, the faster the edge rate, the greater the crosstalk.

The polarities for mutual capacitive coupling are positive for forward and negative for backward. This is the only difference in behavior from that of inductive interference spikes; otherwise, both coupling modes are essentially identical. This coupling spreads out over a period of $2t_p$ where t_p is the time period for signal transmission round-trip.

Backward crosstalk increases linearly with the coupled length. If the coupled length is electrically long, with respect to propagation delay of the round-trip signal, backward crosstalk will show a saturated value and not increase as the coupled length increases.

Under typical operating conditions when a contiguous ground plane is provided, both inductive and capacitive crosstalk coupling voltages approximate the same result. Forward crosstalk cancels while backward crosstalk reinforces. Stripline topology provides a balance between both inductive and capacitive coupling since forward-coupling coefficients are small. Microstrip allows the electric fields generated to partially radiate through free space instead of only through the dielectric material of the PCB. Although less capacitive coupling exists, this coupling can still be present which leads to a small negative forward coupling coefficient.

With an imperfect planar structure, such as a slotted or hashed reference plane, the inductive crosstalk component becomes larger (more inductance in the plane). This inductive crosstalk will become greater than the capacitive component. Forward crosstalk will then be created which is greater than that observed in stripline and is negative in polarity. Forward crosstalk will always be less than backward crosstalk [14].

7.6.1 Units of Measurement—Crosstalk

Crosstalk is measured in units of dB because the reference level is not an absolute power level. The reference is 90 dB loss from the interfering circuit to the victim circuit. As a result, this unit measures how much crosstalk coupling loss is above 90 dB. The relationship describing this is shown in Eq. (7.13) [4].

$$\text{dB} = 90\text{-(crosstalk coupling loss in dB)} \tag{7.13}$$

For example, circuit A couples with circuit B. Circuit A is at a 58 dB lower power level. We calculate the crosstalk from circuit A to B as 32 dB.

Crosstalk is also expressed by Eq. (7.14) when applied to a source and victim circuit. Since we reference voltage against voltage, there is no unit of measurement except that of the basic dB.

$$X_{\text{talk(dB)}} = 20 \log \frac{V_{\text{victim}}}{V_{\text{source}}} \tag{7.14}$$

7.6.2 Design Techniques to Prevent Crosstalk

To prevent crosstalk within a PCB, design and layout techniques listed here are useful within a PCB.

Crosstalk will sometimes increase with a wider trace width. This is not true if the separation distance is held constant as a result of the ratio of self and mutual capacitance being held at a fixed ratio value. If the ratio is not fixed, mutual capacitance, C_m, will increase. With parallel traces, the longer the trace, the greater the mutual inductance, L_m. An increase in impedance, along with mutual capacitance, will increase with faster rise times of a signal transition, thus exacerbating crosstalk. The design and layout techniques are as follows.

1. Group logic families according to functionality. Keep the bus structure tightly controlled.
2. Minimize physical distance between components.
3. Minimize parallel routed trace lengths.
4. Locate components away from I/O interconnects and other areas susceptible to data corruption and coupling.
5. Provide proper terminations on impedance-controlled traces, or traces rich in harmonic energy.
6. Avoid routing of traces parallel to each other. Provide sufficient separation between traces to minimize inductive coupling.
7. Route adjacent layers (microstrip or stripline) orthogonally. This prevents capacitive coupling between the planes.
8. Reduce signal-to-ground reference distance separation.
9. Reduce trace impedance and signal drive level.
10. Isolate routing layers that must be routed in the same axis by a solid planar structure (typical of backplane stackup assignments).
11. Partition or isolate high noise emitters (clocks, I/O, high-speed interconnects, etc.) onto different layers within the PCB stackup assignment.

The best technique to prevent or minimize crosstalk between parallel traces is to maximize separation between the traces or to bring the traces closer to a reference plane. These techniques are preferred for long clock signals and high-speed parallel bus structures. An illustration of various crosstalk configuration is shown in Fig. 7.14.

Because of the current density distribution described in Eq. (7.12), the associated local magnetic field strength drops off with distance. An easy method to calculate trace separation is to use Eq. (7.15). This equation expresses crosstalk as a ratio of measured noise voltage to the driving signal. The constant K depends on the circuit rise time and the length of the interfering traces. This value is always less than one. For most approximations, the value of one is generally used. This equation clearly shows that to minimize crosstalk, we must minimize H and maximize D [3].

$$\text{Crosstalk} \approx \frac{K}{1 + \left(\frac{D}{H}\right)^2} = \frac{K(H)^2}{H^2 + D^2} \tag{7.15}$$

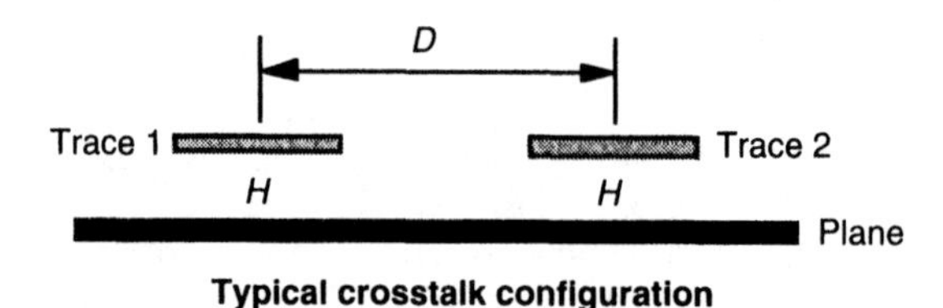

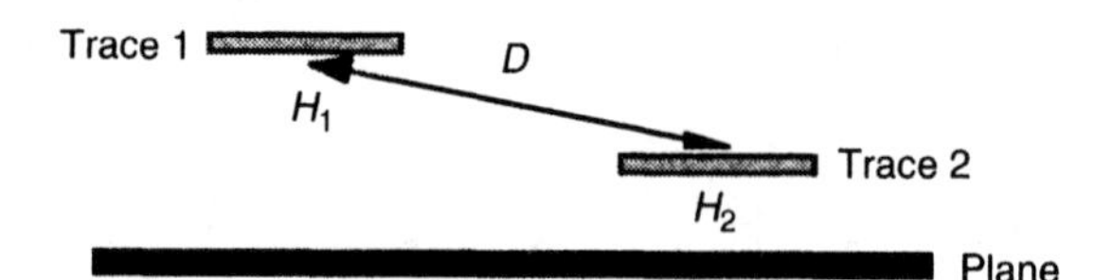

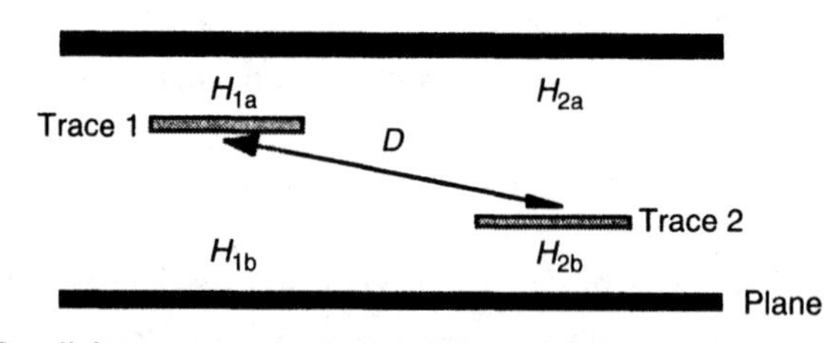

Figure 7.14 Calculating crosstalk separation.

For embedded microstrip, if the parallel traces are at different heights, the H^2 term becomes the product of the two heights as shown in Fig. 7.14 and Eq. (7.16). The dimension D becomes the direct distance between the centerline of the traces [3].

$$\text{Crosstalk} \approx \frac{1}{1 + \left(\frac{D}{H1 * H2}\right)^2} \tag{7.16}$$

If the traces are routed stripline between two reference planes, determine H using a parallel combination of heights to each plane, detailed in Eq. (7.17).

$$H_{\text{total}} = \frac{Hna * Hnb}{Hna + Hnb} \tag{7.17}$$

We can also determine the distance spacing for microstrip traces for eliminating crosstalk by using Table 7.2. When using Table 7.2, special notes are required.

1. PCB trace: $Z_L = 50\ \Omega$ (Z_s & $Z_L = 100\ \Omega$ in parallel) and 1-cm length.
2. Z_s and Z_L are real, not complex values.

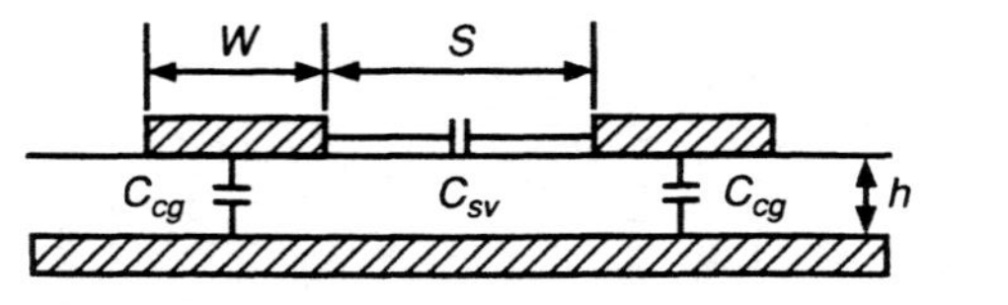

TABLE 7.2 Capacitive Crosstalk Coupling Distance Spacing

S/W	$W/h = 3$ ($C_{cg} \approx 1.2$ pF/cm) $Z_o = 50\ \Omega$				$W/h = 1$ ($C_{cg} \approx 0.5$ pF/cm) $Z_o = 90\ \Omega$				$W/h = 0.3$ ($C_{cg} \approx 0.1$ pF/cm) $Z_o = 120\ \Omega$			
(C_{sv}, pF/cm)	10 (0.003)	3 (0.02)	1 (0.06)	0.3 (0.17)	10 (0.02)	3 (0.06)	1 (0.12)	0.3 (0.28)	10 (0.08)	3 (0.3)	1 (0.45)	0.3 (1.2)
F (ω) = 1 kHz	−174	−158	−148	−140	−158	−148	−142	−136	−146	−134	−130	−122
3 kHz	−164	−148	−138	−130	−148	−138	−132	−126	−136	−124	−120	−112
10 kHz	−154	−138	−128	−120	−138	−128	−122	−116	−126	−114	−110	−102
30 kHz	−144	−128	−118	−110	−128	−118	−112	−106	−116	−104	−100	−92
100 kHz	−134	−118	−108	−100	−118	−108	−102	−96	−106	−94	−90	−82
300 kHz	−124	−108	−98	−90	−108	−98	−92	−86	−96	−84	−80	−72
1 MHz	−114	−98	−88	−80	−98	−88	−82	−76	−86	−74	−70	−62
3 MHz	−104	−88	−78	−70	−88	−78	−72	−66	−76	−64	−60	−52
10 MHz	−94	−78	−68	−60	−78	−68	−62	−56	−66	−54	−50	−42
30 MHz	−84	−68	−58	−50	−68	−58	−52	−46	−56	−44	−40	−32
100 MHz	−74	−58	−48	−40	−58	−48	−42	−36	−46	−34	−30	−22
300 MHz	−64	−48	−38	−30	−48	−38	−32	−26	−36	−24	−20	−12
1 GHz	−56	−40	−30	−22	−38	−30	−22	−18	−28	−18	−14	−8
3 GHz	−52	−36	−26	−20	−32	−24	−18	−14	−24	−14	−10	−4
10 GHz	−52	−36	−26	−18	−30	−22	−16	−10	−24	−14	−10	−4

$$\text{Crosstalk} = 20 \log \frac{R_v\,(C_{sv})\,\omega}{\sqrt{[R_{v\omega}(C_{cg} + C_{sv})]^2 + 1}}$$

where R_v = impedance of victim trace (table uses 100 Ω)
C_{sv} = capacitance between source and victim trace
C_{cg} = capacitance between trace and ground
ω = frequency

Source: M. Mardiguian. *Controlling Radiated Emissions by Design.* New York: Van Nostrand Reinhold, 1992. Reprinted by permission.

3. Crosstalk given per cm of parallel trace run. For other lengths, an approximate correction of 20 log (f_{cm}) may be added, with f_{cm} as parallel trace length in cm. This correction is applicable for frequencies up to 1 GHz.
4. For other lengths and Z_v (impedance of the victim trace), apply the correction factor: 20 log[($Z_v \cdot l$)/100] where l is the length of the trace.
5. Clamp at –4 dB for no ground plane.
   −10 dB for $W/h = 1$
   +4 dB for buried traces
6. If Z_s and Z_L <> 100 Ω, add 20 log ($Z_v/50$), with $Z_v = \dfrac{Z_s Z_L}{Z_s + Z_L}$.

7.7 THE 3-W RULE

Crosstalk can exist between traces on a PCB. This undesirable effect is associated not only with clock or periodic signals, but also with other system critical nets. Data, address, control lines, and I/O all are affected by crosstalk and coupling. Clocks and periodic signals create the majority of problems and can cause functionality problems with other functional sections. Use of the *3-W* rule will allow a designer to comply with PCB layout criteria without having to implement other design techniques. This design technique takes up physical real estate and may make routing more difficult.

The basis for use of the *3-W* rule is to minimize coupling between traces. This rule states that *the distance separation between traces must be three times the width of a single trace, measured from centerline to centerline.* Otherwise stated, *the distance separation between two traces must be greater than two times the width of a single trace.* For example, a clock line is 6 mils wide. No other trace can be routed within a minimum of 2*6 mils of this trace, or 12 mils, edge-to-edge. As observed, much real estate is lost in areas where trace isolation occurs. An example of the *3-W* rule is shown in Fig. 7.15 [1].

Note that the *3-W* rule represents the approximate 70% flux boundary at logic currents. For the approximate 98% boundary, *10-W* should be used.

Use of the *3-W* rule is mandatory for *only* high-threat signals such as clock traces, differential pairs, video, audio, the reset line, or other system critical nets. Not all traces on a PCB have to conform to *3-W* routing. Using this design guideline, before routing the PCB, it is important to determine which traces must be routed *3-W*.

As shown in the middle drawing of Fig. 7.15, a via is located between two traces. This via is usually associated with a third routed net and may contain a signal that is susceptible to electromagnetic disruption. For example, the reset line, a video or audio trace, an analog level control trace, or an I/O interface may pick up electromagnetic energy, either inductively or capacitively. To minimize crosstalk corruption to the via, the distance spacing between adjacent traces must include the angular diameter and clearance of the via. The same requirement exists for this distance spacing between a routed trace rich in RF spectral energy that may couple a component's breakout pin (pin escape) to this routed trace.

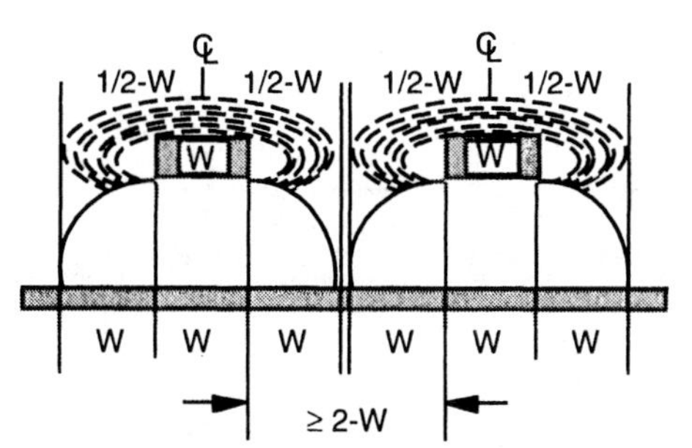

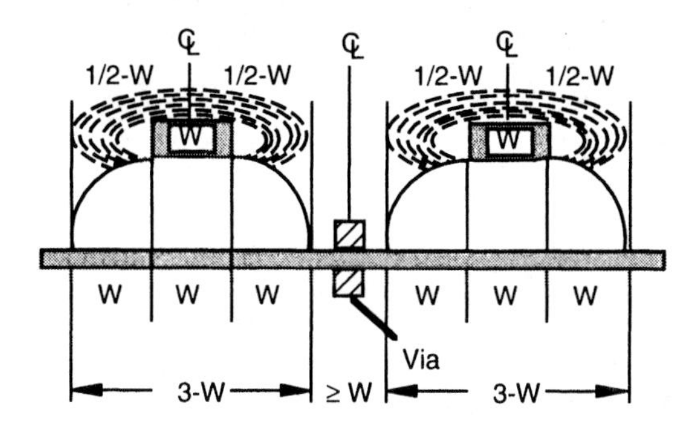

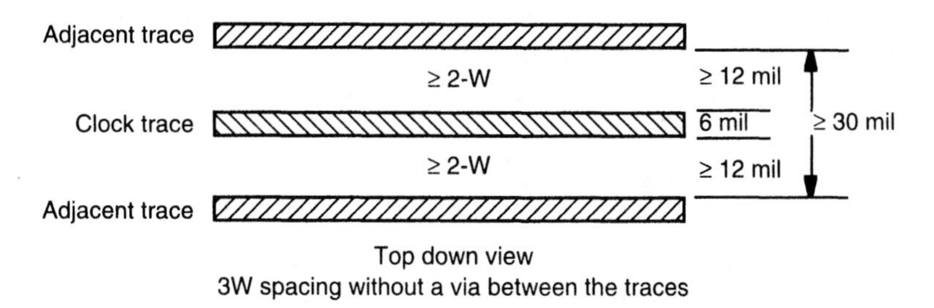

Figure 7.15 Designing with the 3-W rule.

Use of the *3-W* rule should not be restricted to only clock or periodic signal traces; differential pairs (balanced, ECL, and similar sensitive nets) are also prime candidates for *3-W*. The distance between paired traces must be *1-W* for differential traces. For differential traces, power plane noise and single-ended signals can capacitively (or inductively) couple into the paired traces. This can cause data corruption if traces not associated with the differential pair are physically closer than *3-W*.

An example of routing differential pair traces within a PCB structure is shown in Fig. 7.16 [1].

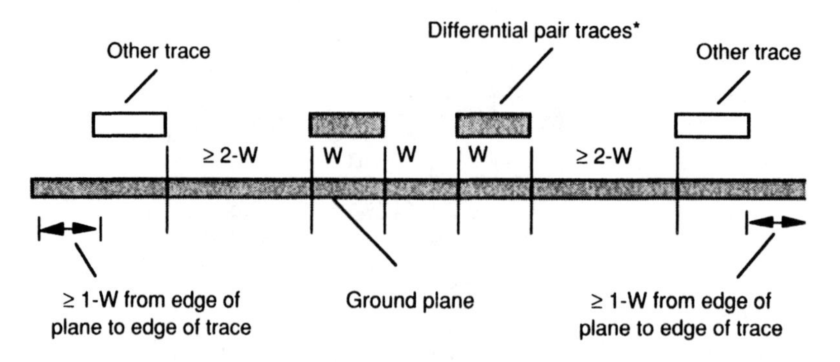

Figure 7.16 Parallel differential pair routing and the 3-W rule.

REFERENCES

[1] Montrose, M. 1996. *Printed Circuit Board Design Techniques for EMC Compliance.* Piscataway, NJ: IEEE Press.

[2] Paul, C. R. 1992. *Introduction to Electromagnetic Compatibility.* New York: John Wiley & Sons.

[3] Johnson, H. W., and M. Graham. 1993. *High Speed Digital Design.* Englewood Cliffs, NJ: Prentice Hall.

[4] Ott, H. 1988. *Noise Reduction Techniques in Electronic Systems.* 2nd ed. New York: John Wiley & Sons.

[5] Motorola, Inc. 1989. *Low Skew Clock Drivers and Their System Design Considerations* (#AN1091).

[6] Motorola, Inc. 1988. *MECL System Design Handbook (#HB205),* Chapters 3 and 7.

[7] Motorola, Inc. 1989. *Transmission Line Effects in PCB Applications (#AN1051/D).*

[8] IPC-D-317A. 1995, January. *Design Guidelines for Electronic Packaging Utilizing High-Speed Techniques.* Institute for Interconnecting and Packaging Electronic Circuits.

[9] IPC-2141. 1996, April. *Controlled Impedance Circuit Boards and High Speed Logic Design.* Institute for Interconnecting and Packaging Electronic Circuits.

[10] IPC-TM-650. 1996, April. *Characteristic Impedance and Time Delay of Lines on Printed Boards by TDR.* Institute for Interconnecting and Packaging Electronic Circuits.

[11] Van Doren, T. 1995. *Circuit Board Layout to Reduce Electromagnetic Emission and Susceptibility.* Seminar Notes.

[12] Mardiguian, M. 1992. *Controlling Radiated Emissions by Design.* New York: Van Nostrand Reinhold.

[13] Goyal, R. 1994, March. *Managing Signal Integrity.* IEEE Spectrum. New York.

[14] Paul, C. R. 1984. *Analysis of Multiconductor Transmission Lines.* New York: John Wiley & Sons.

[15] Ramo, Simon, J. R. Whinnery, and T. Van Duzer. 1984. *Fields and Waves in Communication Electronics.* 2nd ed. New York: John Wiley & Sons.

[16] Witte, Robert. 1991. *Spectrum and Network Measurements.* Englewood Cliffs, NJ: Prentice-Hall.

8

Trace Termination

Before examining use of terminations, a thorough understanding of both the environment and the relationship that traces have to the overall PCB assembly must be achieved.

Trace termination plays an important role in ensuring optimal signal integrity as well as minimizing creation of RF energy. To prevent trace impedance problems and provide higher quality signal transfer between circuits, termination may be required. Transmission line effects in high-speed circuits and traces must always be considered. Even if the clock speed is low, say 4 MHz, and the driver and receiver are in the FCT family (2-ns edge rate), the reflections from a long trace route and fast edge rate can cause the receiver to double clock on a transition. Any signal that clocks a flip-flop is a possible candidate for causing transmission line effects regardless of the actual frequency of operation.

At what point in the design cycle should transmission line effects be considered? The following are some suggestions:

1. Clock lines, FIFO read and write strobe, and any signal that is used to latch or clock a flip-flop either externally in a circuit or internally in an off-the-shelf device (e.g., SIMM modules).
2. When using high-speed logic, such as ACT, FCT, FTTL, ASTTL, BCT, ABT, BTL, GTL, GaAs, or ECL.
3. CMOS components, as CMOS is susceptible to latch-up when a high-to-low transition goes much below the low-voltage transition state (approximately 2 volts).
4. On address and data lines when the designer is pushing the speed of the technology used, along with extremely fast edge rates on device inputs.

Less concern is generally given to transmission line effects with address and data signals, provided the system is synchronous. Designers must consider worst-case setup and hold times with 2–4 ns of safety incorporated into the timing diagram. Designing in a safety margin guarantees that transmission line effects won't typically cause any harm as long as they have settled down by the time the clock or strobe latches them into a register. EMI problems may be increased however, if the signals ring and have overshoot and undershoot.

EXAMPLE

Given a 50-MHz signal (20 ns edge-to-edge), the minimum setup time for flip-flops is 2 ns. The worst-case delay from the source driver to the load is 17 ns. With this example, there is 1 ns to spare after accounting for setup time. If the trace is 6 inches long, transmission line effects will add approximately 2 ns, making the round-trip travel time 21 ns. There is a 1-ns delay over the desired requirements rather than 1 ns under.

8.1 TRANSMISSION LINE EFFECTS

When high-speed, fast edge rate signals are used within a digital design, transmission line effects are observed. Traces must be considered as a transmission line if the round-trip propagation delay of the signal traveling in the trace exceeds the switching-current transition time between logic states. Faster logic devices and their corresponding increase in edge rates are becoming more common in the sub-nanosecond range. A very long trace in a PCB can become an antenna for radiating RF currents or causing functionality problems.

A PCB trace looks very different for high signal speeds than it does at DC levels. For example, a typical PCB trace has a DC resistance of 12 milliohms per inch. When a signal wave propagates down a trace, the trace impedance will range from the few tens of ohms to as much as 100 ohms. Characteristic impedance is identified by the letter Z_o. Characteristic impedance is equal to the square root of L/C where L is inductance and C is capacitance. The ratio of voltage to current is constant only for a matched transmission line. The (x) subscript indicates variations along the line.

$$Z_o = \sqrt{\frac{L_o}{C_o}} = \frac{V_{(x)}}{I_{(x)}} = \frac{V_{(x)}^{+} + V_{(x)}^{-}}{I_{(x)}^{+} - I_{(x)}^{-}} \tag{8.1}$$

where + = forward wave and – = reverse wave.

What mechanism makes transmission lines the preferred choice for data transfer? A transmission line provides a constant impedance path from a source to load without discontinuities. Discontinuities affect signal integrity and may corrupt the voltage levels of the intended signal to a nonfunctional value. Ringing, reflections, overshoot, undershoot, and crosstalk are also problem areas observed in traces that are not routed as a transmission line. If a shield is added to the transmission line, a coax exists. A coax is the best transmission line for signal functionality. This is in addition to minimizing, or preventing RF currents from being created and causing harmful interference to other electronic equipment. A shield around the transmission line also enhances RF immunity protection from externally generated RF sources.

If the load impedance is greater than the characteristic impedance of the trace, Z_o (transmission line), positive reflections will occur. This will result in a higher voltage level at the load than that of the voltage supply, for example, a 6.5V signal with a 5.0V voltage source. If the termination impedance is less than the characteristic impedance, a negative reflection will cause the termination voltage to be less than that of the signal source. Reflections create overshoots. Overshoots affect adjacent lines or traces, as coupling is enhanced by the larger amount of voltage that exists. This coupling may cause induced logic errors, increase edge transition times, and may affect signal timing requirements. When one load is at a logic HI state, and the other end of the transmission line is LOW (typical of TTL), ringing oscillatory resonances occur. The ringing "oscillatory" resonance allows overshoot to alternate in consecutive reflections. The conditions described are illustrated in Fig. 8.1.

Source Z	Load Z	EMI results	Waveform at Load
Z_o	Z_o	None	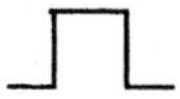
Z_o	High	Trace-trace coupling	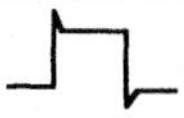
Z_o	Low	Edge rate changes	
Low	High	Trace coupling, EMI and crosstalk	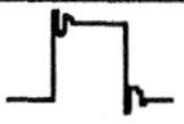

Figure 8.1 Transmission line effects. (*Source:* Oren Hartal, *Electromagnetic Compatibility by Design*, © 1994. Reprinted by permission of R&B Enterprises.)

8.2 TERMINATION METHODOLOGIES

The need to terminate a PCB trace is based on several design criteria. The most important criterion is the existence of an electrically long trace within a PCB. (Electrically long traces are discussed in Chapter 7.) In this chapter, the length of a routed trace on the PCB is discussed before termination is required. When a trace is electrically long, or when the length exceeds one-sixth of the electrical length of the edge rate, the trace requires termination. Even if a trace is short, termination may still be required if the load is capacitive or highly inductive to prevent ringing.

The easiest way to terminate is to use a resistive element. Two basic configurations exist, source and load. Several methodologies are available for these configurations. The five most commonly used termination methods are as follows and are detailed in Fig. 8.2. A summary of termination methods is presented in Table 8.1. Each termination method is discussed in depth in this chapter.

1. Series termination
2. Parallel termination

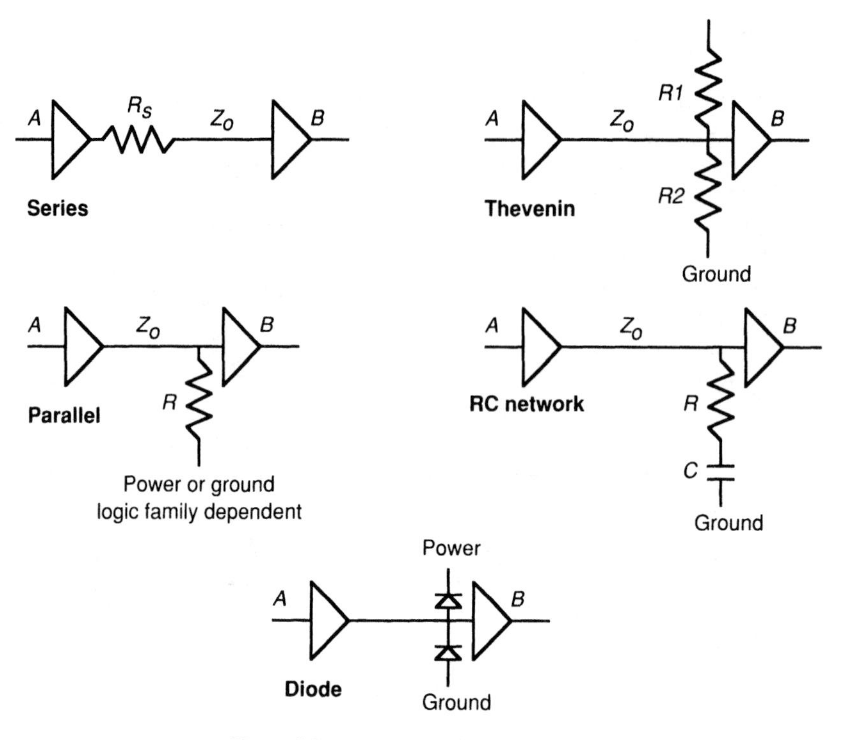

Figure 8.2 Common termination methods.

3. Thevenin network
4. RC network
5. Diode network

Termination not only matches trace impedance and removes (or reduces) ringing and reflections, it may also sometimes slow down the edge rate of the clock signal if incorrect values are applied. Inappropriate termination may degrade signal amplitude and integrity to the point of nonfunctionality. Reducing either *dI/dt* or *dV/dt* within the trace will reduce the creation of RF currents generated by high-amplitude voltage and current levels.

Another way to describe this *dI/dt* and *dV/dt* concern is to relate these functions to Ohm's law, $V = IR$. The following text demonstrates very briefly how to translate Ohm's

TABLE 8.1 Termination Types and Their Properties

Termination Type	Added Parts	Delay Added	Power Required	Parts Values	Comments
Series	1	Yes	Low	$R_s = Z_o - R_o$	Good DC noise margin
Parallel	1	Small	High	$R = Z_o$	Power consumption is a problem
Thevenin	2	Small	High	$R = 2 * Z_o$	High power for CMOS
RC	2	Small	Medium	$R = Z_o$ C = 20–600 pF	Check bandwidth and added capacitance
Diode	2	Small	Low	—	Limits undershoot; some ringing at diodes

© Table and figure reprinted by permission of Motorola, Inc.

law into a "simple" electromagnetic concepts using the equation $V_{ref} = I_{RF} * Z$. If the impedance (Z) of the trace remains constant, then both dV (RF voltage) and dI (RF current) will increase or decrease with the time-variant pulse of the signal. With less RF voltage and RF current, less radiated or conductive RF energy is generated, along with all the EMI undesirable side effects; hence, EMI performance improves. In addition to less RF currents, the edge rate of the signal may also be increased (slower edge rate), along with a reduction of spectral RF energy. However, if the value of Z is too large, then nonfunctionality may occur due to excessive signal degradation. To guarantee proper functionality at all times, Z must be optimally calculated.

Before the different termination methodologies are examined, a baseline network is provided. The following discussions are based on the simple model of Fig. 8.3.

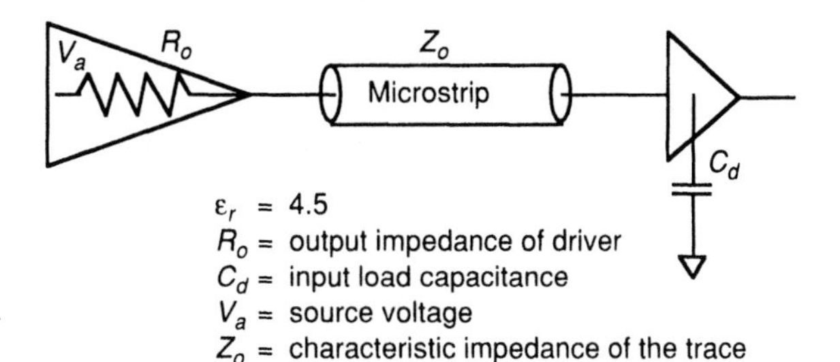

Figure 8.3 Simple baseline model for reference purposes.

Using this baseline model, we can examine the two simulation plots in Fig. 8.4. The edge rate of the driver is 0.8 ns for this 66-MHz, CMOS voltage level. Because components are going through die shrink and edge rates are decreasing, a realistic value of edge rate has been used in this analysis for clock drivers commonly used within high-technology products.

One plot shows the effects of this simple model using a 3-inch (7.6-cm) trace. The other plot shows an 18-inch (45.7-cm) trace, typical of a PCB layout in many designs. Notice that no termination is provided. The electrically short trace, 3 inches, shows a nearly perfect waveform from the source, along with a typical waveform at the load with minor overshoot and undershoot, typical with an improperly designed transmission line. According to Chapter 6, an electrically long trace for an 0.8-ns signal is 5.8 inches, round-trip. The round-trip propagational delay for a 3-inch trace is 6 inches; hence, the value of 3 inches was chosen to approximate the breaking point for a long transmission line. If the trace were less than 3 inches, the signal at the load would be identical to that of the source driver.

The item of interest in Fig. 8.4 is the 18-inch trace. The propagation delay of a microstrip transmission line with an ε_r of 4.5 is 1.72 ns/ft (0.36 ns/cm). At 18 inches, the time it takes for signal propagation is 2.5 ns, one-way travel. As observed in the plot, the signal is received at the load 2.5 ns after the source signal leaves the driver. This is what occurs with propagational delay (path time) within a transmission line. For critical nets where timing skew is important, an engineer must learn what the finished routed length of a trace is in the PCB artwork prior to releasing the artwork for manufacturing. If SPICE or any other transmission path analysis is done on a typical 3-inch trace, and the PCB designer uses an 18-inch trace in a synchronous system operating at frequencies of 100 MHz and above, signal integrity concern becomes mandatory. Details on propagation delay within a PCB trace were presented in Chapter 6. The discussion that follows is based on this baseline data.

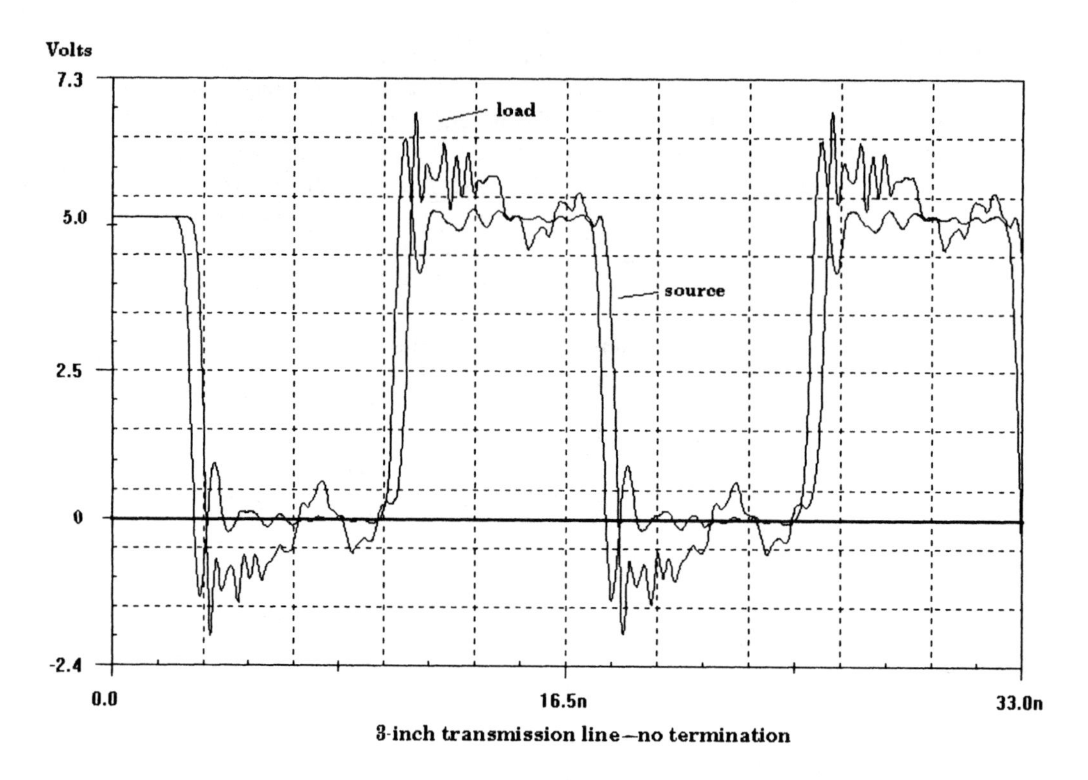

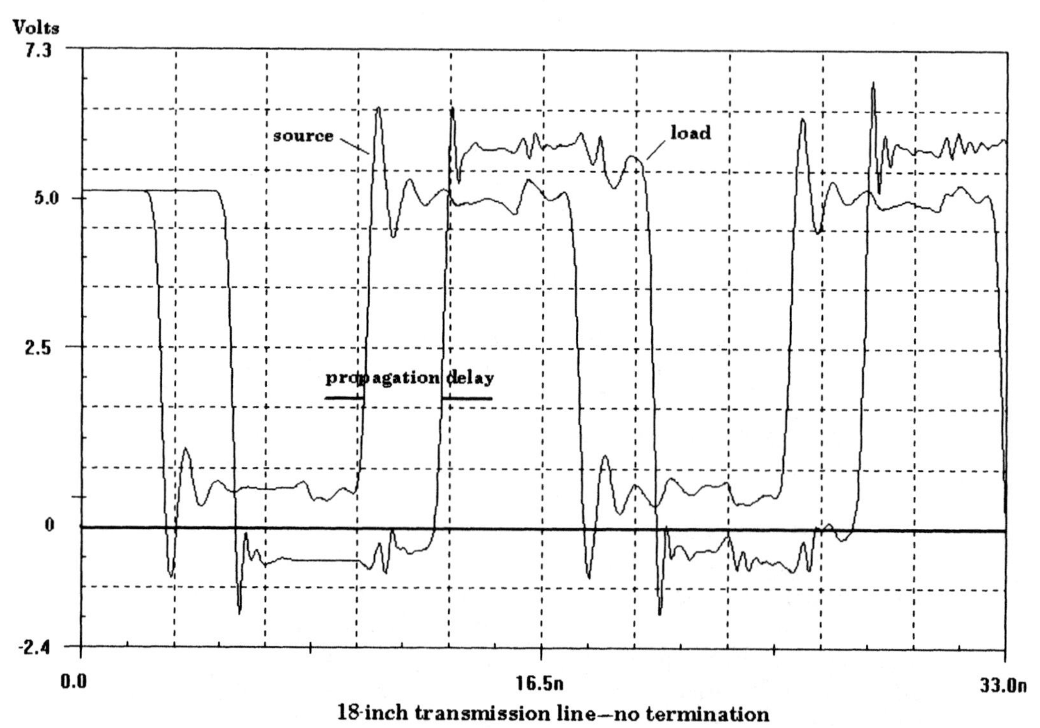

Figure 8.4 Baseline plots for reference purposes.

8.2.1 Source Termination

Source termination [3] provides a mechanism whereby the output impedance of the driver and resistor matches the impedance of the trace. The reflection coefficient at the source will be zero. Thus, a clean signal is observed at the load. In other words, the resistor absorbs the reflections.

The discussion that follows is summarized below.

1. A series termination provides a 50% reduction in drive voltage from the output of the resistor before traveling from source to load.
2. The reflected signal from the load will propagate back to the source at a 50% voltage level.
3. A reflection coefficient of +1 (open circuit) is observed at the far end. This means that the 50% reflected signal, plus the incoming 50% source signal, will add together and provide the full voltage level signal at the load without reflections.

8.2.2 Series Termination

Series termination is optimal when a lumped load or a single component is located at the end of a routed trace. A series resistor should be used when the driving device's output impedance, R_o, is less than Z_o, the loaded characteristic impedance of the trace. This resistor must be located *directly* at the output of the driver without use of a via between the component and resistor (Fig. 8.5). The series resistor, R_s, is calculated by

$$R_s = Z_o - R_o \tag{8.2}$$

where R_o = output resistance of the source driver
Z_o = characteristic impedance of the transmission line
R_s = series resistor

For example, if $R_o = 22\ \Omega$ and trace impedance, $Z_o = 55$ ohms, $R_s = 55 - 22 = 33\ \Omega$. Use of a 33-ohm series resistor is common in today's high-technology products. The series resistor, R_s, can be calculated to be greater than or equal to the source impedance of the driving component and lower than or equal to the line impedance, Z_o. This value is typically between 15 and 75 (usually 33) ohms.

Series terminations minimize the effects of ringing and reflection. Source resistance plays a major role in allowing a signal to travel down a transmission line with maximum quality. If a source resistor does not exist, there will be very little damping. The system will ring for a long time (tens of nanoseconds). PCI drivers are optimal for this function because they have an extremely low-output impedance. A series resistor at the source that

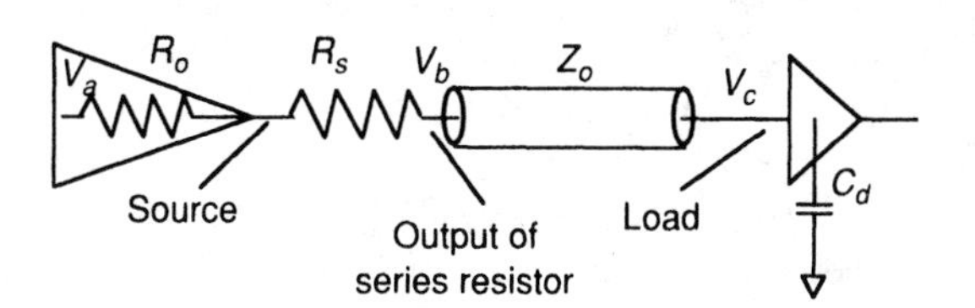

Figure 8.5 Series termination circuit.

is about two-thirds of the transmission line impedance will remove ringing. A target value for a *slightly underdamped system* (to make edges sharper) is to have $R_s = 2/3Z_o$. A wavefront of slightly more than half the power supply voltage proceeds down the transmission line and doubles at the open circuit far end, giving the voltage level desired at the load. The reflected wavefront is almost completely absorbed in the series resistor. Sophisticated drivers will attempt to match the transmission line impedance so that no external components are necessary.

When $R_s + R_o = Z_o$, the voltage waveform at the output of the series resistor is at one-half the voltage level sourced by the driver assuming a perfect voltage divider exists. For example, if the driver provides a 5V output, the output of the series resistor will be 2.5 volts. The reason for this is described by Eq. (8.3). If the receiver has high-input impedance, the full waveform will be observed immediately when received, while the source will receive the reflected waveform at $2 * t_{pd}$ (round-trip travel).

$$\Delta V_b = \Delta V_a \left(\frac{Z_o}{R_o + R_s + Z_o} \right) \tag{8.3}$$

Problems with Impedance Matching

The main problem with series termination is observed when the driving device has different output impedance values in both the LOW and HI states. This problem affects TTL logic and some CMOS devices as well. Both TTL and CMOS have different output impedances in both the logic HI and LOW state. This difference in output impedance, as well as the series resistor, may allow the trace impedance to vary from 55 Ω to 10 Ω depending on the logic state. This impedance mismatch condition may cause poor signal quality and possible nonfunctionality. In addition, certain load devices may have different input and output impedances that are not intuitively known. Hence, use of a series resistor may not be optimal under this condition of varying input/output impedances. Despite the compromises, it can still be effective.

The 1/2 V_{max} plateau can place the signal in an indeterminate logic state that can lead to improper operation should a bus structure be provided with multiple loads at various routed spacings. Signal integrity issues exist for multiple devices located on a routed bus, *except* for the receiver at the end of the net.

A well-designed CMOS clock driver should have approximately the same output impedance in both the HI and LOW logic state. This design requirement prevents many other termination problems from occurring. An additional advantage of using a series resistor is that a DC current path to ground or power is not set up. Without a DC current path, V_{OL} and V_{OH} levels are not degraded. If each clock output is driving only one device, series termination is the optimal choice. If a clock trace must connect to multiple loads or receivers, series termination is not the best choice.

When a device is sending a voltage-level transition down the transmission line, the source driver will see only the input impedance of the load which should be a high value. A series resistor is always located directly in the tranmission path. When the driver sends out a logic high signal, a direct path to ground through a low impedance (e.g., a pull-down resistor) will not occur. For a logic low state, the source driver will not consume power from the voltage source as if a pull-up resistor was in the circuit.

When a series resistor is used, the impedance of the trace is changed as a function of frequency, described by Eq. (8.4) where ω is frequency (in radians), L is series inductance, and C is capacitance of the trace. It is observed that when R exceeds ωL, characteristic impedance becomes inversely proportional to the square root of the frequency available. This occurs at low frequencies when the wavelength is long and the line is electrically short. As a result, transmission line models are not valid for this case, and characteristic impedance is not of much concern. However, at high frequencies, when ωL exceeds R, the characteristic impedance becomes constant.

$$Z_o(\omega) = \sqrt{\frac{R + j\omega L}{j\omega C}} \tag{8.4}$$

Effects of Edge Rate Degradation

When series termination is used, the rise time of the signal can be affected, especially if the value of the resistance is not correctly selected. At any point along the transmission line, looking back toward the source, we see a drive impedance, Z_o. When a capacitive load is provided, a response is observed which appears as a simple RC low-pass filter with a time constant of

$$\tau = RC = Z_o C \tag{8.5}$$

Using Eq. (8.4) for the 10–90% rise time of an RC filter, it is possible to determine the value of the rise time degradation as

$$t_{(10\text{—}90)} = 2.2 Z_o C \tag{8.6}$$

This rise time degradation is twice longer than the rise time of an end-terminated circuit. This condition occurs only if the transmission line impedance and load are the same.

Analysis of Series Termination

To better observe the effects of series termination and signal integrity, examine Fig. 8.6. This figure shows both the source and load voltages of a clock signal on a 3-inch and an 18-inch long trace. Both trace inductance and trace capacitance, in addition to load capacitance, play a role in these plots. The edge rate is 0.8 ns. The source is a clock skew driver with 66-MHz outputs. The 3-inch trace is electrically short, while the 18-inch trace is electrically long. One trace shows the signal directly at the output of the driver; the other trace is the output of the series resistor R_s; and the third trace is at the input of the load. Propagation delay at the load is described by $t_{pd} = 0.7\ (Z_o C_d)$, or the propagational delay equations detailed in Chapter 6 can be used if the values of $Z_o C_d$ are not known.

Interesting results are seen in Fig. 8.6. With an electrically short trace, a nearly perfect waveform is measured along the trace route for the 3-inch trace with typical ringing. The 1/2V at the output of the series resistor is easily observed. For the 18-inch trace, the waveform at the load is acceptable for signal integrity concerns, along with a noticeable 1/2V at the output of the series resistor. Both waveforms are nearly identical in appearance, except for the voltage level. A reduced voltage level at the output of the resistor does not affect the functionality of the signal. The item of interest is what is received at the load's input.

The same discussion regarding propagation delay of a transmission line detailed in the section "baseline" is applicable for series termination.

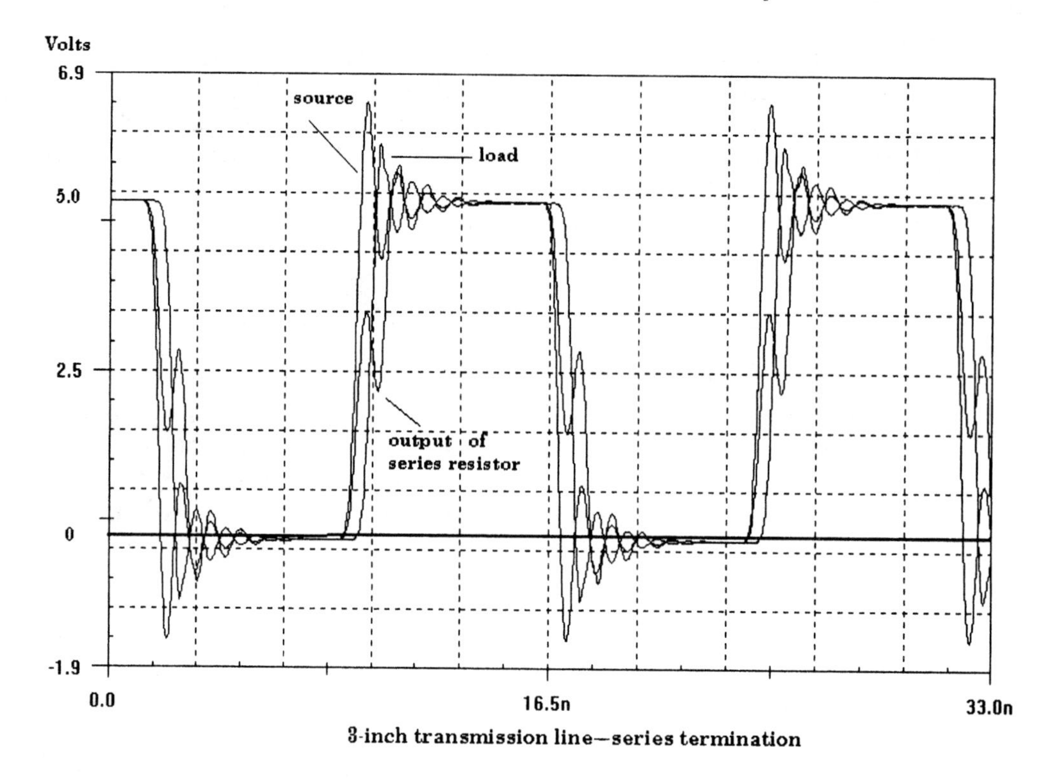

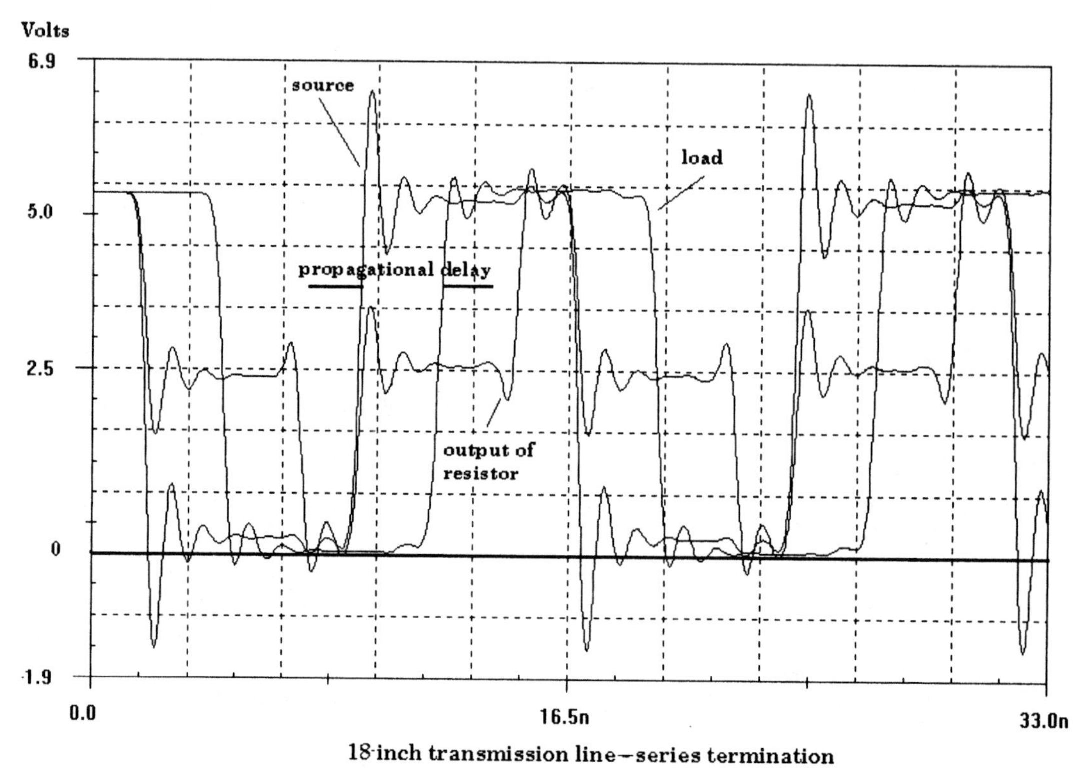

Figure 8.6 Series termination plots.

With a series termination resistor, minimal ringing is observed at the load compared to the unterminated trace shown in Fig. 8.4. This undistorted waveform is desired for signal functionality. An increase in propagation delay, however, is easily observed. The series resistor illustrates the masking of reflections. Both the 3-inch and 18-inch plots look nearly identical at the load because the circuit behaves identically when properly terminated, regardless of trace length. We can ignore the 6.6 V overshoot from the source.

Since $R_s + R_o = Z_o$, the voltage level at V_b (output of the series resistor) is one-half the voltage of V_a(source). The voltage waveform measured is divided evenly, with half of the voltage transmitted to the receiver. If the receiver had a very high-input impedance, the full waveform would have been observed at the load at t_{pd}, while the source would receive the reflected waveform at $2 * t_{pd}$ where t_{pd} is the one-way propagation delay.

When to Use Series Termination

Advantages of Series Termination

1. Series terminators can provide a slower rise time, which results in smaller residual reflections and less EMI.
2. Series resistors help reduce the spectral distribution of RF energy.
3. Series resistors reduce ground bounce.
4. Overshoot is reduced.
5. Signal quality/integrity is enhanced.
6. Minimal power dissipation occurs.
7. Distribution to multiple end-point loads from a common source (Fig. 8.7) is easily implemented.

Disadvantages of Series Termination

1. Series termination does not perform optimally when both TTL and CMOS devices are on the same net.
2. Series termination normally cannot be used when driving distributed loads because in the middle of the trace route, the voltage is only one-half the source voltage. Devices in the middle of a trace route will not get their proper voltage level until much later in the clock cycle.
3. Daisychain topologies are not appropriate with series termination, although a series resistor can be used with a parallel capacitor to slow down the edge time to extend beyond the propagation time of device interval reflections. All loads

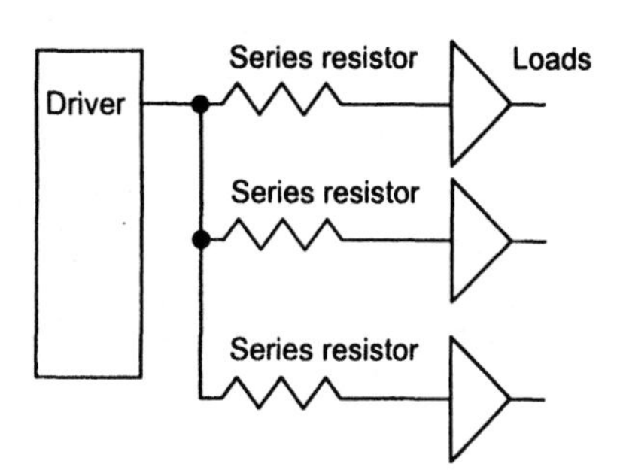

Figure 8.7 Distribution to multiple end-point load from a common source.

must be located at the end of the trace route. If a device is positioned somewhere between source and load, a distorted waveform will occur from improper voltage reference levels, along with possible reflections that may exist in the middle of the signal transmission path.

8.2.3 End Termination

End termination is used when multiple loads exist within a trace route. Multiple-source drivers may be connected to a bus structure or daisychained. The last device on a routed net is where the load termination must be positioned.

To summarize the discussion that follows.

1. The signal of interest travels down the transmission line at full voltage and current level without degradation.
2. The transmitted voltage level is observed at the load.
3. The termination will remove reflections by matching the line, thus damping out the overshoot and ringback.

There is a right way and a wrong way when placing end terminators on a PCB. This difference is shown in Fig. 8.8. Regardless of the method chosen, termination must occur at the "very end of the trace." For purposes of discussion, the RC method is shown in this figure.

Effects of Edge Rate Degradation

An interesting result occurs when termination is provided at the end of a trace route. This observation can describe the effects of edge rate degradation using a simple approach and the circuit of Fig. 8.9. This circuit is modeled as a Thevenin equivalent. The receiver appears as a capacitive load to the transmission line. For this simple circuit, the Thevenin equivalent of the impedance of the network is $Z_o/2$, assuming $R_s = R_L$, which is never the case in actual practice. The capacitor represents input shunt capacitance of the receiver.

When using end termination, the time constant or edge rate degradation is similar to Eq. (8.5) except the impedance of the circuit is $Z_o/2$, as shown in Eq. (8.7).

$$\tau = RC = Z_o C/2 \tag{8.7}$$

Also, similar to using Eq. (8.6) for the 10–90% rise time of an RC filter, the edge time degradation can determine the actual rise time degradation as

$$t_{(10\text{–}90)} = 2.2 Z_o C/2 = 1.1\, Z_o C \tag{8.8}$$

For Eq. (8.8), the difference in edge rate degradation is assumed to be half that of series or source termination. This approximation is due to tolerances and variations within the network (transistors internal to the component plus discrete devices). For system critical nets where timing skew is important and synchronous operation must be assured, end termination may be a better choice.

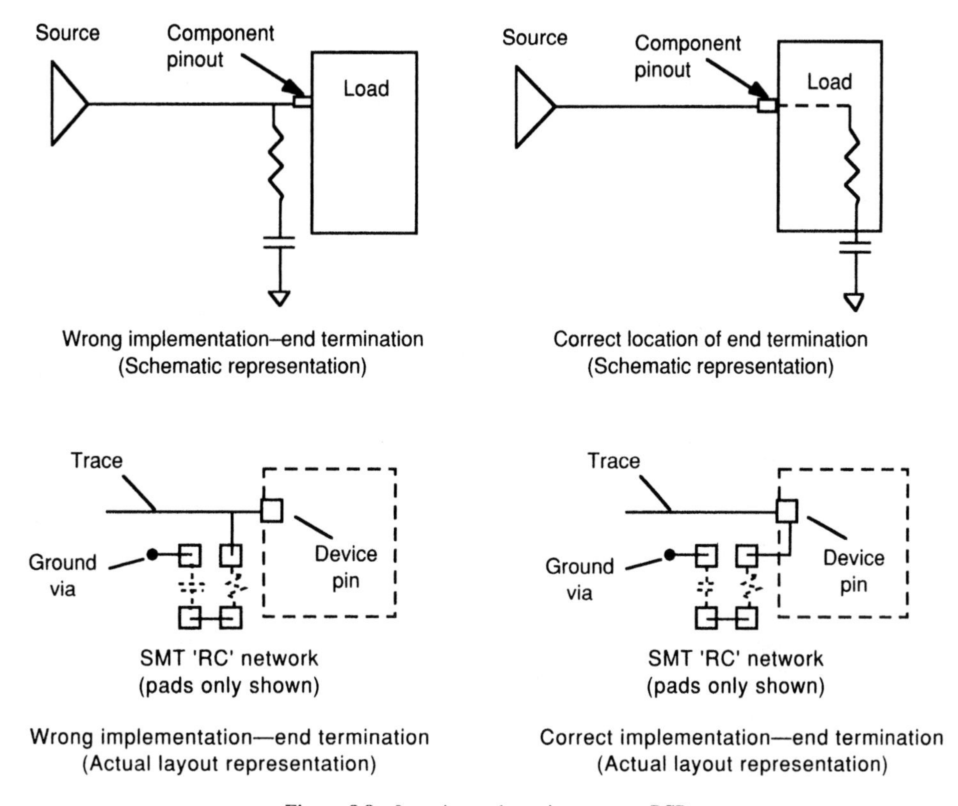

Figure 8.8 Locating end terminators on a PCB.

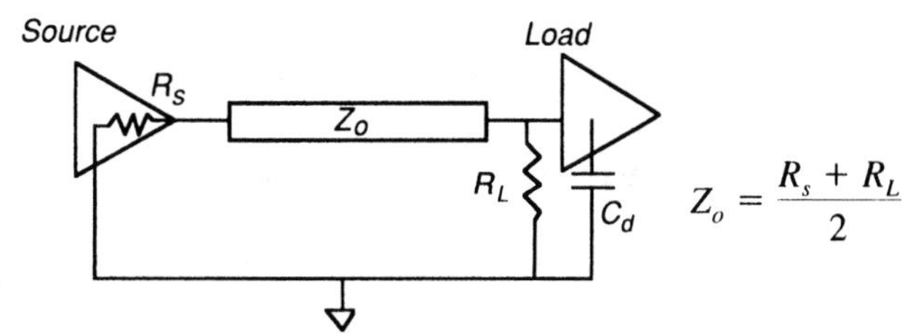

Figure 8.9 Equivalent circuit of end termination.

8.2.4 Parallel Termination

For simple parallel termination, a single resistor is provided at the end of the trace route (Fig. 8.10). This resistor, *R*, must have a value equal to the required impedance of the trace or transmission line. The other end of the resistor is tied to a reference source, generally ground. Parallel termination will add a small propagation delay to the signal due to the addition of the Z_oC time constant that is present in the network, described by Eq. (8.8). This equation includes the total impedance of the network. The termination resistor is only one component of the impedance equation. The total impedance, Z_o, is the result of the termination resistor, line impedance, and source output impedance. The *C* variable in the equation is the input shunt capacitance of the load.

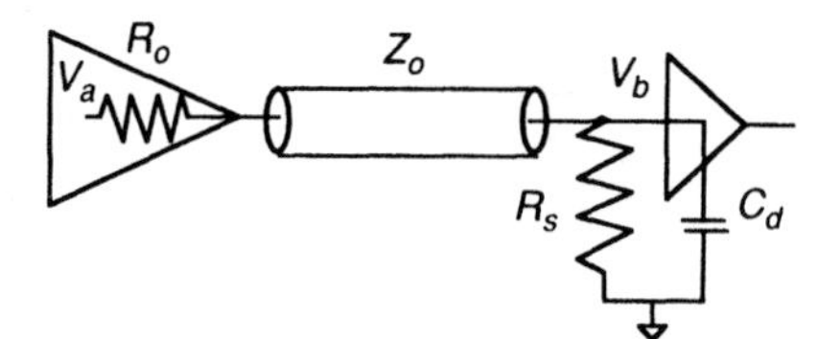

Figure 8.10 Parallel termination circuit.

A disadvantage of parallel termination is that this method consumes DC power, since the resistor is generally in the range of 50 to 150 Ω. In applications of critical device loading or where power consumption is critical, for example, battery-powered products (notebook computers), parallel termination is a poor choice. The driver must source current to the load. An increase in drive current will cause an increase in DC power consumption from the power supply, an undesirable feature in battery-operated products.

Simple parallel termination (resistive only) is rarely used in TTL or CMOS designs. This is because a large drive current is required in the HI logic state. When the source driver switches to V_{cc}, or logic HI, the driver must supply a current of V_{cc}/R to the termination resistor. When in the logic LOW state, no drive current exists. Assuming a 55 Ω transmission line, the current required for a 5V drive signal is 5V/55Ω = 91 mA. Very few drivers can source that much current! The drive requirements of TTL demand more current in the logic LOW state than logic HI. CMOS sources the same amount of current in both the LOW and HI logic states.

Since parallel termination creates a DC current path when the driver is in the HI state, excessive power dissipation and V_{OH} degradation (noise margin) occurs. A driver's output is always switching, thus DC current consumed by the termination resistor must exist. At higher frequencies, the AC switching current becomes the major component of the overall circuit. When using parallel termination, one should consider how much V_{OH} degradation is acceptable by the receivers.

When parallel termination is provided, the net result observed on an oscilloscope should be nearly identical to that of series, Thevenin or RC, since a properly terminated transmission line should respond the same regardless of the termination method used. This effect is observed in the various plots of termination methods provided in this chapter.

When using simple parallel termination, a single pull-down resistor is provided at the load. This allows fast circuit performance when driving distributed loads. This resistor has a Z_o value equal to the characteristic impedance of the trace and source driver. The other end of the resistor is tied to a reference point, usually ground. For ECL logic, the reference is power. The voltage level on the trace is described by Eq. (8.9). On PCB stackups that include Omega layers, parallel termination is commonly found. An Omega layer is a single layer within a multilayer stackup assignment that has resistors built into the copper plane using photo-resist material and laser etched for the desired resistance value. This termination method is extremely expensive and found in only high-technology products where component density is high and large pin-out devices physically leave no room for hundreds or even thousands of discrete termination resistors.

$$\Delta V_a = \Delta V_b \left(\frac{Z_o}{R_o + Z_o} \right) \tag{8.9}$$

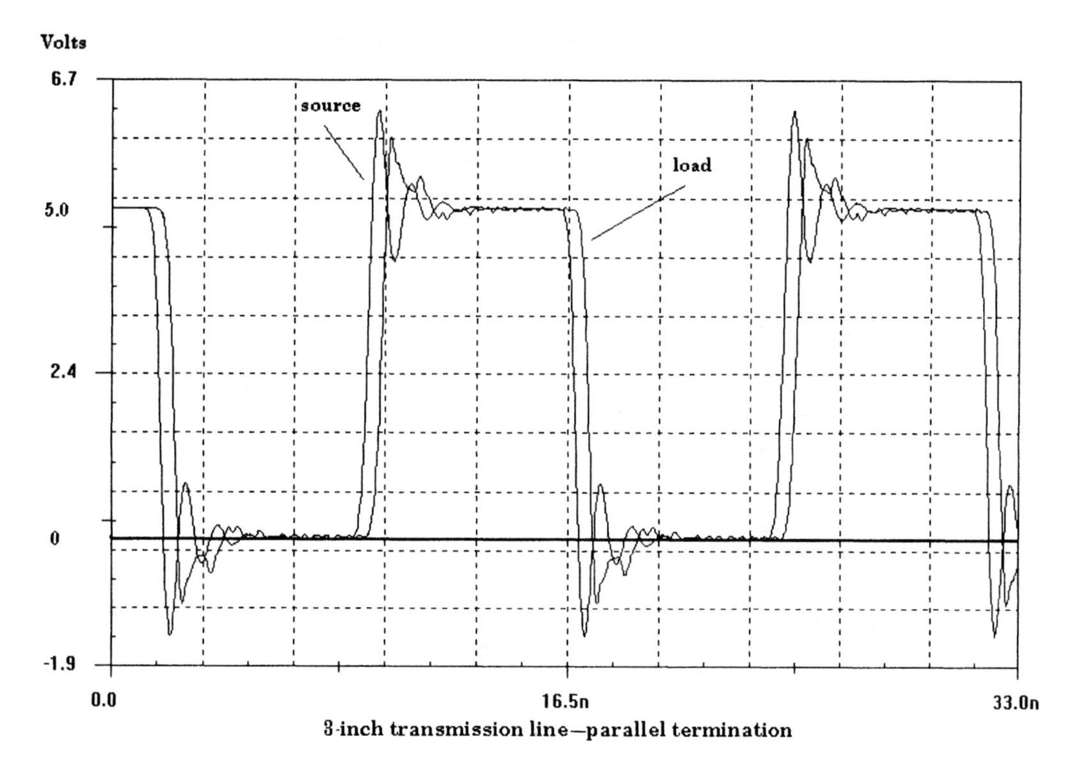

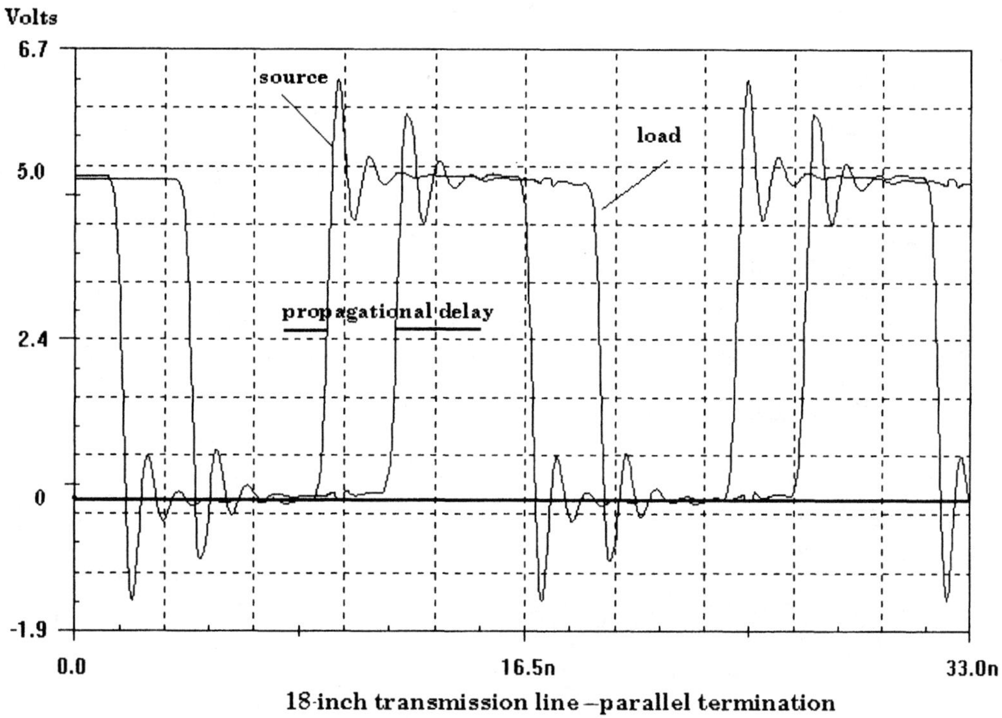

Figure 8.11 Parallel termination plot.

Loading a long trace with additional devices will affect the propagation delay of the source driver's signal, which was discussed in Chapter 7.

Analysis of Parallel Termination

A nearly undistorted waveform will be observed along the full length of the line using parallel termination, similar to that of a coax. If a routed net has multiple receivers and drivers, an increase in propagational delay will be observed owing to additional lumped capacitance provided by all devices connected into the net. If the input shunt capacitance is 5 pF, and six devices are provided on the routed net, a total of 30 pF is presented to the source driver. With the characteristic impedance of the trace and lumped capacitance, the signal will be delayed by the time constant $\tau = 1.1 * Z_o C$. We observe the signal directly at the output of the driver with expected overshoot, and ringing. The signal at the load is acceptable for system performance.

The same discussion regarding propagation delay of a transmission line in the section "baseline" for series termination is applicable. This propagation delay is easily observed in the 18-inch-long trace between source and load. Figure 8.11 shows what parallel termination looks like with both a 3-inch and 18-inch long trace. The results of a properly terminated transmission line will be identical, regardless of termination method chosen. The plot of Fig. 8.11 (18-inch trace) is nearly identical to that of Fig. 8.6 (18-inch trace). The same comments regarding propagational delay of a signal-routed microstrip also applies for parallel termination.

When to Use Parallel Termination

Advantages of Parallel Termination

1. Can be used with distributed loads.
2. Fully absorbs the transmitted wave to eliminate reflections.
3. Sets the line voltage level when nothing is driving the line.
4. Is excellent for busses when distributed loads are available at the end of the trace route.

Disadvantages of Parallel Termination

1. Increased power consumption.
2. Reduced noise margins unless the drivers can source high current circuits.

8.2.5 Thevenin Network

Thevenin termination has one advantage over parallel termination. Thevenin provides a connection that has one resistor to the power rail and the other resistor to ground (Fig. 8.12). Unlike parallel termination, Thevenin permits optimizing the voltage transition points between logic HI and logic LOW. When using Thevenin termination, an important consideration in choosing the resistor values is to avoid improper setting of the voltage reference level of the loads for both the HI and LOW logic transition points. The ratio of $R1/R2$ determines the relative proportions of logic HI and LOW drive current.

Designers commonly, but arbitrarily, use a 220/330 ohm ratio (132 Ohms parallel) for driving bus logic. Determining the resistor ratio value may be difficult to do if the switch point for logic families are different. This is especially true when both TTL and

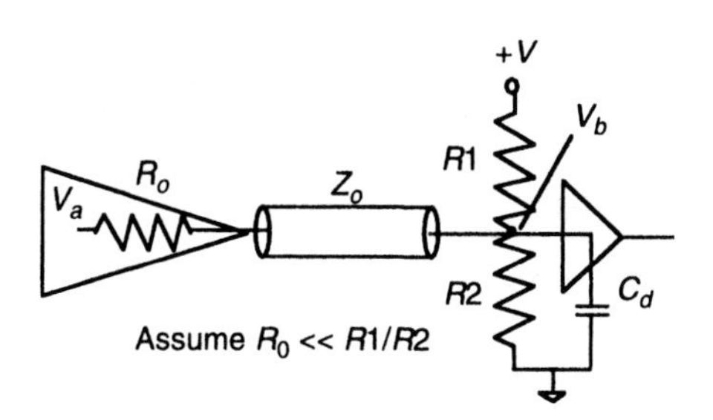

Figure 8.12 Thevenin termination circuit.

CMOS are used. A 1:1 resistor ratio (e.g., 110/110 ohms to create a 55-ohm termination value or characteristic Z_o of the trace) would limit the line voltage at 2.5V, thus allowing an invalid transition level for certain logic devices. Hence, Thevenin termination is optimal for TTL logic, not CMOS.

The Thevenin equivalent resistance must be equal to the characteristic impedance of the trace. Thevenin resistors will provide a voltage division for the signal on the trace. To determine the proper voltage reference desired, one should use Eq. (8.10).

$$V_{ref} = \frac{R2}{R1 + R2} V \tag{8.10}$$

where V_{ref} = desired voltage level to the input of the load
V = voltage source from the power rail
$R1$ = pull-up resistor
$R2$ = pull-down resistor

For the Thevenin termination circuit

$R1 = R2$: The drive requirements for both logic HI and LOW are identical. The setting may be unacceptable for certain logic families.
$R2 > R1$: The LOW current requirements are greater than HI. This setting works well with TTL and CMOS devices.
$R1 > R2$: The HI current requirements are greater than LOW. This is a more appropriate selection for the majority of designs.

With these constraints, $I_{OH\max}$ or $I_{OL\max}$ must never be exceeded. This condition must exist, as TTL and CMOS sinks (positive) current in the LOW state. In the high state, TTL and CMOS sources (negative) current. Positive current refers to current that enters a device, while negative current is the current that leaves the component. ECL logic devices source (negative) current in both logic states.

With a properly chosen termination ratio for the resistors, an optimal DC voltage level will be present for both logic HI and LOW states. The advantage of using parallel termination over Thevenin is parallel's use of one less component. If we compare plots of the effects of parallel compared to Thevenin, we notice that both termination methods provide identical results. A terminated trace will always appear identical regardless of termination method chosen.

Thevenin termination is rarely used because of a large drive current required in the HI state. The results in Fig. 8.13 show a nearly perfect waveform, along with the expected delay of the signal at the load. Figure 8.13 shows the signal directly at the output of the

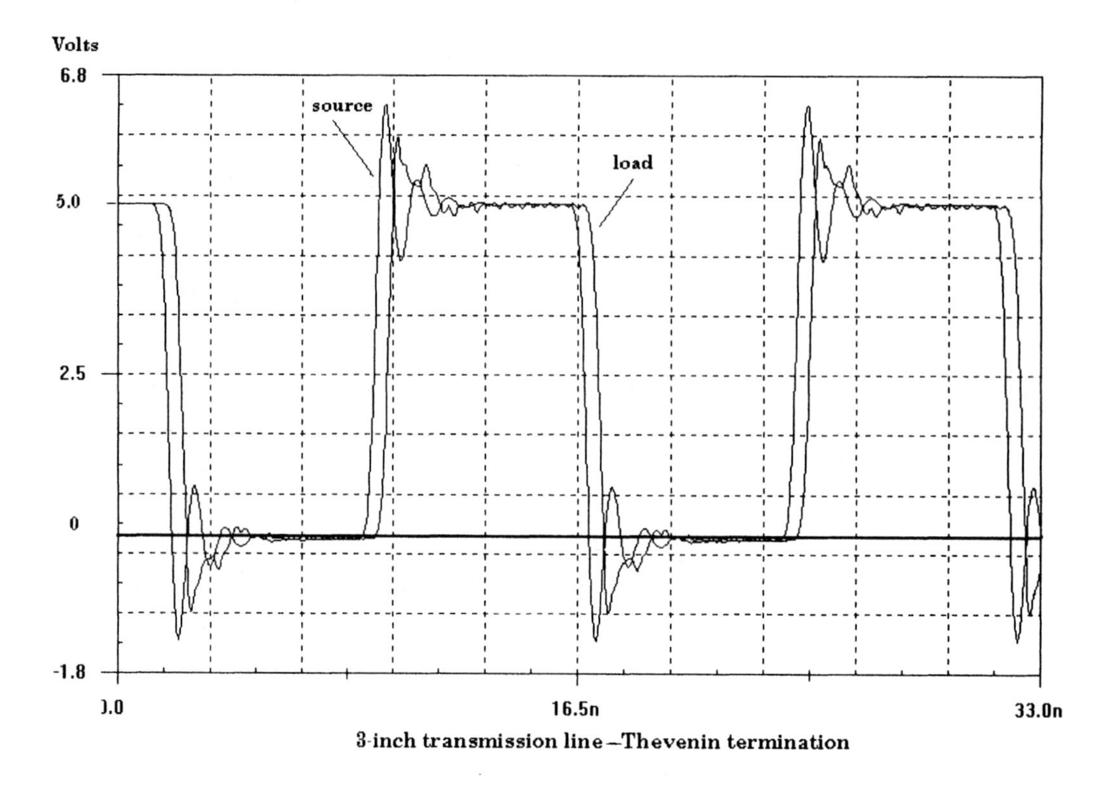

3-inch transmission line—Thevenin termination

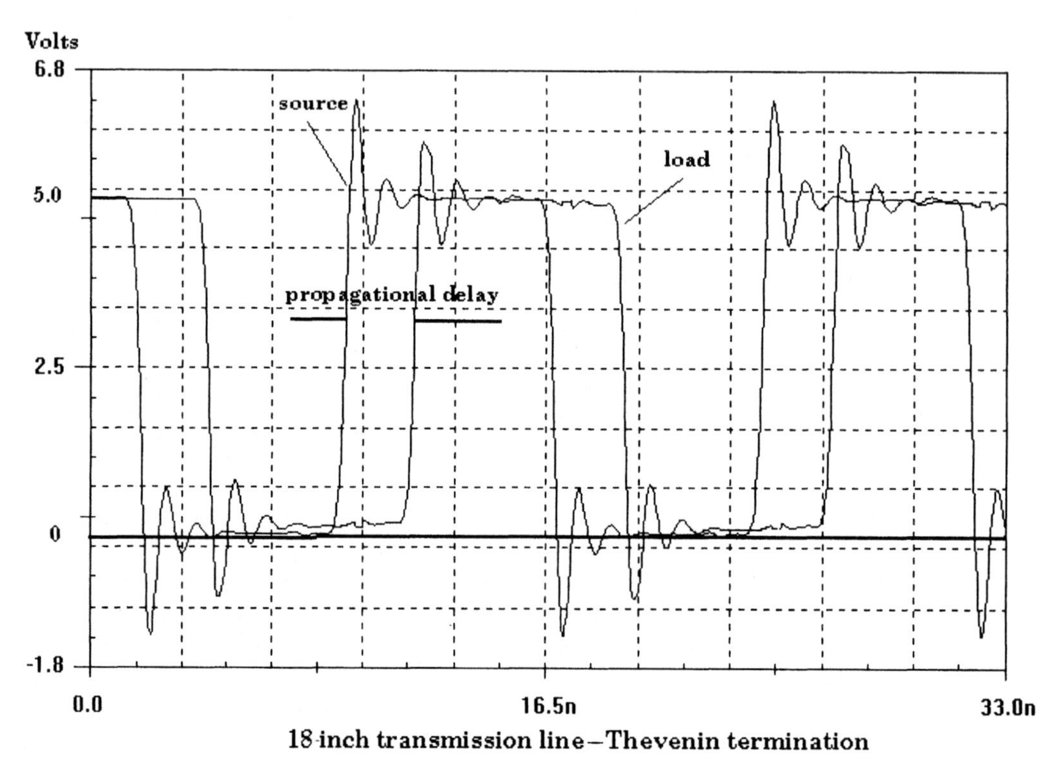

18-inch transmission line—Thevenin termination

Figure 8.13 Thevenin termination plot.

driver with expected overshoot, and ringing, in addition to what the signal would look like at the load for both a 3-inch (electrically short) and 18-inch trace (electrically long).

Analysis of Thevenin Termination [2]

A nearly undistorted waveform will be observed along the full length of the line using Thevenin termination, similar to that of parallel termination. An increase in delay will also be observed on busses with multiple receivers and drivers on the net due to the additional lumped capacitance provided by all devices tied into the routed net. As with parallel termination, the characteristic impedance of the trace, along with the lumped capacitance, will cause the signal to be delayed by the time constant $\tau = 1.1 * Z_o C$. This delay is easily observed in the trace identified as load in Fig. 8.13. We also see what the signal profile looks like directly at the output of the driver along with typical ringing which always occurs. The signal at the load is acceptable for system performance.

Close examination of parallel termination, Fig. 8.11, and Thevenin, Fig. 8.13, indicates identical plots. This is what termination will do to a signal trace. To observe the difference, compare the waveform at the load to the unterminated trace in Figs. 8.4.

The results of a properly terminated transmission line will be identical, regardless of termination method chosen. The plot of Fig. 8.13 (18-inch trace) is nearly identical to those of Fig. 8.6 and 8.11 (18-inch trace). The same comments regarding propagational delay of a signal routed microstrip also apply for parallel termination.

When to Use Thevenin Termination

Advantages of Thevenin Termination

1. Can be used with distributed loads throughout a routed net.
2. Fully absorbs the transmitted wave to eliminate reflections.
3. Sets the line voltage level when nothing is driving the line.
4. Is excellent for busses.

Disadvantages of Thevenin Termination

1. Increases power consumption.
2. Reduces noise margins unless the drivers can source high current circuits.

When the driver is sourcing HI, say 5.0V, and the pull-down resistor is 330 Ω, the resistor must absorb 5.0/330 Ω = 15 mA of current. If the driver cannot source this much current, the HI value of the voltage, V_{OH}, will go down and the noise margin ($V_{IH} - V_{OH}$) will also go down. On busses terminated at both ends with 220/330 Ω resistors, the driver has to source double the current, 30 mA.

EXAMPLE: BACKPLANE IMPLEMENTATION

On a typical TTL-based backplane with traces on the outer layers (microstrip) (Fig. 8.14), the bus lines are terminated at both ends with a 220-Ω resistor to power and a 330-Ω resistor to ground. Both ends must be terminated since the source driver may be located anywhere on the bus. This 220/330 Ω combination provides 132-Ω termination, which is a typical value for PCB traces routed on a backplane assembly. With this termination methodology, the use of high-current bus drivers is required.

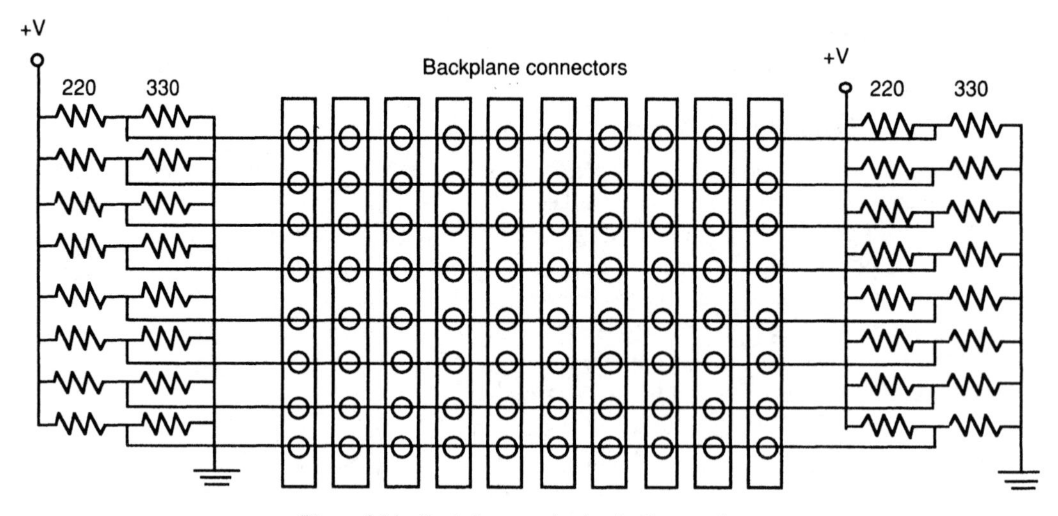

Figure 8.14 Backplane termination implementation.

8.2.6 RC Network

The RC (also known as AC) termination method works well in both TTL and CMOS systems. The resistor matches the characteristic impedance of the trace (identical to parallel). The capacitor holds the DC voltage level of the signal since the source driver does not have to provide current to drive an end terminator. As a result, AC current (RF energy) flows to ground during a switching state since a capacitor will allow RF energy (which is an AC wave, not the DC logic level of the signal) to pass through. Although a minor propagation delay is presented to the signal due to the RC time constant, less power dissipation exists than in parallel or Thevenin termination. From the viewpoint of the circuit, all three end termination methods are identical. The main difference lies in power dissipation, with RC consuming far less power than the other two.

The termination resistor must equal the Z_o of the trace, while the capacitor is generally very small (20–600 pF). The RC time constant must be greater than twice the loaded propagation delay (round trip travel time). This time constant is greater than twice the loaded propagation delay because a signal travels from source to load and returns. It takes one time constant each way for a total of two time constants. If we make the time constant slightly greater than the total propagation delay (x2), reflections will be minimized or eliminated. RC termination finds excellent use in buses containing similar layouts.

To determine the proper value of the resistor and capacitor, Eq. (8.11) provides this simple calculation which includes the round-trip propagation delay $2 * t'_{pd}$.

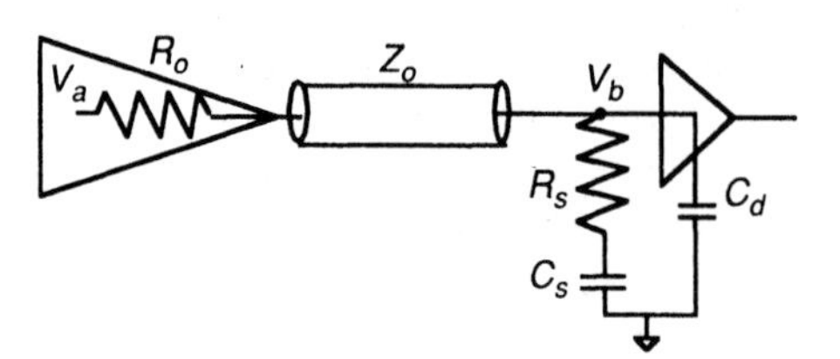

Figure 8.15 RC network circuit.

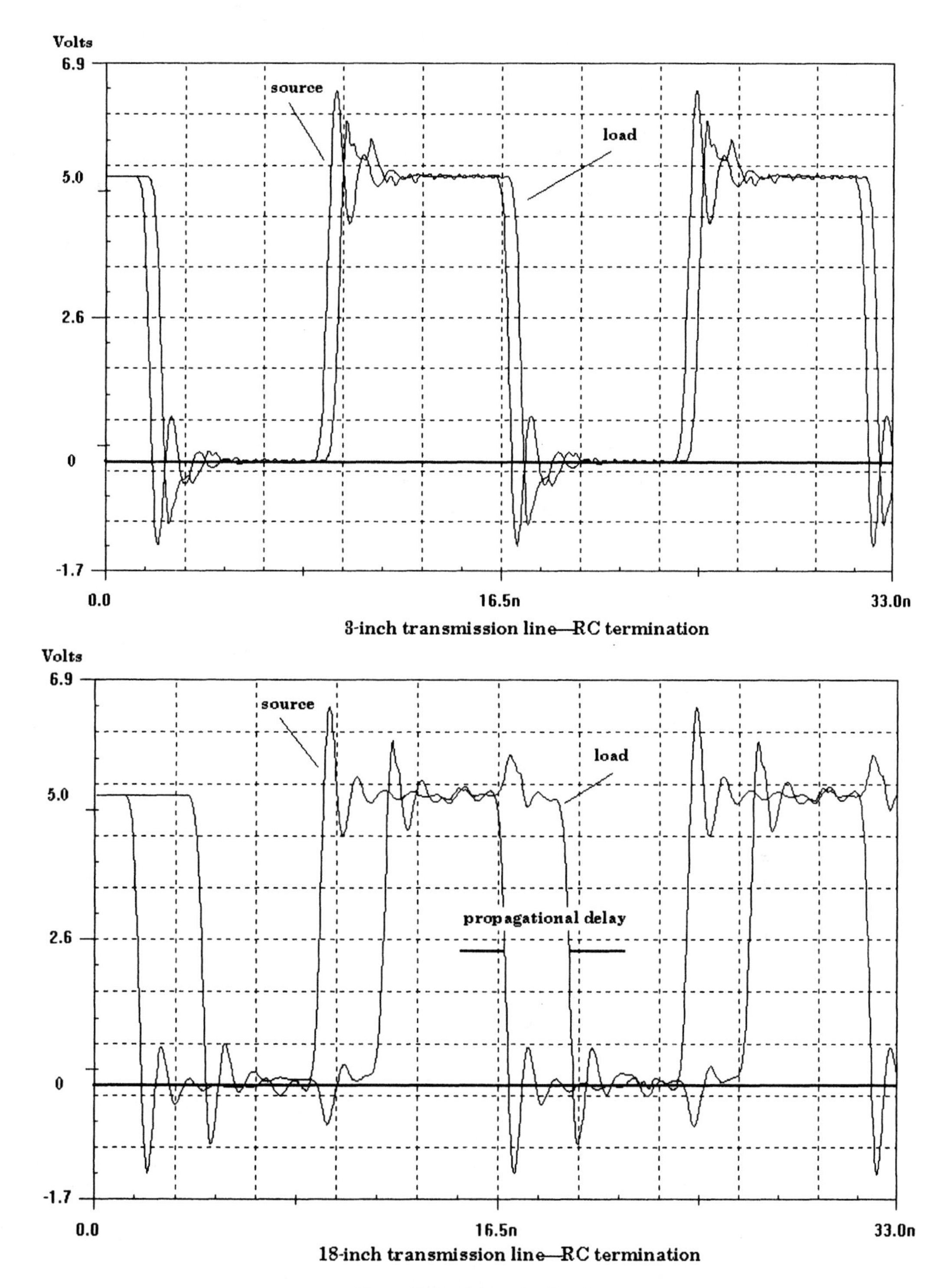

Figure 8.16 RC termination plot.

$$\tau = R_s C_s \quad \text{where } \tau > 2 * t'_{pd} \quad \text{for optimal performance} \tag{8.11}$$

Figure 8.16 shows the results of RC termination. The lumped capacitance (C_d plus C_s) affects the edge rate of the signal, causing a slower signal to be observed by the load.

If the round-trip propagation delay is 4 ns, RC must be > 8 ns. Calculate C_s using the known round-trip propagation delay. Propagation delay is discussed in Chapter 6 using the value of ε_r that is appropriate for the dielectric material provided and for the actual speed of propagation required.

Note: The self-resonant characteristic of the capacitor is critical during evaluation to avoid inserting equivalent series inductance (ESL) into the circuit.

Analysis of RC Network

Again, like parallel and Thevenin, a nearly undistorted waveform will be observed along the full length of the line. Optimal termination will provide a clean signal at the load regardless of which termination method is used.

The capacitor appears to the trace as an AC short to the RF component of the high-speed signal. The reflected wave is fully absorbed with no reflection since the resistor matches the trace impedance. The capacitor blocks the DC current so that no power consumption is traveling through the resistor to ground. The capacitor also prevents the DC noise margins from eroding, since there is no IR drop across the resistor. In addition, the RC network acts as a low-pass filter to remove glitches that may occur on the signal trace.

When to Use the RC Network

Advantages of RC Termination

1. Can be used with distributed loads and bus layouts.
2. Fully absorbs the transmitted wave to eliminate reflections.
3. Has low DC power consumption.

Disadvantages of RC Termination

1. May slow down very high-speed signals.
2. Can produce reflections due to the time constant of the RC network. This is definitely a concern for high-frequency, fast edge rate signals.

When differential paired signals exist, RC termination finds popular use. This termination method is shown in Fig. 8.17. Any termination method that is appropriate for this circuit may be used—RC, parallel, Thevenin, or series. For the example shown, the RC network requires only three components. The capacitor prevents power consumption in addition to maintaining the proper voltage reference to the signals of interest. The components should be close in tolerance to avoid mismatches and to assure equal values for each signal of the pair!

8.2.7 Diode Network

This termination method is commonly used for termination of differential or paired networks. A schematic representation was detailed in Fig. 8.2. Diodes are often used to limit overshoot on traces while providing low-power dissipation. The major disadvantage

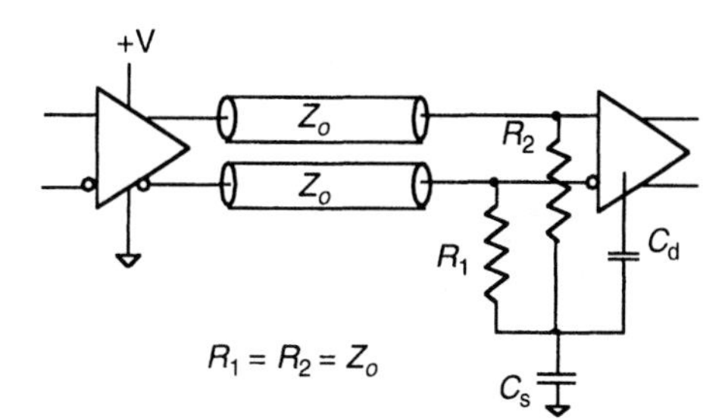

Figure 8.17 RC network of differential paired traces.

of diode networks lies in their frequency response to high-speed signals. Although overshoots are prevented at the receiver's input, reflections will still exist in the trace as diodes do not affect trace impedance or absorb reflections. To gain the benefits of both techniques, diodes may be used in conjunction with the other methods discussed herein to minimize reflection problems. The main disadvantage lies in large current reflections that occurs with this termination network. One should be aware, however, that when a diode clamps a large impulse current, this current can be propagated in the ground plane, thus increasing EMI!

A summary of the various termination methods is provided at the end of this chapter, comparing advantages and disadvantages.

8.3 TERMINATOR NOISE AND CROSSTALK

Multiple terminators may be provided within a single package instead of being used as discrete components. These packages may be Single-In-Line Package (SIPs), Dual-In-Line Package (DIPs), or other configurations with a shared power and/or ground pin. This shared pin may contain undesired lead-length inductance, which may affect signal functionality when a logic transition occurs from high-to-low or low-to-high, in addition to allowing creation of RF currents to exist.

With an inductive effect present, described by $L = dI/dt$, a current surge will be observed by all terminators simultaneously due to this fixed inductance value. This current surge may cause a signal bounce to develop, similar to ground bounce on the power/ground planes. If the bounce is severe enough, functionality concerns exist. To minimize the bounce effect of multiple terminators within the same package, one should use only those package designs with separate power and ground pins and internal decoupling, if possible.

The difference in a typical termination package configuration is shown in Fig. 8.18. When a common ground pin is provided, a common current path will occur depending on the internal manufacturing process used. A common current path introduces a large amount of mutual inductance between the resistors in the package. This common current path will allow crosstalk to be generated internal to the terminator.

Designers should design and lay out a PCB to minimize creation of crosstalk between signal traces and to prevent mutual coupling of RF currents. With crosstalk concerns taking a major role in the layout of traces, one tends to forget that terminators may cause crosstalk.

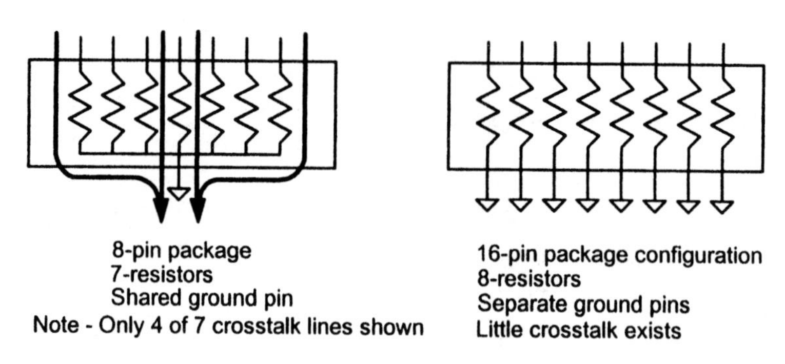

Figure 8.18 Shared pins within a terminator package.

Signal integrity and crosstalk are discussed in Chapter 7. Crosstalk occurs owing to the combination of capacitive and inductive coupling between traces, traces-to-planes, and traces to components. Capacitive and inductive coupling is additive to the overall amount of crosstalk that will exist.

To minimize crosstalk between traces, use of the *3-W* rule is required (discussed at the end of Chapter 7). How is implementation of the *3-W* rule with terminators accomplished? Figure 8.19 illustrates this common design oversight and how to fix it.

Terminators can cross-couple RF energy between circuit traces. This cross-coupling could be much worse than natural crosstalk present between two adjacent transmission lines. Crosstalk in terminations comes from both mutual inductive and capacitive coupling. Inductive coupling usually dominates. The total amount of coupling is the sum of both parts.

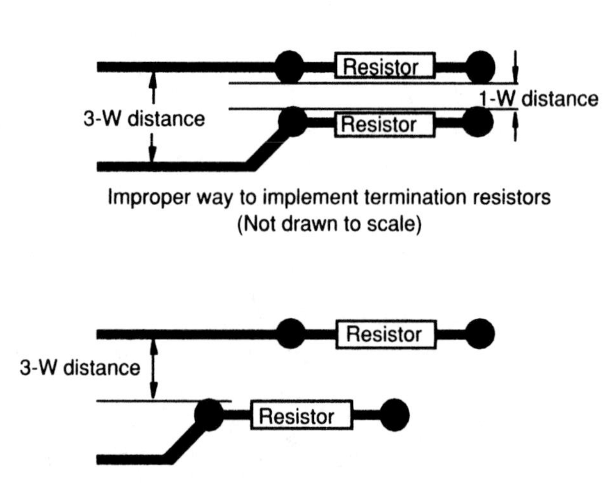

Figure 8.19 Crosstalk between termination resistors.

8.4 EFFECTS OF MULTIPLE TERMINATIONS

Multiple or dual terminations on a PCB trace can present functionality concerns. This is due to the additive effects that a dual termination presents to the circuit. During layout it is sometimes desirable to place component pads onto the PCB; this provides the ability to choose a termination method based on measurements from a prototype build. Although the layout can allow for choosing an optimal termination method, problems can and will occur if careful attention is not taken during installation of the components, or if rework is performed by a person who will install components on pads regardless of whether they are specified in the Bill of Materials. When multiple mounting locations are provided, the designer can use either series, parallel, Thevenin, or RC. An example of how a design engineer may specify optional termination methods for experimentation purposes is presented in Fig. 8.20.

Real-Life Situation—Using Dual Terminations (what can happen). Why would someone dual terminate a trace if poor performance were expected? Let's assume a product is being tested on a remote open field test site with only one attempt to make the product pass radiated emissions requirements. Mounting pads are provided for all possible termination methods. In the rework kit are 0 Ω resistors and a selection of various other resistors and capacitors. These components are used to rework the board (in the field), such as removing series termination or to convert an RC to parallel in order to better match trace impedance and enhance signal functionality. All these components are used in an attempt to make the product work as desired.

For example, component placement pads for a series terminator are provided on the top side of the PCB. During compliance testing, radiated emission is observed. The 0 Ω series resistor is replaced with 33 Ω. Also provided is end termination, which is located on the bottom side of the board unavailable for inspection and unknown to the test engineer. The test engineer may not know what kind of termination is provided, if any, or how terminations even work. The series resistor, 33 Ω, now allows for radiated emissions compliance. This rework is incorporated into the product and shipped without further investigation by signal integrity engineers since product revenue generally takes priority over functionality, especially if a project is behind a scheduled shipment date.

Many engineers will use only a spectrum analyzer when investigating an EMC event. If a sufficiently *high bandwidth oscilloscope* is used in conjunction with a spectrum

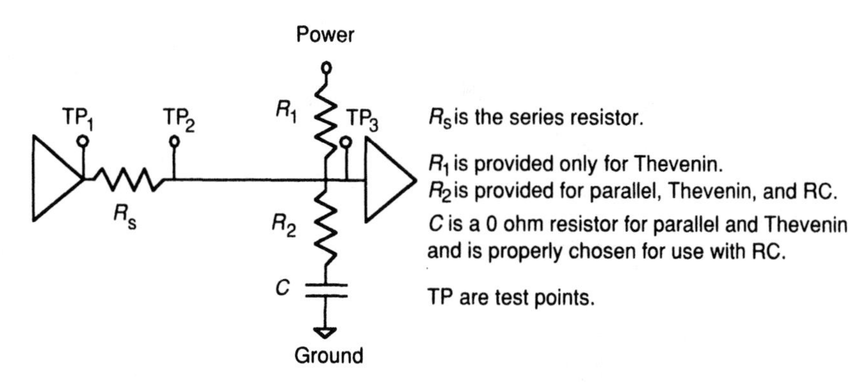

Figure 8.20 Providing for optional termination selection during PCB layout.

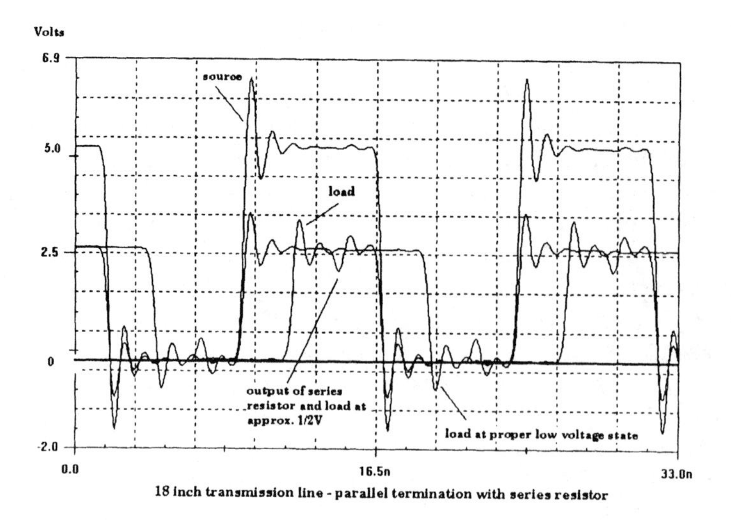

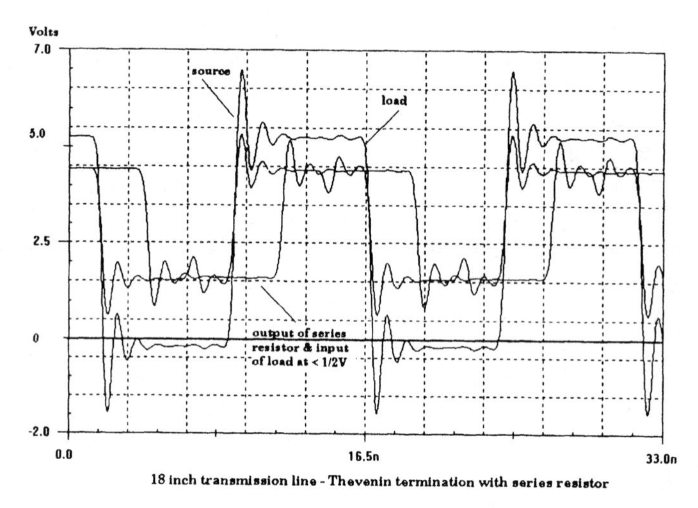

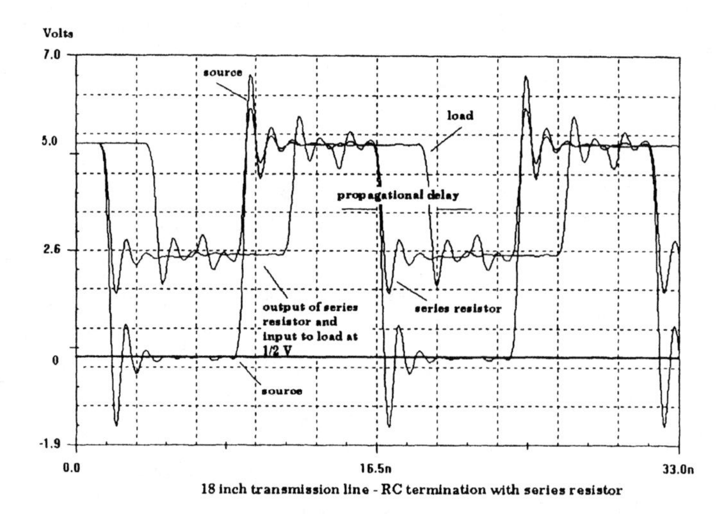

Figure 8.21 Effects of dual terminations.

analyzer, signal integrity problems may be easily observed, with related concerns addressed. Most engineers will use either a spectrum analyzer or an oscilloscope, depending on their comfort level. Few engineers will consider both the time *and* frequency domain aspects of components, a network or circuit. When management demands that a product, which was to ship last week is behind schedule, additional engineering resources to investigate the effects of a termination change become unacceptable, for corporate revenue sometimes takes a higher priority than proper engineering analysis.

To illustrate actual results that can occur with dual terminations, an example is provided with multiple components daisychained on an 18-inch (45.7-cm) trace, 0.8-ns edge rate, with components dispersed through the trace route. Figure 8.20 illustrates a schematic drawing of this circuit. In Fig. 8.21, we observe the effects of dual terminations on the PCB using SPICE simulation. The trace is electrically long and requires termination per the definition provided in Chapter 6 for an electrically long trace.

For parallel and Thevenin, one will observe a 1/2V level, logic level state dependent. This 1/2V level is caused by the series resistor. The ringing that occurs at the load is minimal and is a nearly perfect signal at the load. However, an interesting situation is noted. For parallel termination with a series resistor, the voltage level observed at the load in the low-to-high transition state is at 1/2V. It becomes obvious that the circuit cannot work as designed, especially if mixed logic is provided on the net. The same results occur with Thevenin except in the high-to-low transition state.

For RC networks, a more dramatic effect is observed. As a result of the terminator, the series resistor appears at the 1/2V level as expected. At this point, the output of the resistor looks identical with smooth rounding of the signal. This rounding occurs from $\tau = RC$ where τ is the rounding time constant, R is the termination resistance, and C is the total shunt capacitance of the network. The load receives a clean signal in the HI logic state but at the expense of a much slower edge rate. If the timing requirements of the load are not critical, rounding of the signal will enhance EMI performance significantly without affecting signal integrity. Since the capacitor holds the DC voltage level of the signal on the trace, the voltage degradation that was present when both parallel and Thevenin termination are dual terminated with a series resistor will not be seen.

As observed, providing dual termination can, and will, cause functionality concerns to exist. Dual termination must never be used in any design without a thorough understanding of what will happen to signal integrity along with creation of EMI.

8.5 TRACE ROUTING

Engineers and designers will sometimes daisychain periodic signal and clock traces for ease of routing. Unless the distance is small between loads (with respect to propagation length of the signal rise time), reflections may occur from daisychained traces. Sometimes daisychaining impacts signal quality and EMI spectral energy distribution to the point of nonfunctionality or noncompliance. Therefore, radial connections for fast edge signals and clocks are preferred over daisychaining for nets with a single, common drive source. Each component should have its respective trace terminated in its characteristic impedance as shown in Fig. 8.22. Parallel and Thevenin termination at the end of the trace route is rarely feasible because the drivers usually cannot tolerate the total current sink of the terminated loads.

To prevent undesired effects of unmatched loads, termination may be required. Five common termination methods are available (discussed earlier in this chapter). Each

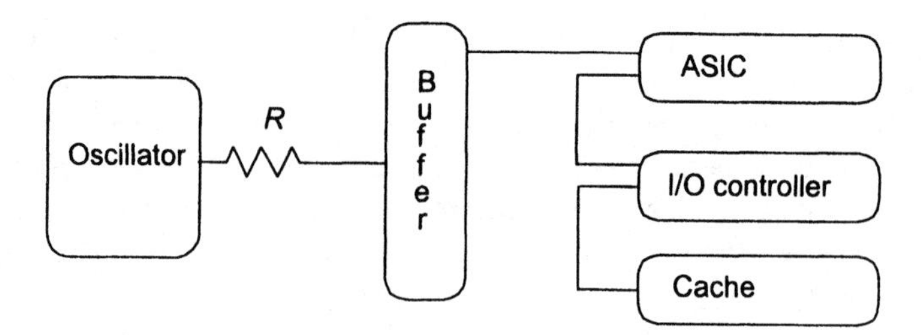

Poor trace routing for clock signals
(Note daisychaining of clock signal)

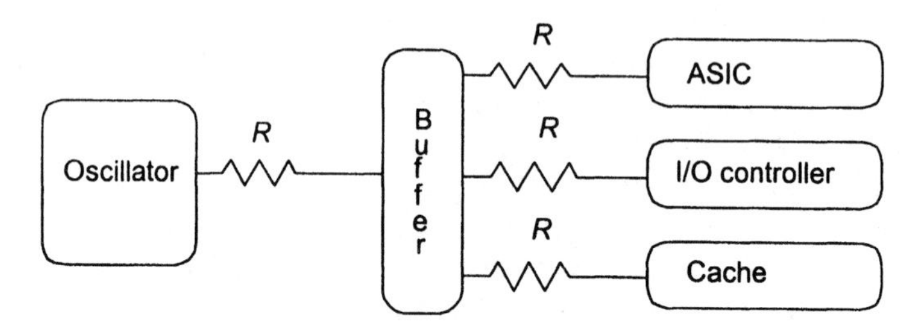

Optimal trace routing for clock signals with series termination

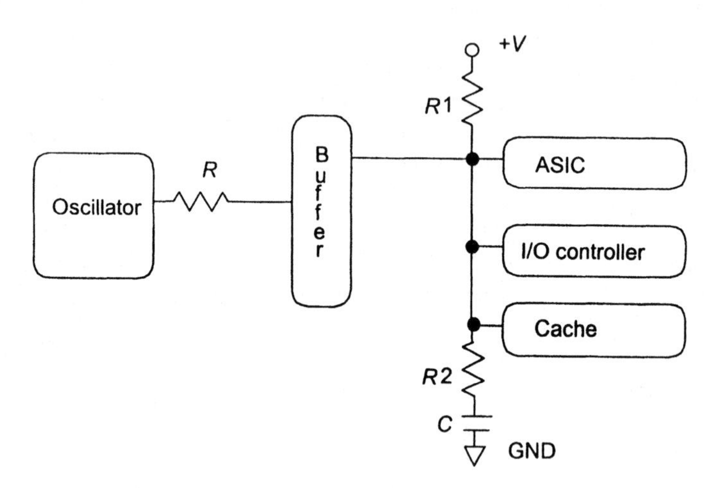

Figure 8.22 Termination of clock traces.

method is dependent on the complexities of layout geometry, component count, and power consumption. When a driver is overloaded, termination can degrade the trace if incorrectly specified or implemented.

If an electrically long signal trace route exists, this trace *must be correctly terminated!* Long lines generally require use of high-current driving components. One should calculate the terminating resistor value at the Thevenin equivalent or the characteristic impedance of the trace. Use of "T-stubs" or bifurcated lines is generally not allowed. If a

T-stub has to be used, the maximum permissible stub length cannot exceed $T = L_d^{tr/10}$, where L_d is the routed length of the trace. The length of each "T" from the center leg must be identical. In T-stub lines, the capacitance and load characteristics of the devices at the end of each T-arm should be exactly equal.

If a T-stub is required because of problems with layout or routing, it must be as short as possible. The measurement feature of the CAD system should be used to measure routing lengths. If necessary, one should serpentine route the shorter trace until it equals its counter trace length exactly.

A potential or fatal drawback of using T-stubs lies in future changes to the artwork. If a different design engineer or PCB designer makes a change to the layout or routing to implement rework or a redesign, knowledge of this T-stub implementation may not be known, and accidental changes to the trace may occur, posing potential EMI or functionality problems.

8.6 BIFURCATED LINES

Bifurcated lines is another term for T-stubs. This condition occurs when a trace is split into multiple trace routes from a driver. An example of a bifurcated topology is shown in Fig. 8.23, along with a recommended termination method.

The impedance of a bifurcated line is not constant throughout the trace route. Let us assume, for example, that the trace from point A to X is 50 ohms. The impedance of the bifurcated lines must be $2Z_o$ from point X to point B, or twice that of the desired impedance characteristics. This is because, for this example, the two traces are running in parallel, each with an impedance of 100 ohms. The parallel impedance of the two traces will be 50 ohms, the desired impedance of the network. Since the individual bifurcated traces are now 100 ohms, use of end termination of 100 ohms each is the only practicable termination methodology. To use parallel termination, the driver must be able to source sufficient current to multiple loads on the net.

If proper termination is provided, reflection and ringing that may occur will be prevented. If termination is not provided, the return current will see an impedance discontinuity at point X. This discontinuity will be observed as ringing at the source driver, point A. The termination must be provided at the end of the trace route as was shown in Fig. 8.8.

If bifurcated routing must occur, how can an optimal design be set up within a PCB when split traces must be $2Z_o$? The easy layout technique available to the PCB designer is to make the trace width smaller than the primary trace from point A to X. This becomes almost impossible for most applications, for the line widths of traces are already approaching a very small dimension. To increase trace impedance, it is necessary to make the traces even smaller, which in many situations is smaller than the manufacturing process a PCB fabricator is capable of performing.

Another problem with bifurcated lines lies with RF loop currents created within the network. If one of the bifurcated signal traces is routed on a different routing plane than the other bifurcated trace, RF return current will try to reference both traces to different 0V return planes. When this occurs, a potentially large loop area is created, with a corresponding increase in radiated emissions.

To summarize, use of bifurcated lines, or T-stubs is not desirable for both signal integrity and EMI compliance reasons.

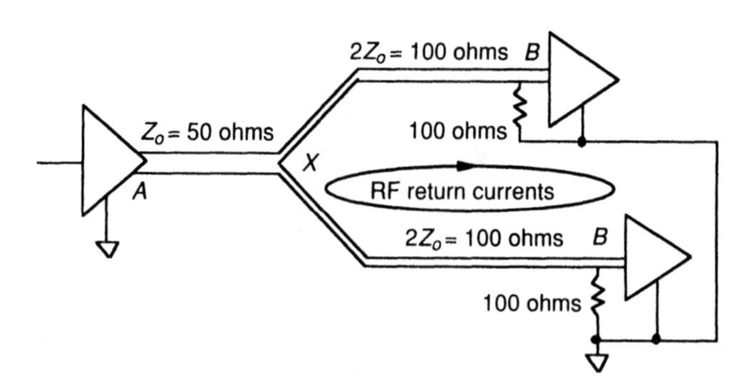

Figure 8.23 Example of bifurcated lines.

8.7 SUMMARY—TERMINATION METHODS

Different termination methods are available, each for a specific application along with advantages and disadvantages. The termination method that provides optimal performance for most designs (CMOS or TTL) is dependent on what the circuit designer requires. The following is a brief summary of termination methods presented earlier in this chapter.

1. Series is excellent for point-to-point trace routes (one load on the net). In addition, series termination works well for those traces that are electrically short (small propagation time from source-to-load and return from load-to-source) with respect to the clock frequency (t_{pd}). Series termination may also be used to slow down edge times so that the effect of propagation discontinuities in the signal path is minimized. In addition, it becomes a simple process to allow for the fanout of multiple load radials from a common source to occur using separate transmission lines that do not corrupt other circuits in the network.
2. Parallel is preferred for busses and point-to-point nets with fast clock/pulses (frequencies).
3. Thevenin networks are difficult to implement owing to the reduced voltage level that exists in both the HI and LOW state if a combination of both CMOS and TTL exists on the same net.
4. The RC network provides good signal quality but at the expense of added components. Drawbacks exist at high frequencies and for long trace lengths owing to limited damping that occurs with poor impedance matching and edge rate degradation.
5. Dual terminations degrade signal functionality and should not be used without fully understanding the consequences.

What happens when a trace on a PCB cannot be terminated?

1. There may be 100% positive signal reflection, resulting in signal doubling in amplitude at the source driver. This may cause destruction of the component owing to excessive voltage levels.
2. The signal may become 70–80% negative at the source driver, which will erode signal functional and noise margin levels of the circuit.

3. Loads located at a distance from the driver may be triggered by false signal transitions, resulting from ringing that will occur at trigger threshold levels.
4. Load-sensitive components should be located at the end of a routed net to minimize reflections between these sensitive components and to prevent false edge triggering.
5. Edge-sensitive signals should be prevented. Voltage-level sensitive components should be used with a sufficiently long setup time to guarantee proper triggering.

REFERENCES

[1] Montrose, M. 1996. *Printed Circuit Board Design Techniques for EMC Compliance.* Piscataway, NJ: IEEE Press.

[2] Montrose, M. 1996. "Analysis of the Effectiveness of Clock Trace Termination Methods and Trace Lengths on a Printed Circuit Board." *Proceedings of the IEEE EMC Symposium.* Piscataway, NJ: IEEE.

[3] Johnson, H. W., and M. Graham. 1993. *High Speed Digital Design.* Englewood Cliffs, NJ: Prentice Hall.

[4] IPC-D-317A. 1965, January. *Design Guidelines for Electronic Packaging Utilizing High-Speed Techniques,* Institute for Interconnecting and Packaging Electronic Circuits (IPC).

[5] IPC-2141. 1966, April. *Controlled Impedance Circuit Boards and High Speed Logic Design,* Institute for Interconnecting and Packaging Electronic Circuits.

[6] Hartal, O. 1994, *Electromagnetic Compatibility by Design.* West Conshohocken, PA: R&B Enterprises.

9

Grounding

9.1 REASONS FOR GROUNDING—AN OVERVIEW

Grounding is required within most products. Although this ground may be fully connected, isolated, or floating, a ground structure must still be present. Grounding is often confused with providing a current return path for signals. In reality, only a few grounding issues are related to PCBs. These concerns relate to providing a reference connection between analog and digital circuits and a high-frequency connection between the PCB return plane and an external metal chassis.

Grounding, though probably the most important aspect of a design, is least understood by many engineers. It is not easy to understand intuitively and does not usually allow for straightforward definition, modeling, or analysis since many uncontrolled factors affect its performance. Every circuit is ultimately referenced to a ground source and cannot be left to chance; it must be designed in from the very beginning. One cannot assume that because a ground system is present, for example, a metal enclosure, optimal performance will be achieved. Desired performance is not easily achieved if no thought is given to its design.

Grounding is one primary method of minimizing unwanted noise pickup and partitioning circuit segments. Proper implementation of PCB ground methods and cable shields will prevent a majority of noise problems. One advantage of a well-designed ground system is protection against unwanted interference and emissions for basically zero cost in material usage.

9.2 DEFINITIONS

The word *grounding* is vague and means different things to different people. For logic designers, it refers to a reference level for logic circuits and components. For system and mechanical engineers, ground is the metal housing or chassis that connects circuits. For electricians, it refers to the third wire safety ground as mandated by their respective National Electric Codes.

To prevent confusion, the following words used in this chapter are defined as follows.

Bonding. Making a low-impedance electrical connection between two metal surfaces.

Circuit. Multiple devices with a source impedance, load impedance, and interconnects. For digital circuits, multiple sources and loads may be part of one circuit where all devices are referenced to the same point or use a common signal return conductor. For EMC, circuits usually originate in one location and terminate in another.

Circuit Referencing. The process of providing a common 0V reference voltage for multiple circuits to allow communication between the two. Circuit referencing is one of the most important reasons for providing a ground reference. This reference point is not intended to carry functional current.

Earthing (British term). The connection of the safety ground wire to earth at the service entrance of a building.

Equipotential Ground Plane. A piece of metal used as a common connection point for power and signal referencing. This plane may not be at equipotential levels for RF frequencies owing to its electrically large size.

Ground Loop. A circuit that includes a conducting element (plane, trace, wire) assumed to be at ground potential where return currents pass through. At least one ground loop will be present within a circuit. Although a ground loop is acceptable, the severity of the problem for currents flowing through the loop depends on the unwanted signals that may be present, which can cause system malfunction.

Ground Stitch Location. The process of making a solid ground connection from a PCB to a metallic structure for the purposes of providing systemwide ground referencing regardless of which grounding methodology is used.

Grounding. A generic term with as many definitions as there are engineers. This word must be preceded by an adjective.

Grounding Methodology. A chosen method for directing return currents in an optimal manner appropriate for the intended application.

Holy Ground. Sometimes referred to as the actual location used. *See also* Single-point ground.

Hybrid Ground. A grounding methodology that combines single-point and multipoint grounding simultaneously, depending on the functionality of the circuit and the frequencies present.

Multipoint Ground. A method of referencing different circuits together to a common equipotential or reference point. Connection may be made by any means possible in as many locations as required.

Referencing. The process of making an electrical connection or bond between two circuits that allows the 0V reference from both circuits to be identical.

RF Ground. The process of providing a ground reference point using a specific methodology to allow a product to comply with both emissions and immunity requirements.

Safety Ground. The process of providing a return path to earth ground to prevent the hazard of electric shock through proper connection and routing of a permanent, continuous, low-impedance, adequate fault capacity conductor that runs from a power source to a load.

Shield Ground. The process of providing a 0V reference or electromagnetic shield for both interconnect cables and main chassis housing.

Single-point Ground. A method of referencing many circuits together at a single location to allow communication between different points. All signals will thus be referenced to the same location.

When discussing grounding concepts, it is best to use an adjective to better define what we want, such as signal ground, chassis ground, safety ground, and analog ground.

9.3 FUNDAMENTAL GROUNDING CONCEPTS

The two primary areas related to grounding are

1. Safety ground (including protection against the effects of lightning and electrostatic discharge).
2. Signal voltage referencing ground.

If a ground is connected by a low-impedance path to earth, this method is identified as a safety ground. Signal grounds may or may not be connected to earth potential. Connection of the two ground methods may be unsuitable for a particular application and may exacerbate EMC problems.

Safety grounding minimizes or prevents a voltage difference between exposed conducting surfaces. The more conductors we make available to reduce the voltage difference to extremely low levels, the less chance of electric shock may be present to harm someone or cause death. The more ground connections, the less chance of harm to the operator.

Signal voltage referencing ground provides for all parts of an electrical system to be referenced to a common source. For signal referencing, the voltage difference typically must be less than a few millivolts. The implementation of signal voltage referencing, the number of ground connections, and their location must be chosen carefully. Few critically located connections between signal voltage points can allow a product to be either compliant or noncompliant to EMC regulations.

Common misconceptions exist regarding grounding. Most analysts believe that ground is a current return path and that a good ground reduces circuit noise. This belief causes many to assume that we can sink noisy RF currents into the earth, generally through a building's main grounding structure. This is valid if we are discussing safety grounding, not signal voltage referencing.

Current requires a return path to complete a closed-loop circuit. We usually only consider AC or DC supply current and not RF current. Although an RF return path is mandatory, it need not be at ground potential. Free space is not at ground potential. Analog ground is isolated from digital or chassis ground to prevent disruption to sensitive circuits. Not all currents within a system require a safety ground or a signal voltage reference. For example, low-voltage battery-operated devices do not require any external safety ground connection, for no shock hazard exists.

To guarantee that a system works within a specific design requirement, signal ground may not be the same as current return. Signal currents should not flow on grounding conductors except under certain conditions. Regardless of application, for both safety and signal referencing, we must either reduce the ground voltage difference between two circuits or avoid having a voltage potential difference at all.

Why is safety ground discussed in a book about EMC and PCBs? The reasons are obvious. Many PCBs contain hazardous voltages. These include power supply assemblies, telecommunication circuits, relay-driven instrumentation control units, power switching modules, and the like. User safety cannot be separated from EMC. The field of regulatory compliance includes both product safety (meeting essential safety requirements based on National Electric Codes or governmental mandated legislation) and EMC limits for emissions and immunity. Product safety standards mandate the amount of creepage and clearance distance between traces to prevent electric shock to the user. For example, the distance spacing that must be used for a product safety requirement between hazardous voltages on traces may prevent optimal flux cancellation from a voltage or signal trace to a ground fill or ground trace for single- or double-sided PCBs.

Creepage and clearance is of concern because AC or high-voltage traces may be subject to an abnormal failure condition. Failures include primary-to-secondary, primary-to-ground, or primary-to-primary. To prevent a shock hazard due to an abnormal failure, traces must be routed with a specific amount of spacing (distance) between high-energy (voltage) traces and secondary or ground circuits. This requirement is especially critical in power supplies and related circuitry.

When routing AC voltage traces, one should use sufficient trace width and spacing to comply with legally mandated creepage and clearance requirements. The following definition of creepage and clearance is extracted from, and is identical to, all international product safety standards.

- Creepage is the shortest path between two conductive parts, or between a conductive part and the bounding surface of the equipment, measured along the surface of the insulation.
- Clearance is the shortest distance between two conductive parts, or between a conductive part and the bounding surface of the equipment, measured through air.
- Bounding surface is the outer surface of the electrical enclosure considered as though metal foil were pressed into contact with the accessible surface of insulation material.

When dealing with ground currents, several fundamental concepts must be remembered.

- Whenever a current flows across a finite impedance, a finite voltage drop occurs (Ohm's law). As stated in Ohm's law, there can never be "zero volt potential" in the real world. The units may be in the pico range (voltage or current). Still, a finite value will exist.
- Current must always return to its source. This return may consist of numerous paths with various amplitudes provided for each return current proportional to

the finite impedance within each and every path (Kirchhoff's law). Unintended currents can travel in alternate return paths which may not be designed to handle these currents.

To illustrate these fundamental concepts, Fig. 9.1 shows two subsystems connected to a metallic plate, chassis, or other item identified as ground. These subsystems may be analog, digital, or another defined source. If digital, the power (+Vcc) current returns to its source, the power supply through a return system. Current is constantly changing when devices switch logic states, consuming power. In analog circuits, the return current may contain low-frequency or high-frequency narrowband or broadband signals. Analog signals generally have dedicated return or "grounds" that are different from digital logic.

As can be observed in Fig. 9.1a, the return current path of Subsystem #2 travels through the same return line as Subsystem #1. The two currents add up at the power supply source. Since a return path will have a finite impedance, either resistive or inductive, currents within the return structure will cause a voltage potential to be developed between the two subsystems. The ground point of Subsystem #1 is varying at a rate proportional to the signals in Subsystem #2. By virtue of this coupling through a common impedance, the power source now sees two separate voltage potentials simultaneously.

So far, this has been a discussion about ground-noise voltage. What about the voltage that is observed at the load? The voltage of the ground point for Subsystem #2 is $Z_{g1}I1 + (Z_{g1} + Z_{g2})I2$. Subsystem #2 contains the signals of Subsystem #1 through Z_{g1} in addition to its own signal. This situation is identified as common-impedance coupling.

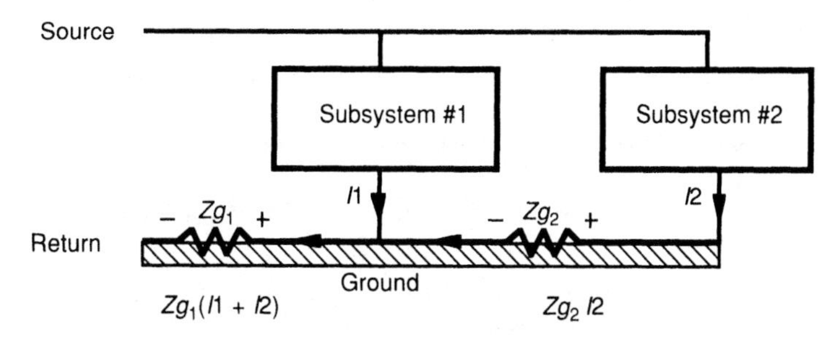

Figure 9.1a Common-impedance coupling in a ground structure. (*Source:* Clayton Paul, *Introduction to Electromagnetic Compatibility*, © 1992. Reprinted by permission of John Wiley & Sons.)

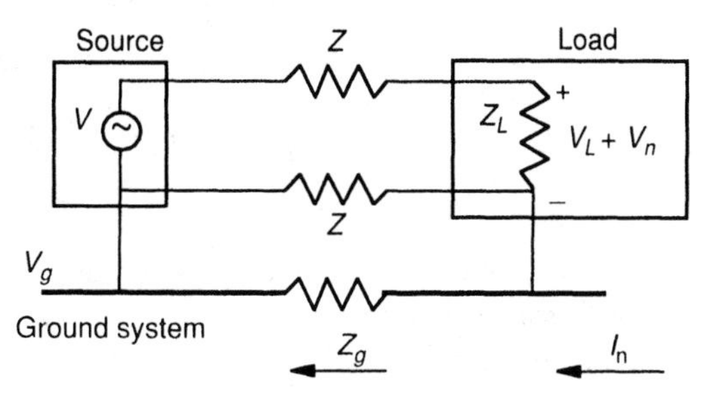

Figure 9.1b Conductive coupling of ground noise with interconnecting cables.

Figure 9.1b illustrates a connection commonly found in data communications where a signal (e.g., RS-232) and its return path are provided. Let's assume the source is a computer and the load is a monitor or modem located some distance away. The ground system common to both devices is the third wire safety ground within their respective power cord. This power cord ground has a high impedance at RF frequencies. In this case, the noise source from the external system drives a noise current I_n into the safety ground wire, which is common to all devices. Let's assume load Z_L is greater than the return wire impedance Z and the power wire impedance Z_g. The ground noise developed on the return wire adds to the signal voltage of the load, described by

$$V_n = \frac{Z - Z_g}{Z + Z_g} * (V_g) \tag{9.1}$$

A misconception regarding ground impedance is the type of impedance that exists. Most engineers assume that ground impedance is at DC potential or has low-frequency resistance. At high-frequency of operation, 30 MHz and above, the primary impedance component that is observed is inductive, not *resistance* or *skin effect*. Resistance and skin effect are negligible compared to the inductance. As presented in earlier chapters, inductance is approximately 15 nH/inch for a 0.020-inch trace. Using $X_L = 2\pi fL$, at 100 MHz, the inductive reactance is 9.43 Ω/inch. A #28AWG wire (radius of 6.3 mils) has an inductive reactance of 65.9×10^{-3}Ω/inch. As observed, there is a significant magnitude of difference between resistance and inductance at 100 MHz. This is why resistance is not a concern at RF frequencies.

When designing a product, minimal or zero cost may be incurred during the design cycle when grounding is taken into consideration. A well-designed ground system, not only on the PCB, but systemwide, will offer both improved emissions and immunity protection. A grounding system that was not thought about during a design cycle, or reimplemented from a design on a different product (because it once worked on that product, so why redesign?), is a sign of system failure related to system functionality or EMC compliance.

The important areas of concern include the following.

- Minimize or reduce current loops by careful layout of high-frequency components.
- Partition areas of the PCB, or system, to keep high-bandwidth noise circuits from low-frequency circuits.
- Design the PCB, or system, to keep interfering currents from affecting other circuits through a common ground return path.
- Carefully select ground points to minimize loop currents, ground impedance, and transfer impedance of the circuit.
- Consider the current flow through the ground system as it relates to noise being injected into or from a circuit.
- Connect very sensitive (low noise-margin) circuits to a stable ground reference source.

The next section examines various ground systems and how they apply to a product's overall design. Following this examination, there is a description of how to implement grounding methods in an optimal manner.

9.4 SAFETY GROUND

The primary concern associated with a safety ground is the protection of people, animals, and other living creatures from the hazard of electric shock. When a product is at a hazardous voltage potential, serious injury or death may occur.

If the system is powered by AC voltage above certain levels (defined below), exposed metal must be bonded to a "green or green/yellow stripe wire" safety ground provided within the AC mains power cord. This requirement also applies to battery-operated devices if the battery charger is built into the module unit or built onto the PCB, powered by AC mains voltage. If the unit operates from DC voltage, then only the remote power charger unit needs to comply. If a conflict occurs between EMC compliance and product safety, safety takes precedence. No exception to this requirement exists.

Electric shock occurs when current passes through the human body. Currents on the order of a milliampere can cause a reaction in persons of good health and may cause indirect danger due to involuntary reaction. Higher currents can have more damaging effects. Voltages up to 42.4 VAC peak, or 60 Vdc, are not generally regarded as dangerous under dry conditions. Electrical parts that have to be touched or handled should be at earth potential or properly insulated to prevent electric shock.[1]

Under normal conditions, any voltage (absolute value) greater than 42.4 VAC peak, or 60 VDC, that may exist on a PCB (or system) is considered hazardous and requires special attention by a product safety compliance engineer.

How does all this discussion about hazardous voltages affect PCBs? Telecommunication circuits operate at –48Vac. Power supplies are sometimes provided on a PCB connected to AC mains voltage. Solenoids drive 115V or 230V motors. Process control equipment generally uses voltages above 42.4 VAC peak. These are only a few examples of PCBs that may contain hazardous voltages; hence, this chapter requires a discussion on safety grounds within PCBs.

In Fig. 9.2, the stray impedance between voltage potential (PCB) at point V_1 and chassis is identified as Z_1. The stray impedance between chassis and ground is identified as Z_2. The potential of the chassis is the impedance of Z_1 and Z_2 acting as a voltage divider. The chassis potential, relative to the PCB, is

$$V_{\text{chassis}} = \left(\frac{Z_2}{Z_1 + Z_2}\right) \tag{9.2}$$

This potential could reach hazardous levels, enough to cause a shock hazard to exist.

It must never be assumed that as long as everything is connected to earth ground through an appropriate means (green/yellow wire, braid strap, and the like), then all is well. This ground wire will have a high impedance at RF frequencies which varies as the frequency varies. In general, safety earth ground is not required for EMC compliance. Examples are battery-powered units. A good low-impedance connection to an RF reference point provided by a local chassis, frame, or other metallic structure is necessary and in many instances must be provided in parallel with safety earth ground for those devices connected to an AC mains source.

[1]This description of electric shock hazard is extracted from the international product safety standard, EN 60950, *Specification for safety of information technology equipment, including electrical business equipment.*

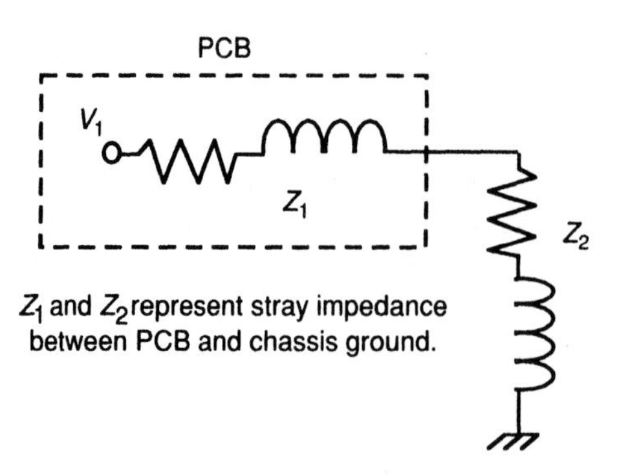

Figure 9.2 Stray impedance from PCB to chassis ground.

If we observe common-mode emissions emanating from a power cord, a safety earth ground connection may be required. A line filter can be installed at the mains power inlet which places the line filter in series between the mains wall receptacle and the system. Internal to the line filter are capacitors from line to ground ("Y" capacitors) which shunt the RF currents to ground. For this application, the ground wire is a return path for RF currents.

At times it is beneficial that the safety earth ground path be removed from the RF generation circuit [8]. This is best accomplished by inserting a choke (RF conductor) in series with the earth return. This choke provides an alternative path for interference currents to remain within the system. These currents are hopefully prevented from radiating to the external environment by a Faraday shield or Gaussian structure (sheet metal covers).

To summarize, a voltage potential at hazardous levels must not exist. Under an abnormal fault condition, such as the PCB shorting out and energizing a metal chassis housing, the housing can assume full-voltage potential and create a shock hazard.

9.5 SIGNAL VOLTAGE REFERENCING GROUND

The majority of design concerns related to EMC compliance lies in signal ground and referencing one circuit to another. As discussed earlier, both source and load must be at the same voltage reference level for proper functionality. Logic circuits base their voltage transition states at a 0V reference level. If the reference level between two circuits is not the same, functionality concerns occur such as noise margin erosion and threshold levels for logic switching (in addition to the creation of a ground-noise voltage). This ground-noise voltage will cause common-mode currents to be developed, which is exactly what is not wanted.

A *ground* is usually defined as an equipotential point that serves as a reference potential between two or more items. This term is not representative of actual applications, since digital ground may be completely different from analog ground, which may also be different from chassis ground. This term also does not emphasize the return path that currents at RF frequencies take. There may be less inductance between a noisy circuit and a ground point than the connection to an equipotential point. RF current will always take the path of "least impedance." At low frequencies, where $R >> \omega L$, the current will take

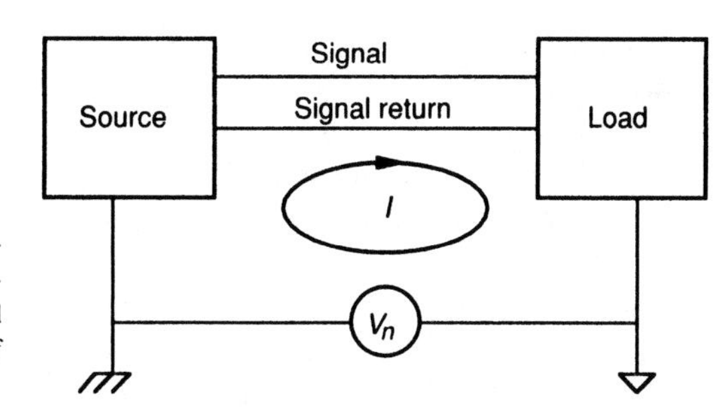

Figure 9.3 Typical grounding observed between two circuits. (*Source:* H. Ott, *Noise Reduction Techniques in Electronics Systems*, 2nd edition © 1988. Reprinted by permission of John Wiley & Sons, Inc.)

the path of least resistance, as resistance dominates the impedance. At high frequencies, $R << \omega L$, inductance dominates.

A better definition of signal ground is a low-impedance path for signal current to return to its source. This is applicable to any application or environment. Current is the item of concern, not voltage. If a voltage difference exists between two circuit points through a finite impedance, current will be created (Ohm's law). The current path in the ground structure determines the magnetic coupling between circuits. Since a closed-loop path is present, with current flowing in the loop, a magnetic field is developed (see Chapter 2). The physical size of the loop area determines the frequency of the radiated emissions. The current level determines the amplitude of the radiated noise.

Designers must always keep in mind the path that RF current will take during a product design. They cannot concern themselves only with functionality and with how well their chosen logic devices work based on simulation data. The design engineer and PCB designer must work together to ascertain the anticipated path through which the return currents will flow during component placement. The question to ask is, "Where will the current flow?" Any conductor carrying current will have a voltage drop associated with it, along with its corresponding current. This current is usually at RF potentials.

The signal ground system is determined by the type of product design, frequency of operation, logic devices used, I/O interconnects, analog and digital circuits, and product safety (electrical shock hazard).

A typical grounding scheme used to describe the signal ground concept is shown in Fig. 9.3 where the load is connected to one ground reference point and the source is connected to another reference. Ground-noise voltage, V_n, is caused by losses in the return path.

In implementing a grounding methodology, two basic categories exist, single-point and multipoint. Within each methodology, hybrid combinations may exist. The signal ground methodology that is best for a particular application is dependent on the design. Several different methods may be used at the same time, only if the designer understands the concept of current flow and return paths.

9.6 GROUNDING METHODS

Many grounding methods and terms have been devised, including digital, analog, safety, signal, noisy, quiet, earth, single-point, and multipoint. Grounding methods must be specified and designed into a product and not be left to chance. Designing a good grounding system is also cost effective in the long run. In any PCB, a choice must be made between two basic concepts of grounding; single versus multipoint. Interactions with other grounding methods can exist if they are planned for in advance. The choice of grounding is dependent on product application. It must be remembered, if single-point grounding is used, to be consistent in its application. The same rule exists for multipoint grounding. A multipoint ground should not be mixed with single-point ground unless the design allows for isolation or partitioning between planes and functional subsections!

The discussion that follows is divided into three main grounding concepts. These concepts are single-point, multipoint, and hybrid.

9.6.1 Single-point Grounding

A single-point ground connection is one in which ground returns are tied to a single reference point within a product design. The intent of this "holy" ground location is to prevent currents from two different subsystems (at different reference levels) from sharing the same or common return path for RF currents, thus producing common-impedance coupling.

Single-point grounding is best when the speed of components, circuits, interconnects, and the like is in the range of 1 MHz or less, which means that the effect of distributive transfer impedances is minimal. At higher frequencies, the inductance of the return path will start to become noticeable. At still higher frequencies, the impedance of the power planes and interconnect traces becomes noticeable. These impedances can be very high if the trace lengths coincide with odd multiples of a quarter-wavelength based on the edge rate of the periodic signals. With a finite impedance in the current return path, a voltage drop is developed, along with creation of unwanted RF currents.

Owing to the significant impedance at RF frequencies, these traces and ground conductors will act as loop antennas and radiate RF energy based on the physical size of the loop. A convoluted loop is still a loop, regardless of shape. At frequencies above 1 MHz, a single-point ground generally is not used for this reason. However, exceptions do exist if the design engineer recognizes the pitfalls and designs the product using highly specialized and advanced grounding techniques.

In Fig. 9.4, two methods are shown for single-point grounding: series and parallel connection. The series connection is in a daisychain fashion. This type of configuration allows common-impedance coupling between the ground reference of each subsystem, which is undesirable at frequencies above 1 MHz. This figure only shows the inductance

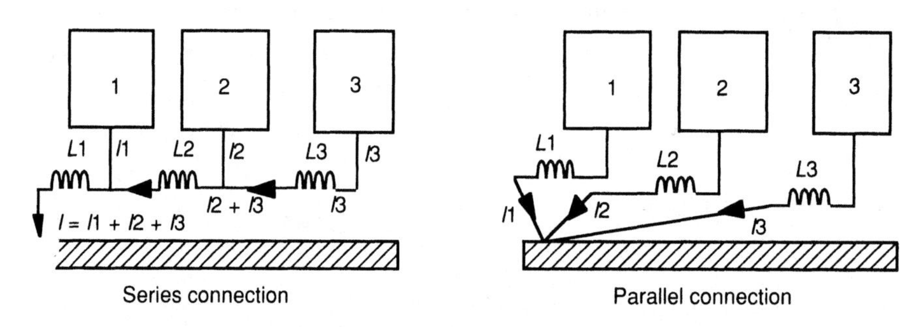

Figure 9.4 Single-point grounding methods. *Note:* Inappropriate for high-frequency operation.

within the ground path. Distributed capacitance is also present among the three circuits to ground. When both inductance and capacitance are present, a resonance will occur. For this configuration, three different resonances are possible.

For series connection, the total amount of current that is observed across the final return path L_1 is the summation of $I_1 + I_2 + I_3$. The voltage potential at I_1 (V_A) and I_3 (V_C) is also not at zero potential, and is described by

$$V_A = (I_1 + I_2 + I_3)\,\omega L_l \qquad (9.3)$$

$$V_C = (I_1 + I_2 + I_3)\omega L_1 + (I_2 + I_3)\omega L_2 + (I_3)\omega L_3 \qquad (9.4)$$

With this widely used configuration, a large amount of current across this finite impedance will produce a voltage drop. The voltage reference between circuits and the reference structure may be sufficient to cause the system to fail to work as desired. During the design cycle, one must be aware of the pitfalls of using series connection for single-point grounding. This grounding method should not be used when widely different power levels are present, since high-power consuming circuits produce large ground currents, which in turn will affect low-level components and circuits. If this method must be used, the most sensitive circuits must be located immediately at the input power location and as far away from low-level components and circuits.

A more optimal single-point ground method is parallel. Using this method has a disadvantage, however, in that each current return path may be at a different impedance value, thus exacerbating ground-noise voltage. If multiple PCBs are provided within an assembly, or if various subassemblies are combined within an end-use product, a particular return path may be physically long, especially if wires are used as the interconnect method. The ground wires may also possess a large impedance that will negate the desired effect of a low-impedance ground connection. Many products fail emissions testing when multiple PCBs are tied together in a parallel fashion, believing that a "holy" ground connection will solve their problems. Like series connection, distributive capacitance is also present from each circuit to ground. The designer should maintain the inductance value from each circuit to ground using this configuration to be approximately the same, but rarely is this the case. As a result, the resonance between each circuit to ground should be approximately the same and may not affect circuit operation to the extent multiple unique resonances will.

Another problem associated with using single-point grounding with wires is radiated coupling, which may occur between the wires, the wire and PCB, or the wire and chassis housing. (Internal radiated noise coupling is discussed later in this chapter.) In addition to RF radiated coupling, crosstalk may occur depending on the physical distance spacing between the current return paths. This coupling may occur by either capacitive or inductive means. The amount of crosstalk that may be present is dependent on the spectral content of the return signal. Higher frequency components will radiate more than lower frequency components.

Single-point grounds are usually found in audio circuits, analog instrumentation, 60-Hz and DC power systems, along with products packaged in plastic enclosures. Although single-point grounding is commonly used in low-frequency applications, it is occasionally found in extremely high-frequency circuits and systems. This application is permitted when a design team understands all the problems that exist with inductance in different ground return structures.

Use of single-point grounding on a CPU-motherboard or adapter (daughter) card allows loop currents to be present between the ground planes and chassis housing if metal is used as the chassis. Loop currents create magnetic fields. Ground loops are examined in greater detail later in this chapter.

Magnetic fields create electric fields, which will radiate RF currents. It is nearly impossible to effectively implement single-point grounding in personal computers and similar devices because different subassemblies and peripherals are grounded directly to the metal chassis in different locations. A distributed transfer impedance exists between the chassis and the PCB that inherently develops loop structures. Multipoint grounding places these loops in regions where they are least likely to cause problems (e.g., they can be controlled and directed rather than allowed to transfer energy inadvertently).

An example of poor implementation of single-point grounding is shown in Fig. 9.5. In this example, the A/D 0V reference is isolated within both the digital and analog section under the assumption that the open connection point (bridge) will provide optimal single-point connection as long as the analog section is not bonded to any other ground location. Single-ended signals are routed across the gap in the area of the converter's moat or isolated area. If low-frequency (kHz) noise frequencies are a problem, the 0V reference connection should be placed as near to the A/D device as possible. Analog and digital power must be isolated from each other referenced by an appropriate filter.

In Fig. 9.5, various current and voltage sources are present. RF return current travels through the bridge. The bridge provides a low-impedance RF return path for all signals that travel to the analog section by either crossing the moat or traveling through the bridge. Since a closed-loop path must be present for signal functionality, any RF energy crossing the moat must complete its return through the bridge.

Since a moat is present, a common-mode voltage potential will be developed at the point furthest from the bridge. The impedance between the two power sources will be different based on the inductance of the power and ground plane structure. With this common-mode voltage, a common-mode RF current loop current is developed, which travels through both the digital and analog sections. Once a loop is created with RF currents, a magnetic field structure exists, causing possible RF emissions.

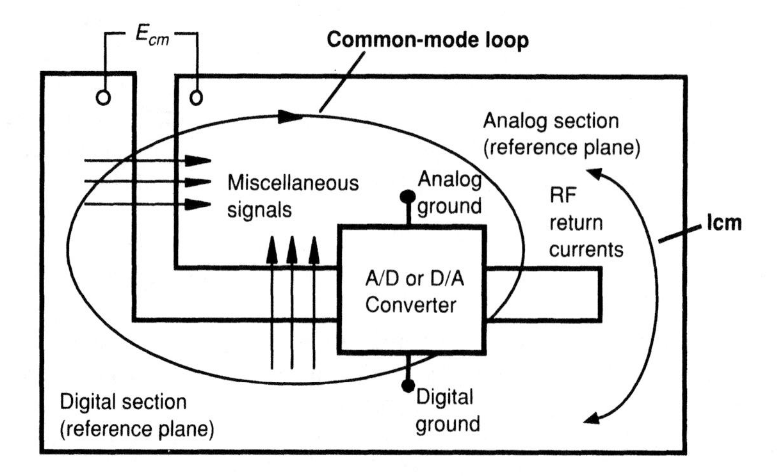

Figure 9.5 Bad implementation of single-point (connection) grounding.

Another bad implementation of single-point grounding in a multicircuit system is detailed in Fig. 9.6. The ground wires are noncurrent-carrying conductors (RF return path). The interconnect wires connected between circuits (Circuit #1 to Circuit #2 and Circuit #1 to Circuit #3) are identified as "GND" by the circuit designer or component manufacturer. These ground connections become part of the signal return path for currents that travel between circuits. These ground traces create an RF current loop, increase self-inductance of the traces, and develop a stray magnetic field between circuits and the 0V reference point. In addition, parasitic capacitance, C, between Circuit #1 and ground and Circuit #3 and ground is shown, along with inductance, L, from all circuits to the single-point ground connection. This small amount of LC may create a resonance that occurs at a frequency or harmonic of an oscillator, thus exacerbating systemwide problem.

To summarize, single-point grounding is not ideal when dealing with products operating above 1 MHz.

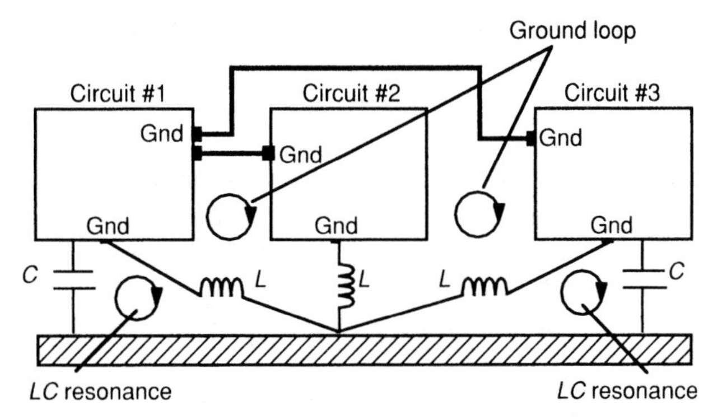

Figure 9.6 Another bad implementation of single-point grounding.

9.6.2 Multipoint Grounding

High-frequency designs generally require use of multiple chassis ground connections to a common reference point in order to minimize ground impedance. Multipoint grounding minimizes ground impedance present in the RF current return path because there are more low-impedance paths to take. Low planar impedance is caused primarily by the lower inductance characteristic of solid power and ground planes or by additional low-impedance ground stitch connection to the chassis reference point.

When a low-impedance ground plane is provided in a multilayer PCB, or a chassis ground stitch connection is provided between the PCB and metal chassis, it becomes important, like single-point grounding, that trace length (or wire length) be kept as short as possible to minimize lead-length inductance. In very high-frequency circuits, the length of the ground leads must be kept to a small fraction of an inch (cm). When using low-frequency circuits, multipoint grounds should be avoided since ground currents from all circuits flow through a common ground impedance, the ground plane. The common impedance of the ground plane can be reduced by using a different plating process on the surface of the material [2]. Increasing the thickness of the plane has no effect on minimizing plane impedance, for RF currents travel on the skin surface layer of the material.

A general rule of thumb is that for frequencies less than 1 MHz, single-point grounding is preferred. Between 1 MHz and 10 MHz, single-point grounding may be used only if the longest length trace or ground stitch connection is less than 1/20 of a wavelength, assuming long edge times and low-frequency spectra. Each and every trace must be considered.

Multipoint grounding minimizes inductance between noise generation circuits and a 0V reference point. This minimization occurs because many parallel RF current return paths exist in parallel, as illustrated in Fig. 9.7. Even with many parallel connections to a 0V reference, ground loops may still be created between each ground stitch location physically distant from other ground connections. These ground loops are prone to magnetic field pickup of ESD energy or creation of radiated EMI. To prevent loop currents between ground locations, it is important to measure the physical distance spacing between the ground connections and to implement the design technique identified in the section "Aspect Ratio" in Chapter 4, where the physical distance between two ground stitch connections should not exceed 1/20 of a wavelength of the highest frequency present within the functional subsection being grounded.

In very high-frequency circuits, lengths of ground leads from components must also be kept as short as possible. Trace lengths as long as 0.020 inch (0.005 mm) add inductance to a circuit of approximately 15–20 nH per inch (depending on trace width). This inductance may permit a resonance to occur when the distributed capacitance between the ground planes and chassis ground forms a tuned resonant circuit. The capacitance value, C, in Eq. (9.5) can be determined through knowledge of the impedance of copper planes. Impedance of cooper planes is discussed in Chapter 4.

$$Z = \frac{1}{2\pi f \sqrt{LC}} \tag{9.5}$$

where Z = impedance (ohms)
f = resonant frequency (Hz)
L = inductance of the circuit (henries)
C = capacitance of the circuit (farads)

Equation (9.5) describes most aspects of frequency domain concerns. This equation, though simple in form, requires knowledge of how to calculate both L and C, which by themselves are not easy to determine, use, and implement.

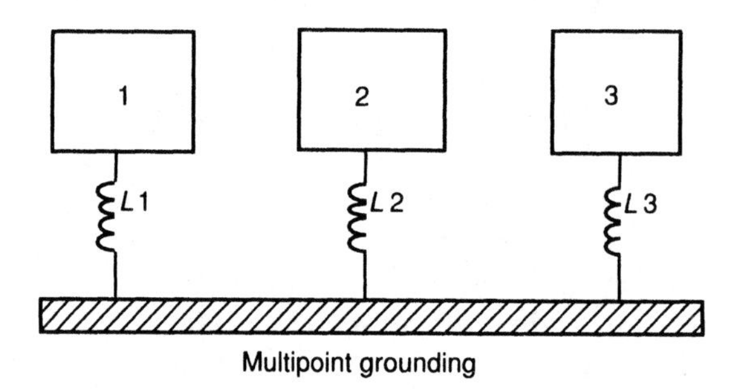

Figure 9.7 Multipoint grounding.

9.6.3 Hybrid or Selective Grounding

A hybrid ground structure is a combination of both single- and multipoint grounding. This configuration is used when mixed frequencies are present within a PCB. Figure 9.8 shows two hybrid ground methods. For the capacitive coupling version at low frequencies, the single-point configuration is dominant, whereas the multipoint configuration works at high frequencies. This is because the capacitor shunts high-frequency RF currents to ground after the single-point connection goes inductive. The key to success is understanding both the frequency present and desired direction of ground current flow.

The inductive coupling version is used when multiple ground stitch locations must be connected to a chassis ground reference for safety reasons and low-frequency connections. The chokes, *L*, prevent RF currents from entering the chassis ground, while allowing low-frequency AC or DC voltages to be referenced to their respective 0V point. The choke keeps the RF current internal to the PCB and forces the return currents to travel through the lowest impedance path to ground the single-point connection (wire), which is at a much lower impedance level than the chokes.

Using capacitors or inductors in a ground topology allows us to steer RF currents in a manner that is optimal for our design. One can take control of the PCB layout by defining the path that the RF currents will take. Failure to recognize the RF current return path may result in either emissions or susceptibility problems.

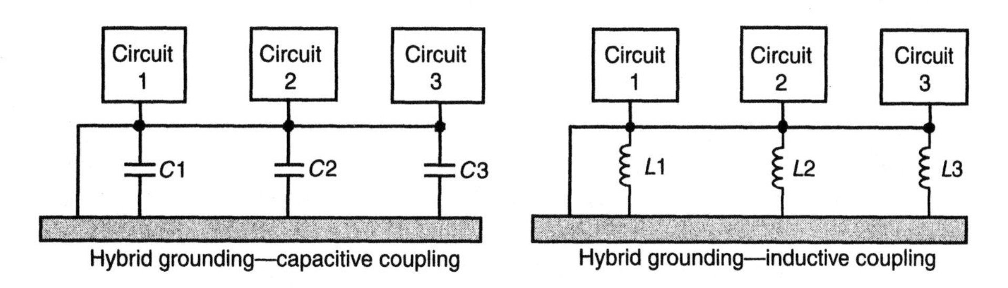

Figure 9.8 Hybrid grounding. (*Source:* H. Ott, *Noise Reduction Techniques in Electronics Systems* © 1988. Reprinted by permission of John Wiley & Sons.)

9.6.4 Grounding Analog Circuits

Many analog circuits are low frequency in operation. Single-point grounding is best for these sensitive circuits but only at the "bridge" between digital and analog. The primary objective is to prevent large ground currents from other noisy components (digital logic, motors, power sources, relays) from sharing a sensitive analog ground path. Ground loops must also be avoided with all sensitive low-frequency analog circuits. With low-frequency analog circuits, it is easy to control both intended and unintended currents.

The degree of quiet required of the analog ground depends on the sensitivity of the analog inputs. The signal-to-noise ratio determines how much interference is allowed to exist before functionality concerns arise. For example, a low-level analog amplifier that requires a 10 μV input is more susceptible to disruption than a 10V input signal. Therefore, a very clean ground system must be present for the 10 μV input amplifier. For higher level analog circuits, ground requirements are less stringent.

Digital circuits affect analog components owing to switching noise from the logic gates internal to digital devices. Usually, there are many significant levels of ground bounce within the power and ground distribution in digital systems. High-speed CMOS components inject more noise into the ground reference than TTL because of higher peak switching currents. CMOS also creates more radiated emissions for the same reasons.

Separate ground references should be provided for both digital and analog, especially if sensitive analog circuits are present. A common reference point must still exist for D/A and A/D converters. This is best achieved at only one point on the PCB; two locations are not permitted at any time. It may sometimes be required to provide a passive filter, such as a ferrite bead between digital and analog circuits. These filters are effective at higher frequencies where parasitic capacitances will attempt to form a ground loop.

Occasionally, complete isolation must occur between analog and digital sections. This is best accomplished by use of optical isolators or isolation transformers, especially when extremely sensitive analog circuits are used alongside digital components.

9.6.5 Grounding Digital Circuits

With higher speed digital circuits, multipoint grounding is preferred because high-frequency currents are developed based on ground-noise voltage and the voltage drop across the layout field of the digital devices. The primary design objective is to acquire a uniform potential common-mode reference system. Single-point grounding does not work well for this reason as parasitics will alter the ground paths desired. Ground loops are usually not a digital problem, as long as a low ground reference impedance is maintained. Ground loops are discussed later in this chapter.

Many digital circuits do not require a ground reference source with filtering. Digital circuits have noise margins in the hundreds of mV and can typically withstand a ground-noise gradient of tens to hundreds of millivolts. Ground "image" planes within the multilayer board are optimal for signal currents, whereas multipoint grounding to chassis is desired to control common-mode return losses.

9.7 CONTROLLING COMMON-IMPEDANCE COUPLING BETWEEN TRACES

A concern associated with common-impedance coupling is to minimize the effects that occur when two metallic structures share a common return path. Two main concepts are used to control common-impedance coupling.

- Lowering the common impedance to a minimum value.
- Avoiding having a common-impedance path.

9.7.1 Lowering the Common-Impedance Path

A ground system requires a metal conductor: trace, wire, strap, chassis frame, PCB planes, and the like. All conductors have a frequency response dependent on the material and geometry. Any conductor will have a DC resistance by

$$R = \rho l/A \text{ (ohms)} \tag{9.6}$$

where R = DC resistance

l = length of the conductor in the direction of current flow (m)

A = cross-sectional area of the conductor perpendicular to the current flow (mm^2)

ρ = resistivity of the material (ohms • mm^2/m)

Resistivities, ρ, of various materials are

copper	$1.7 * 10^{-3}\ \Omega \cdot mm^2/m$
aluminum	$2.8 * 10^{-3}\ \Omega \cdot mm^2/m$
steel	$1.7 * 10^{-2}\ \Omega \cdot mm^2/m$

With common-impedance coupling, skin effect becomes a major factor. (Skin effect was briefly examined in Chapter 2.) As the frequency increases, current through a conductor will migrate toward the edge of the conductor identified as the skin. The area of the conductor available for current flow decreases while resistance increases. For a round conductor, skin effect is illustrated in Fig. 9.9.

Conductors have an intrinsic inductance value that is different from overall inductance. Overall inductance is also identified as external inductance, which is a function of loop area enclosed by the conductor. Internal inductance is not a function of this loop area. For a round conductor, internal inductance is

$$L = 0.2 \cdot l\left[ln\left(\frac{4l}{d}\right) - 1\right] \tag{9.7}$$

where L = internal inductance (μH)

l = conductor length (m)

d = conductor diameter (m)

This equation shows that inductance, L, increases linearly with length, l, while an increase in diameter, d, will reduce the total inductance logarithmically (a proportionally small degree only).

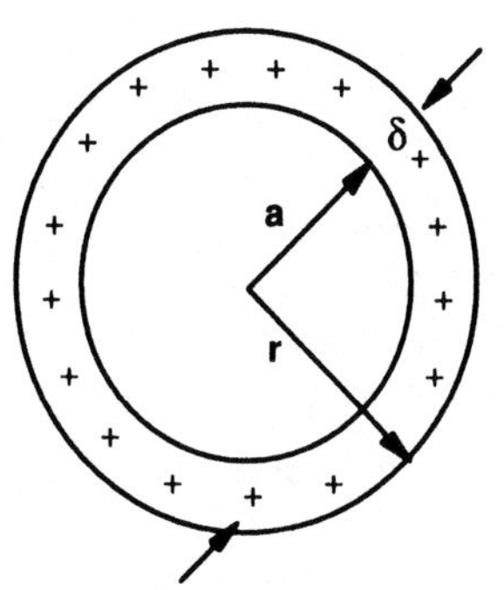

Figure 9.9 Current flow in a conductor—skin effect. (*Source:* Oren Hartal. *Electromagnetic Compatibility by Design* © 1994. Reprinted by permission of R&B Enterprises.)

Rectangular straps and multilayer power planes have a smaller inductance per unit length than that of round wire. The reason for this difference is that a flat strap (and extending this to a ground plane) has a larger perimeter than a round wire with the same cross-sectional area. Inductance of a ground strap is calculated as

$$L = 0.2 \cdot s\left(ln \frac{2s}{w} + 0.5 + 0.2 \frac{w}{s} \right) \tag{9.8}$$

where L = inductance of the ground strap (μH)
s = strap length (m)
w = strap width (m) [must be larger than the thickness by a factor of 10 or more].

When $s/w > 4$ (length-to-width ratio), Eq. (9.8) can be approximated by

$$L = 0.2 \cdot s \cdot ln \frac{2s}{w} \tag{9.9}$$

Equation (9.9) shows that a strap has lower inductance than a round wire and is more useful as a method of providing a low-impedance ground connection at high frequencies. Extending this analysis to a solid plane internal to a PCB, we find that the plane has an impedance that is extremely small compared to a wire or strap except for the perturbations to the planes caused by annular anti-pads around vias. This is the primary reason why ground planes work as well as they do at high frequencies while minimizing common-impedance coupling.

9.7.2 Avoiding a Common-Impedance Path

To reduce common-impedance ground coupling, care must be taken to identify all return paths. This is best achieved when all reference connections from different system circuits follow a dedicated and separate path to a single-point ground connection.

Figure 9.10 illustrates a star configuration for providing power and ground to various subsystems. This implementation technique requires additional wiring and interconnect hardware, not to mention cost.

To help implement an improved method of common impedance grounding, functional circuits must be separated by the power distribution network for each area in addi-

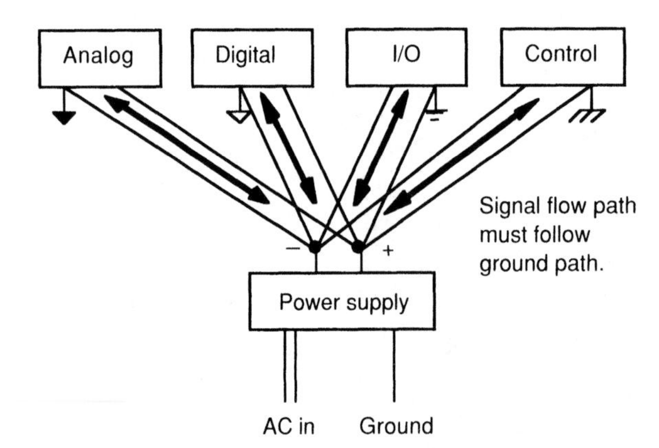

Figure 9.10 Separation of grounds to avoid common-impedance coupling. (*Source:* Oren Hartal, *Electromagnetic Compatibility by Design* © 1994. Reprinted by permission of R&B Enterprises.)

tion to the 0V reference required by logic circuitry. What this means is that we segregate circuits by logical function. Logical function includes the following list and does not include special circuitry that may be required or used, application dependent [4].

- Digital
- Analog
- Audio
- Video
- I/O
- Control logic
- Power supply

By separating noise-generating circuitry to prevent common-impedance coupling, increased noise immunity occurs. Each area must be connected by itself to the main 0V reference (as shown in Fig. 9.10), usually a safety wire ground connection.

As observed, preventing common-impedance coupling is best implemented with single-point grounding, which realistically may not be an option during the design cycle. As examined in the next section, single-point grounding is best when the signal within the circuit is 1 MHz or less containing low-frequency Fourier spectra, while multipoint is preferred for higher frequency signals.

What happens when a product must be multipoint grounded and when common-impedance coupling is to be avoided? For Fig. 9.11, a system must operate in a low-

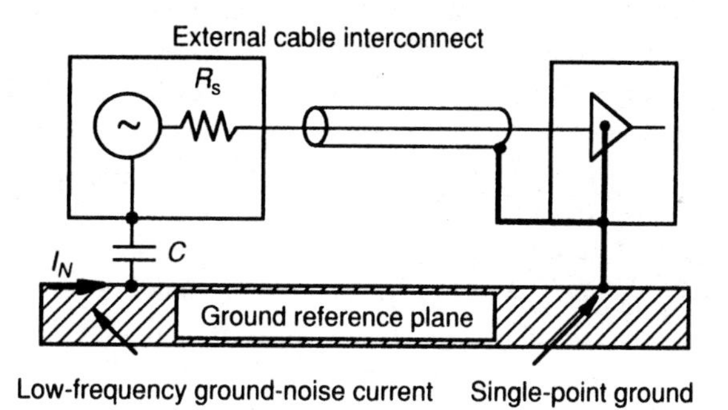

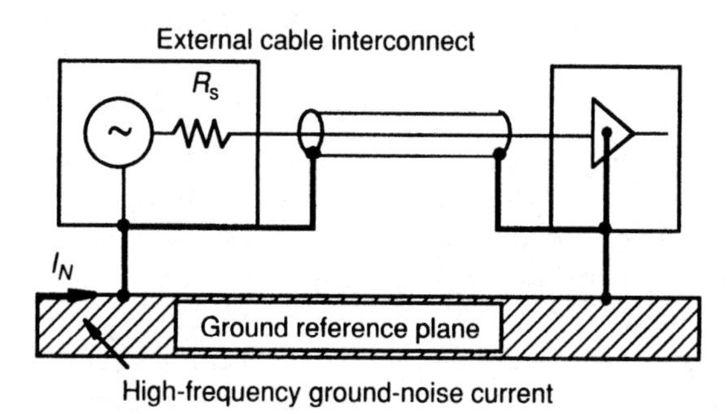

Figure 9.11 Single-point grounding for low frequencies and multipoint for higher frequencies.

frequency environment that requires single-point grounding. An I/O interconnect has high-frequency noise on the cable shield as a result of exposure to externally induced high-energy radiated fields. This cable shield must be single-point grounded if the frequency of operation is less than 1 MHz. For higher frequency signals, *both* ends of the cable shield must be connected to the 0V reference plane.

To solve this problem of attempting a single-point ground connection for low-frequency circuits, a bypass capacitor must be optimally selected for the frequency range of interest, and installed at the end of the cable shield, which is not DC connected to ground. (Optimal selection of this capacitor was described in Chapter 5.)

Other methods of avoiding common-impedance coupling, in addition to using single-point grounding, are available. These are use of an isolation transformer, common-mode choke, optical isolator, or balanced circuitry. These options are examined later in this chapter in the section "Ground Loops."

9.8 CONTROLLING COMMON-IMPEDANCE COUPLING IN POWER AND GROUND

When there are many circuits switching simultaneously, with widely different voltage and current swings (logic family dependent) and all powered from the same power distribution system, coupling of RF energy will probably occur between devices. The power distribution system will always have a finite impedance by virtue of its existence. With an impedance in the planes and with current being consumed by active logic devices, a voltage drop will occur. This voltage drop develops common-mode ground-noise voltage. (Ground-noise voltage is discussed in detail in Chapter 3.) Because a plane structure exists for an entire PCB assembly, ground-noise voltage that is present on one section of the board may be transmitted to other sections, causing both signal quality and EMC problems.

Figure 9.12 illustrates the concept of common-impedance coupling in the power and ground planes. The noise on Device 1's ground reference is described by

$$V_{\text{noise1}} = (I_1 + I_2)(R_{p1} + R_{g1} + Z) \tag{9.10}$$

If Device 2 consumes more current than device 1, with output impedance of the source negligible, we can determine the total amount of ground-noise voltage impressed across device 1. If device 1 is susceptible to disruption, serious concerns develop.

$$V_{\text{noise1}} = I_2(R_{p1} + R_{g1}) \tag{9.11}$$

When investigating a design to minimize common-impedance coupling within a power distribution system, one should take into account the impedance that is presented by that power distribution network. Depending on the design, the supplied voltage and return (ground) may be provided through use of round conductors or flat straps. Equation (9.12) illustrates the amount of inductance that will exist for various configurations. Knowledge of this inductance for these configurations will help the designer understand why RF noise created from one device causes harmful interference to another device. Table 9.1 provides information on the inductance of these conductors operating at 1 MHz [4].

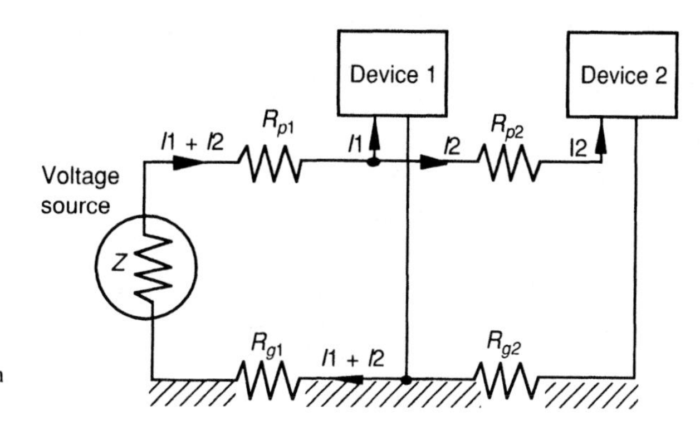

Figure 9.12 Common-impedance coupling in a power and ground structure.

$$Lo_{(\text{round})} = \frac{\mu_o s}{2\pi} \cdot \left[ln\left(\frac{4s}{d}\right) - 1 \right] \quad \text{round conductor}$$

$$Lo'_{(\text{round})} = \frac{\mu_o s}{2\pi} \cdot ln\left(\frac{4h}{d}\right) \quad \text{round conductor over a plane} \tag{9.12}$$

$$Lo_{(\text{flat})} = \frac{\mu_o s}{2\pi} \cdot \left[ln\left(\frac{8s}{w}\right) - 1 \right] \quad \text{flat strap}$$

$$Lo'_{(\text{flat})} = \frac{\mu_o s}{2\pi} \cdot ln\left(\frac{2\pi h}{w}\right) \quad \text{flat strap over a plane}$$

where s = conductor length (meters)
w = width of the conductor (mm)
h = height above ground plane (cm)
d = diameter of conductor (mm)
L = inductance (henry)
$\mu_o = 4\pi * 10^{-7}$

Inductance increases with the length of the conductor and decreases with width. With this increase, it becomes important to keep the length of the conductor as short as possible. Also, the wider the trace, the lower the impedance.

The best way to minimize common-impedance coupling within a power distribution system is to provide separate power and ground sources to specific switching devices. This works well with single- and double-sided PCBs. When separate power and ground

TABLE 9.1 Inductance of Various Conductors at 1 MHz

Conductor Type	Width (mm)	Length (m)	Diameter (mm)	Height (cm)	Inductance (μH)	Reactance (Ω)
Round	—	1	1		1.7	11
Round above a plane	—	1	1	1	0.7	4
Flat strap	10	1			1.3	8
Flat strap above a plane	10	1		1	0.37	2

Source: Oren Hartal, *Electromagnetic Compatibility by Design* © 1994. Reprinted by permission of R&B Enterprises.

planes exist in a multilayer stackup, common-impedance coupling is minimized due to the low impedance of the power distribution system.

9.9 GROUND LOOPS

Ground loops are a primary source of RF noise. RF noise is effectively produced when the physical distance between multipoint ground locations are significant (>1/20 of a wavelength) and connection is made to the main reference ground, usually at AC or chassis potential. In addition, low-level analog circuits can also create ground loops. When a ground loop occurs, it is necessary to isolate or prevent RF energy transference from one circuit corrupting other circuits. A ground loop consists of part signal path and part grounding structure.

Figure 9.13 illustrates what a ground loop looks like within a PCB that is mounted in a chassis where V_n represents common-mode ground loss within the PCB. I_{cm} represents the shunt of current V_n through the chassis. Two separate ground locations are provided, one for each circuit. A difference in ground reference exists between the two circuits due to its finite impedance that occurs between the common reference trace. Unwanted noise from one circuit may be injected into the other circuit. The magnitude of the ground-noise voltage, compared to the signal level in the circuit, is of prime importance. If the signal-to-noise margin is affected, design techniques must be implemented to ensure optimal circuit functionality. All components must have a reference point to determine where the 0V circuit reference is located such that the voltage-level transition is appropriate for the logic family used.

How does one avoid ground loops when a difference in 0V reference exists? Two primary design techniques may be used during the design and layout stages of the PCB.

- Remove one of the grounds (convert to a single-point system)
- Isolate the two circuits using any of the following:
  Transformer
  Common-mode choke
  Optic isolator, or
  Balanced circuitry

Figure 9.14 illustrates the circuit of Fig. 9.13 with modifications to reduce I_{cm}. The first modification provides ground-loop isolation using a transformer. When using a trans-

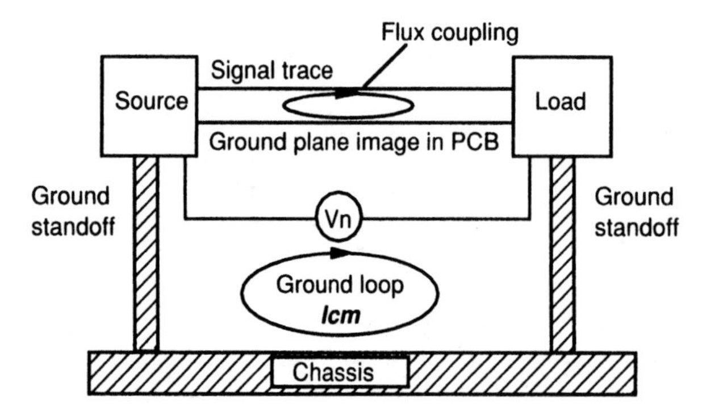

Figure 9.13 Ground loop between two circuits.

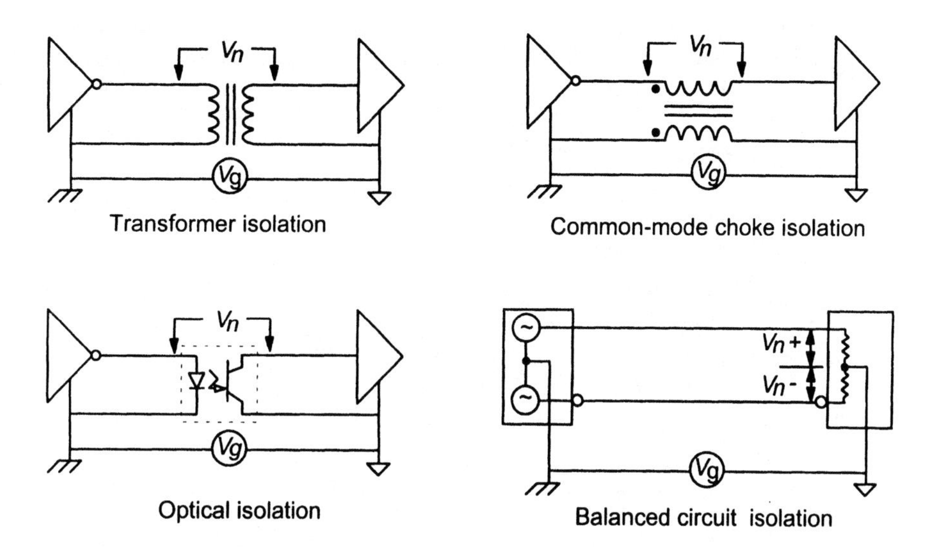

Figure 9.14 Breaking up ground loops between two circuits. (*Source:* H. Ott, *Noise Reduction Techniques in Electronics Systems* © 1988. Reprinted by permission of John Wiley & Sons.)

former, ground-noise voltage will be observed only at the input terminals of the transformer. Any noise coupling that occurs is a result of parasitic capacitance between the input and output windings of the transformer. To prevent parasitic capacitance, the use of a shield may be provided between the primary and secondary windings, connected to the main AC reference point or chassis ground. A disadvantage of using a transformer is physical size, amount of PCB real estate required, and additional cost. In addition, if multiple signals are to travel between isolated areas, a transformer is required for each signal.

Common-mode chokes are also shown in Fig. 9.14 as another technique. The advantage of using common-mode chokes is to remove common-mode currents. If the 0V reference between two components is not at the same reference level due to a finite impedance with the return path, the voltage drop observed will create common-mode noise. A common-mode choke will pass the DC level of the signal while attenuating the high-frequency AC component that is also present within the transmission line. The common-mode choke has no effect on the differential-mode signal of interest. It is the differential-mode signal we want, not common-mode currents. Multiple windings may be wrapped around the same core structure, increasing the density or number of signal lines that the choke can handle.

Optical isolation is another technique used to prevent ground loops and minimize I_{cm}. An optical isolator breaks the transmission path completely. A continuous metallic connection cannot occur between two circuits. This metallic connection is required for the propagation of an electromagnetic field down a PCB trace or wire. These isolators are best suited when a large voltage reference potential exists between circuits. Ground-noise voltage appears across the input of the optical transmitter. These optical isolators are best suited for digital logic designs owing to the nonlinearity of the device when used with analog circuitry.

Balanced circuits include using differential pairs to transmit a signal from source to load. By using differential transmission paths, the currents in both lines are equal. This balance causes a rejection of common-mode currents that may be present within the network.

Many differential-input components manufacturers provide a Common-Mode-Rejection-Ratio (CMRR) number within their data sheets. CMRR is defined as the ratio of

$$\frac{\text{common-mode voltage, } V_{cm}\text{, applied to both inputs required to generate output voltage, } V_o}{\text{differential voltage, } V_{dm}\text{, applied between the inputs to generate output } V_o}$$

CMRR identifies how much common-mode noise will be rejected from entering the device. The better the balance between the differential pairs, the greater the amount of common-mode rejection. At high frequencies, achieving a large CMRR value may be difficult to accomplish.

Common-Mode-Rejection-Ratio (CMRR) is mathematically defined as

$$\text{CMRR} = 20 \log \left| \frac{V_{cm}}{V_{dm}} \right| \text{dB} \quad (V_0 = \text{constant}) \tag{9.13}$$

Using the circuit of Fig. 9.15, we can calculate CMRR as

$$\text{CMRR} = -20 \log \left| \frac{R_1(Z_b - Z_a)}{(Z_a + R_1)(Z_b + R_1)} \right| \text{dB} \tag{9.14}$$

One item to note in Fig. 9.15 is the location of the image plane and chassis plane. The differential-mode transmission line system is referenced from the 0V plane, not the chassis plane. Any I_{cm} that is developed between source and load must flow in the 0V (ground) reference. This distinction must be noted when using differential-mode components. The termination resistors, Z_a and Z_b, must be chosen with a tight tolerance value to assure impedance matching between the two traces. If an impedance imbalance is present, I_{cm} is increased. The development of I_{cm} is described in Chapter 4.

When using Eq. (9.14), the tolerance rating of the resistors is the critical parameter concerned, whereas the R_s resistors are provided to match transmission line impedance to ensure the functionality of the circuit.

When dealing with differential-mode circuits to minimize I_{cm}

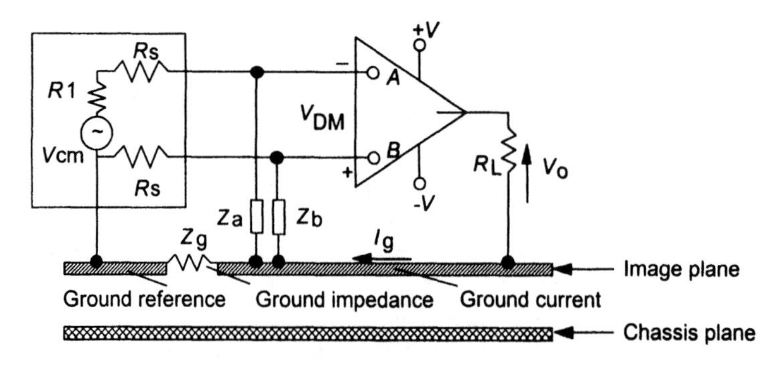

Figure 9.15 Circuit representing common-mode-rejection-ratio.

1. The impedance control of the signal lies only in the image plane.
2. Signal flux is bound to the internal image plane, not the chassis. The chassis is too far away to be of any significant value.
3. The chassis plane only shorts out the common-mode loss that occurs across the image plane.

9.10 RESONANCE IN MULTIPOINT GROUNDING

Problems that arise in PCBs using multipoint grounds are resonances that occur between ground stitch locations and the AC reference or chassis plane. While the AC reference or chassis plane may be at 0V potential referenced to a particular ground structure, this AC reference may be completely different from the 0V reference of the digital or analog circuitry. This difference in reference levels is more apparent when high-frequency, high edge rate signals are present.

Depending on the distance spacing between ground stitch locations, a resonance can occur, depending also on spectral excitation. This resonance exists because parasitic capacitance and inductance are also present between the power and ground planes, in addition to capacitance and inductance induced by the mounting ground stitch standoff mounting posts as shown in Fig. 9.16.

Figure 9.16 illustrates a PCB's image plane secured to a metal mounting plate. In this figure, we see that both capacitance and inductance are present. Capacitance exists between the power and ground planes internal to the PCB. The planes themselves have a finite impedance between ground stitch locations. Using Eq. (9.5), we can determine the self-resonant frequency of the power and ground plane structure, which is difficult to do mathematically. Use of a network analyzer will provide a quick way of determining the *actual* self-resonant frequency between ground points. Multiple measurements are required since the self-resonant frequency of the PCB is dependent on the inductance of the planes, based on the distance spacing of the network analyzer and ground locations for the test probe. Capacitance will, however, remain fixed between the power and ground planes.

Since the PCB's power and ground plane structure is self-resonant at various frequencies, the same analysis for self-resonance is applied to the metallic structure that is used to secure the PCB. This metallic structure may be a chassis for a motherboard, a mounting plate used in a cardcage with a backplane, a shield partition between two boards, or other application not identified herein. Again inductance will occur between the mounting standoffs relative to the actual location of the PCB. Now that we have identified inductance, what about the capacitance?

Because there is a finite distance between the PCB and the metallic structure, capacitance and transfer impedance exist. For example, the PCB can be considered as the positive plate of a capacitor, at voltage potential, and the metal structure as the negative plate, with air as the dielectric medium.

In addition to the overall inductance of the metallic material (which is extremely small), and the parasitic capacitance between the PCB and mounting plate, the standoffs used to secure the board to the chassis (generally pemstuds) are extremely inductive, as described below. These mountings are sometimes the cause of EMI failure.

The explanations of why the standoffs are inductive does not primarily have to do with the standoff itself, but with the metal screw used with the standoff. The screw con-

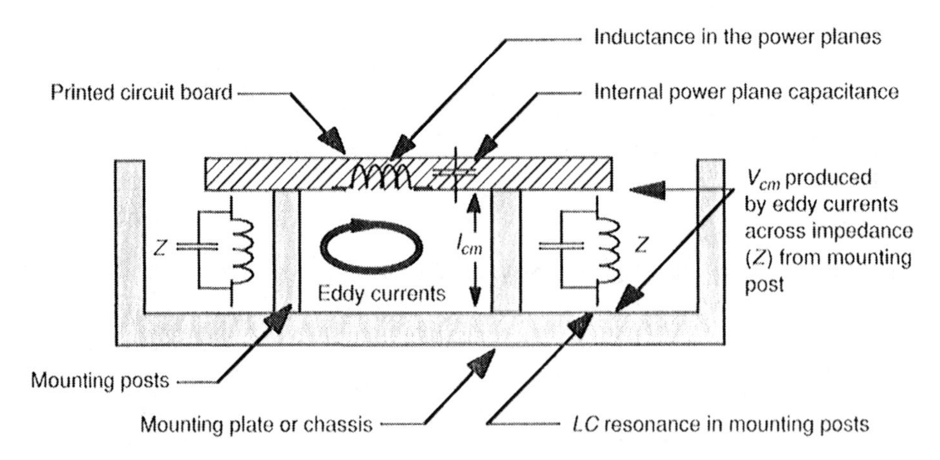

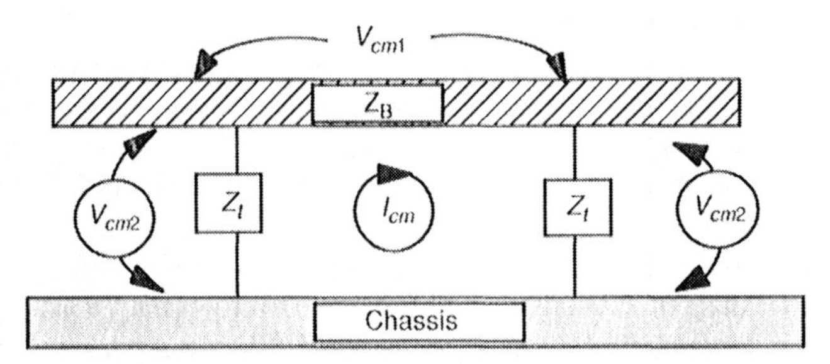

Figure 9.16 Resonance in a multipoint ground to chassis.

tains inductance that may be several orders of magnitude greater than the inductance of the PCB or parasitic inductance of the overall assembly. The reasons why screws are inductive is best illustrated in Fig. 9.17. It is difficult to model screw inductance because of the large number of parameters that cause this inductance to exist. Some of these parameters include material composition, the number of threads that makes contact with the standoff, thread spacing, pitch of the threads, plating material provided, compression strength, and length of the screw from top to bottom.

A screw contains a helical thread, and the edge of the screw thread is the part that mates with the standoff. We cannot guarantee that all threads will make 100% solid bonding contact with the standoff. The standoff must be physically larger in diameter than the screw diameter to allow the screw to be inserted. As a result, we will always have incidental contact between *some* of the threads, not the entire length of the screw. This is observed in Fig. 9.17.

A helical thread performs the same function as a helical antenna when a RF current travels through the screw. This current is located on an extremely thin surface of the helical thread because of skin effect. A voltage potential is developed between the bottom and top of the screw. This voltage reference difference exacerbates creation of RF current.

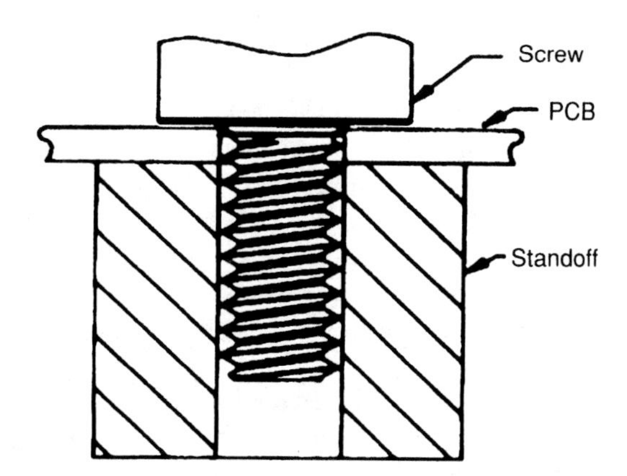

Figure 9.17 Problems grounding the PCB by screws to a standoff.

In addition to the helical thread, a coating of plating material is provided on the screw by its manufacturer. When metal-to-metal contact and rubbing occur between the screw and standoff, the plating can get scraped off, thus exposing the screw to the external environment and pollution based on intended application. Galvanic corrosion can develop, making the screw nonconductive (an insulator) in extreme conditions. If the intended application is to allow a low-impedance, common-mode ground reference path, one cannot exist if corrosion occurs. As such, a screw must be used only for compression between the PCB and metallic structure, and must not be relied upon to transfer RF currents to the 0V reference or ground system. Large mounting pads provided on the bottom of the PCB that overlaps the standoff walls help make the desired low-impedance ground connection, not the screw. The mounting pads of the PCB must be secured against the walls of the standoff. The standoff is usually installed in the chassis with a good bond connection by the sheet metal fabricator. Thus, if a low-impedance path to ground is required for the PCB, this is done through the walls of the standoff, not the screw threads!

Digital circuits must be treated as high-frequency analog circuits. A good low-inductive, 0V reference return is necessary on any PCB containing many digital circuits. The ground planes internal to the PCB (more so than the power planes) generally provide a lower inductive ground-image reference for the power supply and signal return currents. This allows use of constant impedance transmission lines for signal interconnects. When making a ground plane (0V reference) to chassis plane connection, it is necessary to provide for high-frequency decoupling of RF currents.

These high-frequency RF currents are created by the self-resonance of the power and ground plane structure losses caused by via anti-pad holes and the switching noise from digital circuits. High-quality decoupling capacitors should be used at each and every ground connection between the power and ground plane. Optimal selection of decoupling capacitors is detailed in Chapter 5.

9.11 FIELD TRANSFER COUPLING OF DAUGHTER CARDS TO CARD CAGE

RF fields generated from a PCB (components, ground loops, interconnect cables, and the like) will couple to a metallic structure. As a result, RF eddy currents will develop in the structure and will circulate within the unit creating a field distribution. This field distribution may then couple to other circuits, subsystems, interconnect cables, peripherals, and power supplies. One of the most significant ramifications of this field distribution is to develop a common-mode potential between a backplane and the metallic card cage. This potential will exhibit the spectral energy signature not only of the backplane, but the daughter cards as well. In addition, this field will be observed during radiated testing in the near field ($< \lambda/4$) or as a plane wave at a distance greater than $\lambda/4$ at the frequency of concern. Proper implementation of suppression techniques on a PCB, along with proper referencing of the backplane to the card cage to short out the distributively derived potentials, will minimize field transfer coupling between the boards to the backplane and card cage assembly.

The *proper referencing* of the backplane to the card cage noted above takes the form of establishing a very low-impedance RF reference between the backplane and the card cage. This reference method is mandatory to short out the potentials caused by eddy currents developed at and by the daughter cards coupling to the sheet metal. These currents are coupled to the card cage through distributive transfer impedances (often in the low tens of ohms) and then through attempts to *close the loop* by coupling to the backplane. If the common-mode reference impedance between the backplane and the card cage is not significantly lower than the distributive "driving source" (of the eddy currents), an RF voltage will be developed between the backplane and the card cage. This voltage will have the spectral energy profile signature not only of the backplane but also of the daughter cards. This voltage will cause any interconnects that are provided on the backplane to radiate the spectral profile-even DC wire. The spectral voltage developed in this mechanism may contribute to interboard coupling using the backplane-to-card cage relationship as an intermediary![2]

Simply put, the common-mode spectral potential between the backplane and card cage must be shorted out. This may take the form of frequently connecting the backplane ground plane to the card cage (chassis) at regular intervals around the perimeter of the backplane. Alternatively, an "AC chassis plane" can be configured internal to the backplane, positioned immediately adjacent to a logic return plane. A distributive transfer impedance will thus be established between both the AC chassis and the return plane. The chassis plane may also serve as a Faraday partition within the assembly. The location of an AC chassis plane within the backplane must be such that it is never used as an image return reference for signal traces. That is, it must be "capped" by logic ground planes. Generally, to be reasonably effective, the RF transfer impedance between the logic ground planes and the AC chassis plane must be equal to or less than 1 Ω, thereby shorting out the common-mode potential between the daughter cards-card cage-backplane-to-card cage.

The reader is cautioned that the best EMI and system performance will be gained when the signal impedances are well controlled and referenced to ground planes (or 0V reference) rather than voltage planes. In addition, the intrinsic parallel-plane power impedance distribution must be established at as low a value as is reasonably possible.

In Fig. 9.18, if the top and bottom layer of the backplane or daughter card is a solid AC chassis plane, a lower impedance connection to chassis ground is available to both the backplane connector(s) and faceplate screw securement of the PCB. This low-impedance path will now source RF currents to chassis ground, thus preventing ground loops L_1 and L_3 from producing RF potentials between the faceplate to PCB and backplane to PCB, re-

[2]The propagational mechanisms and solutions were derived and modeled by W. Michael King.

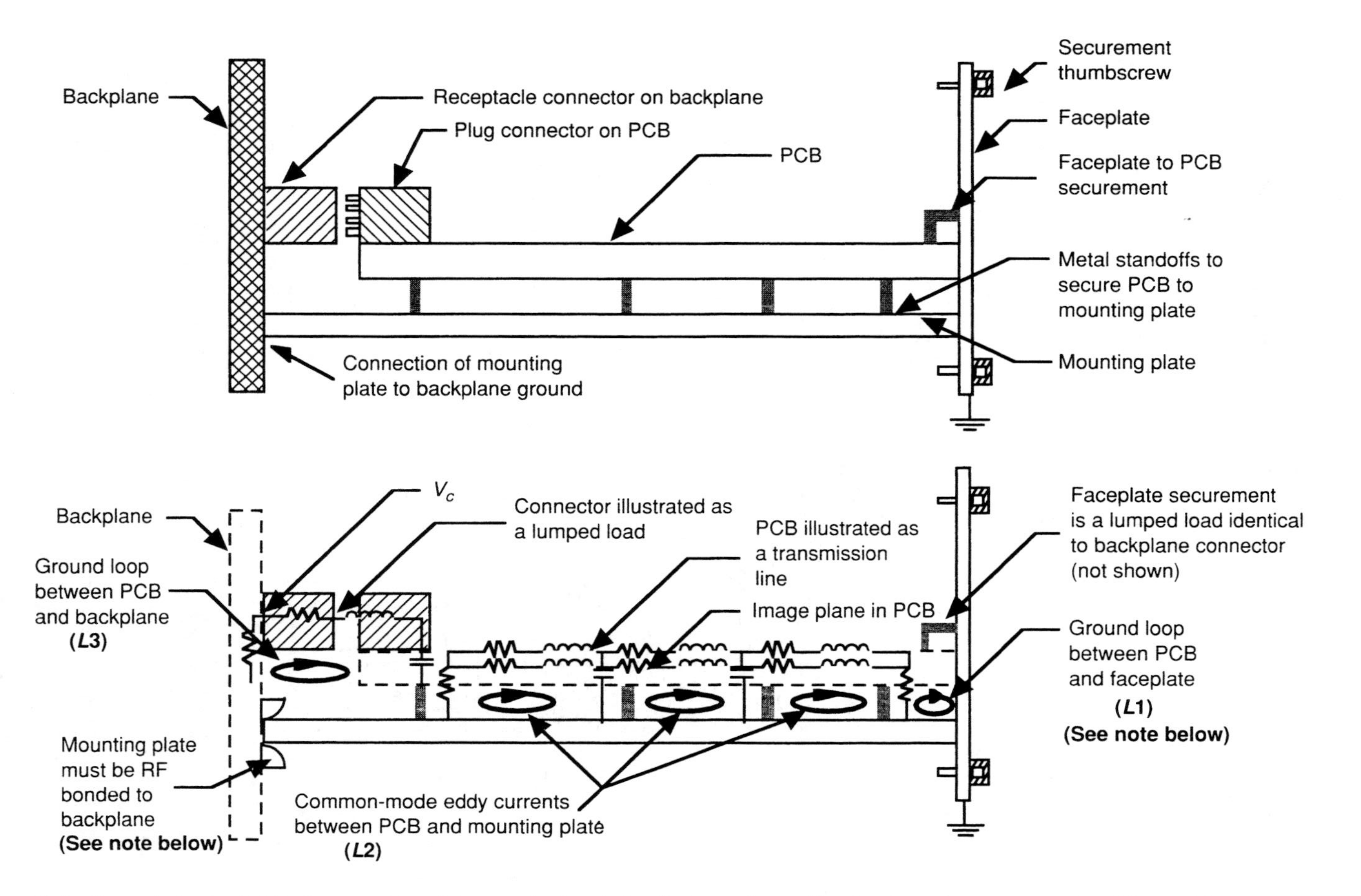

Note: To control potential V_C, the mounting plate must be bonded to the backplane with a low-impedance connection. Without this connection, the mounting plate will couple (or transfer) common-mode eddy currents to the chassis or adjacent printed circuit boards. Common-mode eddy currents are to be avoided at all times if EMC compliance is required.

To control loop $L1$, the faceplate must be RF bonded to the mounting plate.

Figure 9.18 Backplane interconnect impedance considerations.

spectively. With solid bonding of logic ground to chassis ground, potentials from ground loop L_2 are also minimized.

All routing layers must be internal (stripline) to the backplane, with both top and bottom layers as solid AC chassis planes. With the outer layers an AC plane, direct chassis connection from logic ground to chassis ground can easily be achieved using bypass capacitors between the planes.

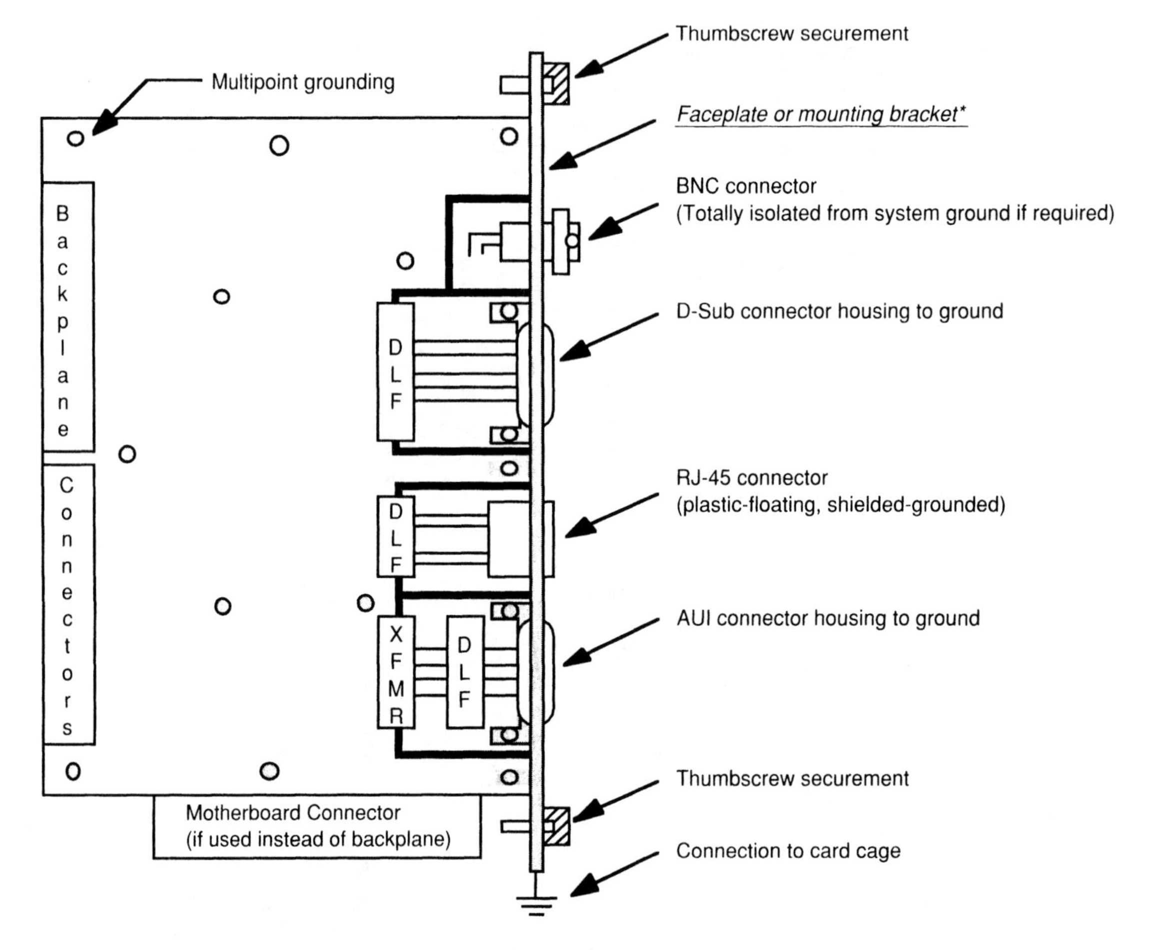

* The mounting bracket is bonded (grounded) to the PCB as indicated in multiple locations.
The faceplate is also secured to the main chassis ground via thumbscrews or by other means.
Note the location of the ground points on the board to minimize ground loops.

DLF refers to Data Line Filter.

Figure 9.19 Multipoint grounding of I/O faceplate or bracket.

9.12 GROUNDING (I/O CONNECTOR)

For products that are low-frequency and that may use single-point grounding, this section is generally not applicable. For low-frequency products, low-impedance connection between logic ground and chassis ground not only can cause electromagnetic interference but can also prevent proper functionality. This is especially true for audio circuits that are devoid of digital processing. For a circuit at "low-frequency," to the extent that it qualifies for single-point grounding, the combination of signal levels, packaging techniques, and all operating frequencies must be such that transfer currents to the case (or external surfaces) through distributive transfer impedance is insignificant in comparison to the operative signal levels *or* the desired EMC criteria.

For products using *multipoint grounding*, this section is applicable whenever an I/O interface is used. Most modular PCBs contain a mounting bracket, faceplate, bulkhead connector, or securement means between control logic and the outside world. This securement may contain various I/O connectors, or it may be a blank panel (e.g., EISA/ISA/PCI adapter bracket). This bracket must be RF bonded by a low-impedance metal path directly to chassis ground. This bracket grounding may also be bonded to logic ground for functionality reasons.

Multiple ground connections must be provided from the ground planes of the PCB to the I/O bracket. Multiple ground points in the appropriate locations redirect RF ground loops between grounding locations on the bracket, distributive transfers to the case, and the opposite end of the PCB. The better the grounding, the more sourcing of RF currents to chassis ground. Figure 9.19 shows how to properly ground a mounting bracket to both chassis and logic ground. All I/O areas are isolated from control logic by a moat, which is also commonly identified as a partition cut, split plane, gap, or isolated area.

REFERENCES

[1] Montrose, M. 1996. *Printed Circuit Board Design Techniques for EMC Compliance*. Piscataway, NJ: IEEE Press.

[2] Coombs, C. F. 1996. *Printed Circuits Handbook*. New York: McGraw-Hill.

[3] Gerke, D., and W. Kimmel. 1994, January 20. "The Designers Guide to Electromagnetic Compatibility." EDN.

[4] Hartal, O. 1994. *Electromagnetic Compatibility by Design*. W. Conshohocken, PA: R&B Enterprises. (Material reprinted by permission.)

[5] Ott, H. 1988. *Noise Reduction Techniques in Electronic Systems*. 2nd ed. New York: John Wiley & Sons. (Material reprinted by permission.)

[6] Paul, C. R. 1992. *Introduction to Electromagnetic Compatibility*. New York: John Wiley & Sons. (Material reprinted by permission.)

[7] Van Doren, T. 1995. *Circuit Board Layout to Reduce Electromagnetic Emission and Susceptibility*. Seminar notes.

[8] William M. King, United States Patent #4,145,674.

Glossary

ELECTROMAGNETIC COMPATIBILITY

Note: Some of these definitions are defined within a particular chapter and repeated for quick reference herein.

Attenuation A reduction in energy measured in units of decibels (dBs).

Bonding Making a low-impedance electrical connection between two metallic surfaces.

Circuit Multiple devices with a source impedance, load impedance, and interconnects. For digital circuits, multiple sources and loads may be part of one circuit where all devices are referenced to the same point, or may use a common signal return conductor. Circuits usually originate in one location and terminate in another.

Circuit referencing The process of providing a common 0V reference voltage for multiple circuits to allow communication between the two. Circuit referencing is the most important reason for providing a ground reference. This reference point is not intended to carry functional current.

Common-mode Signals identical in amplitude and phase at both inputs of a device; the potential or voltage that exists between neutral and ground.

Common-mode current The component of a signal current that creates electric and magnetic fields that do not cancel each other. For example, a circuit with one signal conductor and one ground conductor will have common-mode current as the summation of the total signal current flowing in the same direction on both conductors. Common-mode currents are the primary source of EMI.

Common-mode interference The interference that appears between signal leads or terminals of a circuit referenced to ground.

Common-mode rejection ratio The ratio of common-mode interference voltage at the input of a device to the corresponding interference voltage at the output of the same component. The higher the ratio, the better the performance. This ratio describes the capability of the device to reject the effects of a voltage applied simultaneously to both inputs.

Conducted emissions The component of RF energy that is transmitted through a medium as a propagating wave, generally through a wire or interconnect cable. LCI (Line Conducted Interference) refers to RF energy in a power cord or AC mains input cable. Conducted signals do not propagate as fields but propagate as conducted waves.

Conducted immunity The relative ability of a product to withstand electromagnetic energy that penetrates it through external cables, power cords, and I/O interconnects.

Conducted susceptibility EMI that couples from outside of the equipment to the inside through I/O interconnect cables, power lines, or signal cables.

Containment Preventing RF energy from leaving an enclosure, generally by shielding a product within a metal box (Faraday cage or Gaussian structure), or by using a plastic housing with RF conductive paint. By reciprocity, we can also speak of containment as preventing RF energy from entering the unit.

Decibel Logarithm of a ratio measurement. The basic unit for measuring the power or strength of a signal. Increases or reductions of 6 dB doubles or halves of the power level within the circuit.

Earthing (British term) The connection of the safety ground wire to earth at the service entrance of a building.

Electromagnetic compatibility The capability of electrical and electronic systems, equipment, and devices to operate in their intended electromagnetic environment within a defined margin of safety, and at design levels or performance, without suffering or causing unacceptable degradation as a result of electromagnetic interference.

Electromagnetic interference The lack of EMC, since the essence of interference is the lack of compatibility. EMI is the process by which disruptive electromagnetic energy is transmitted from one electronic device to another via radiated or conducted paths (or both). In common usage, the term refers particularly to RF signals, but EMI can occur in the frequency range from "DC to daylight."

Electrostatic discharge A transfer of electric charge between bodies of different electrostatic potentials in proximity to each other or through direct contact. This event is observed as a high-voltage pulse that may cause damage or loss of functionality to susceptible circuits. Although lighting qualifies as a high-voltage pulse, the term *ESD* is generally applied to events of lesser amperage and more specifically to events that are triggered by human beings. However, for the purposes of discussion, lightning is included in the ESD category because the protection techniques are very similar, though differing in magnitude.

EMI filter A circuit or device containing components that provide a low-impedance path for high-frequency RF energy to be removed. The filter may also be used to protect a particular circuit from electromagnetic field disturbance.

Equipotential ground plane A piece of metal used as a common connection point for power and signal referencing. This plane may not be at equipotential levels for RF frequencies due to its electrically large size.

Faraday shield A term referring to conductive shielding used to contain or control an electric field. This shield may be located between the primary and secondary windings of a transformer or may completely surround a circuit (or system) to provide electrostatic shielding. No ground is necessary. (*Note:* A Faraday shield is in reality a Gaussian structure. Gauss's law describes the functional purpose of this shield, while Faraday's law describes the creation of electric fields from time-varying magnetic fields. Faraday was the first person to validate or prove the validity of Gauss's law; hence his name is attributed to this function of shielding.)

Ferrite components Powered magnetic (permeable) material in various shapes used to absorb conducted interference on wires, cables and harnesses. Acting as a lossy resistance and increased self-inductance, ferrites convert an EMI magnetic-flux density field into heat (an exothermic process). One benefit of this, in contrast to filters that perform by reflecting EMI in their stopbands, is that ferrites do not reflect EMI, which otherwise could enhance radiation and disturb other victim components or circuits.

Ferrite material A combination of metal oxides sintered into a particular ceramic shape with iron as the main ingredient. Ferrites provide two key features: (1) high magnetic permeability that concentrates and reinforces a magnetic field, and (2) high electrical resistivity that limits the amount of electric current flow. Ferrites exhibit low energy losses, are efficient, and function at high frequencies (1 MHz to 1 GHz).

Filter A device that blocks the flow of RF current, for example, 50/60/400 Hz, while passing a desired frequency. For communication or higher frequency circuits, a filter suppresses unwanted frequencies and noise, or separates channels from each other.

Ground loop A circuit that includes a conducting element (plane, trace, wire) assumed to be at ground potential where return currents pass through. At least one ground loop will exist in a circuit. Although a ground loop is acceptable, the severity of the problem of currents flowing through the loop depends on the unwanted signals that may be present which can cause system malfunction.

Ground stitch location The process of making a solid ground connection from a PCB to a metallic structure for the purposes of providing systemwide ground referencing, regardless of which grounding methodology used.

Grounding A generic term with as many definitions as there are engineers. This word must be preceded by an adjective.

Grounding methodology A chosen method for directing return currents in an optimal manner appropriate for the intended application.

Holy ground Sometimes referred to as the actual location used. *See also* Single-point ground.

Hybrid ground A grounding methodology that combines single-point and multipoint grounding simultaneously, depending on the functionality of the circuit and the frequencies present.

Immunity A relative measure of a device or system's ability to withstand EMI exposure while maintaining a predefined performance level.

Insertion loss The ratio between the power received at a load before and after the insertion of the filter at a given frequency, or how much loss a filter provides for its intended function.

Multipoint ground A method of referencing different circuits together to a common equipotential or reference point. Connection may be made by any means possible in as many locations as required.

Parasitic capacitance The capacitive leakage across a component (resistor, inductor, filter, isolation transformer, optical isolator, etc.) that adversely affects high-frequency performance. Parasitic capacitance is also observed between active components (or a PCB) and the sheet metal mounting plate or enclosure.

Permeability The extent to which a material can be magnetized; often expressed as the parameter relating magnetic-flux density induced by an applied magnetic-field.

Radiated emissions The component of RF energy that is transmitted through a medium as an electromagnetic field. RF energy is usually transmitted through free space, however, other modes of field transmission may occur.

Radiated immunity The relative ability of a product to withstand electromagnetic energy that arrives via free-space propagation.

Radiated susceptibility Undesired EMI radiating through free space into equipment from externally induced electromagnetic sources.

Radio frequency A frequency range containing coherent electromagnetic radiation of energy useful for communication purposes; roughly the range from 10 kHz to 100 GHz. This energy may be transmitted as a byproduct of an electronic device's operation. RF is transmitted through two basic modes: radiated and conductive.

Referencing The process of making an electrical connection, or bond, between two circuits, allowing the 0V reference from both circuits to be identical.

RF ground Providing a ground reference point using a specific methodology in order to allow a product to comply with both emissions and immunity requirements; radiated or conducted.

Safety ground The process of providing a return path to earth ground to prevent the hazard of electric shock through proper connection and routing of a permanent, continuous, low-impedance, adequate fault capacity conductor that runs from a power source to a load.

Shield ground Providing a 0V reference or electromagnetic shield for both interconnect cables or main chassis housing.

Single-point ground A method of referencing many circuits together at a single location to allow communication between different points. All signals will thus be referenced to the same location.

Suppression The process of reducing or eliminating RF energy that exists without relying on a secondary method, such as a metal housing or chassis.

Susceptibility A relative measure of a device or system's propensity to be disrupted or damaged by EMI exposure. It is the lack of immunity.

SIGNAL INTEGRITY TERMS*

AC impedance The combination of resistance, capacitive reactance, and inductive reactance seen by AC or time-varying voltage.

Alternating current (AC) A current that varies with time. This label is commonly applied to a power source that switches polarity many times per second, such as the

*Glossary—Signal Integrity Terms provided by ICP-2141, *Controlled Impedance Circuit Boards and High Speed Logic Design*, and IPC-D-317A, *Design Guidelines for Electronic Packaging Utilizing High-Speed Techniques*. © Reprinted by permission of the Institute for Interconnecting and Packaging Electronic Circuits (IPC).

power supplied by utility companies. It may take a sinusoidal shape but could be a square or triangular wave shape.

Amplitude The height or magnitude of a signal measured with respect to a reference, such as signal ground.

Attenuation Reduction in the amplitude of a signal due to losses in the media through which it is transmitted.

Backporching A term used to describe the reflections that follow a fast rise or fall time signal traveling down a long transmission line that has not been properly terminated. Looks like a stair-step function.

Backward crosstalk Noise injected into a quiet line placed next to an active line as seen at the end of the quiet line at the signal source.

Busbar A large copper or brass bar used to carry high power supply current onto a PCB or backplane.

Capacitance A measure of the ability of two adjacent conductors separated by an insulator (a dielectric material) hold a charge when a voltage is impressed between them. Measured in farads.

Characteristic impedance The impedance of a parallel conductive structure to the flow of AC current. Usually applied to transmission lines in PCBs and cables carrying high-speed signals. Normally a constant value over a wide range of frequencies.

Coaxial A term used to describe conductors that are concentric about a central axis. Takes the form of a central wire surrounded by a conductor tube that serves as a shield and ground. May have a dielectric other than air between the conductors.

Crossover Intersection of two conductors separated by insulation.

Current Electrons traveling in a conductor as the result of a voltage difference between two points.

Decoupling Preventing noise pulses injected in the power supply lines by switching (digital) logic, from disturbing other logic on the same power supply circuit by providing a localized point source of charge. Usually done with capacitors.

Dielectric constant *See* Permittivity.

Differential pair Parallel routed signals exhibiting a mutual impedance between both lines, typically 50 to 100 ohms.

Direct current (DC) A current produced by a voltage source that does not vary with time. Normally provided by power supplies to electronic circuits.

Edge rate The rate of change in voltage with time of a logic signal transition. Usually expressed in volts per nanosecond.

Edge transition attenuation The loss in sharpness of a switching edge caused by absorption of the highest frequency component of the transmission line.

Effective relative permittivity (ε_r') The relative permittivity that is experienced by an electrical signal transmitted through a conducted path.

Flat conductor A rectangular conductor that is wider than it is high. Usually refers to signal conductors or traces in a PCB.

Forward crosstalk Noise induced into a quiet line placed next to an active line as seen at the end of the quiet line farthest from the signal source.

Ground A term used to describe the terminal of a voltage source that serves as a measurement reference for all voltages in the system. Often, the negative terminal of the power source, but sometimes the positive terminal.

Impedance The resistance to the flow of current represented by an electrical network. May be resistive, reactive, or both.

Inductance The property of a conductor that allows it to store energy in a magnetic field induced by a current flowing through it. Units of measure—henry.

Line coupling Coupling between two transmission lines caused by mutual inductance and capacitance between them.

Load capacitance The capacitance seen by the output of a logic circuit or other signal source. Usually the sum of the distributed line capacitance and input capacitance of all load circuits.

Logic A general term used to describe functional circuits that perform computational functions.

Noise budget/noise margin The allowance for a change in the system's DC and/or AC voltage which allows a device to operate within specific limits. There are two primary components of the noise budgets: The DC power supply of each integrated circuit, and the logic signal AC noise budget.

Overshoot The effect of an excessive voltage level above the power rail, or below ground reference as observed at a component or device.

Permeability A general term used to express various relationships between magnetic induction and a magnetizing force.

Permittivity (dielectric constant, ε_r) The ratio of the permittivity of the material to that of free space. This term is preferred to the term *dielectric constant*. Permittivity is not constant but varies with several parameters, including electrical frequency at which the measurement is made, temperature, and extent of water absorption in the material.

Power distribution The DC and AC characteristics for defining power distribution may be grouped into two major categories: conductive losses (DC) and dielectric (AC).

DC power distribution Encompasses the output from the power supply to the input of a device.

AC power distribution Divided into three contributing elements of impedance and considered as a power distribution impedance network. These are

(1) Switching transient impedance The impedance between the decoupling capacitor and a device. This element is also referred to as ground or power bounce. It is the highest frequency component of the circuit.

(2) Impedance due to the charging of bulk IC decoupling capacitors The current in this impedance is at a lower frequency and higher amplitude than the current in the first element. This voltage drop will be less than the switching transient impedance. This lower voltage drop is due to a lower impedance resulting from a lower frequency of operation.

(3) Decoupling capacitor impedance Created when the decoupling capacitors supply current to the ICs. Decoupling must provide sufficient current for the devices. This includes high peak current requirements during device switching. The PCB power system must provide this current without lowering the supply voltage below the required minimum device level.

Power/ground bounce Simultaneous switched outputs may be inductively coupled between the power and ground reference. This coupling delay may cause an edge rate transition (change rise and fall times). Without an accurate model of the power and ground structure, which includes package models, accurate simulation is difficult. Provision for low-inductance connections using wider conductors helps to reduce inductive effects.

Propagation delay The time required for a signal to travel through a transmission line, or the time required for a logic device to perform its desired function from input to output.

Pulse A logic signal that switches from one state to the other and back in a short period of time and remains in one state most of the time. Generally used as a clock for logic devices.

Reflections Energy from a high-speed signal edge that is sent back toward the source as a result of encountering a change in impedance in the transmission line on which it is traveling.

Relative permittivity The amount of energy stored in a dielectric insulator per unit electric field, and hence a measure of the capacitance between a pair of conductors in the vicinity of the dielectric insulator, as compared to the capacitance of the same conductor pair in a vacuum.

Ringback The effect of the rising edge of a logic transition meeting or exceeding logic requirements, then recrossing the threshold before settling. Can be caused by a mismatch of logic drivers and receivers, poor termination techniques, and impedance mismatch of the net to the devices.

Ringing The effect within a transmission line that contains overshoot (going past the maximum voltage level of the circuit and below the low-voltage reference level) before stabilizing to a quiescent level.

Rise time Time required for a logic signal to switch from a low state to its high state. Commonly measured between the 10% and 90% voltage levels.

Signal line Any conductor used to transmit a logic signal from one circuit to another.

Skew The effect of a signal being delayed with respect to another signal due to different path lengths, or a delay during a transmission state that may cause timing errors in the design. Skew can be affected by conductor impedance, differing conductor lengths, power supply variations, device tolerances, and load capacitance of inputs.

Stub A branch of the main line of a signal net generally used to reach a load that is not on the direct signal path.

Switching noise When devices are switching, current is either drawn from or passed to the power supply through the power/ground paths. When this current contains high-frequency components, the self-inductance of the package leads and traces becomes significant with respect to transients or switching. These transients are caused by the inductance of the power/ground loop. The layout must be designed to reduce this inductance as much as possible.

Threshold Threshold violations are caused when a rising pulse edge does not reach the voltage threshold of the device input. Weak drivers or poor terminations are often the cause, although it can also be created by device drivers with a large rise time versus pulse width time.

Transmission line Any form of conductor used to carry a signal from a source to a load. The transmission time is usually long compared to the speed or rise time of the signals, so that coupling, impedance, and terminators are important in preserving signal integrity.

Undershoot A condition in which the voltage level does not reach the desired amplitude for both maximum and minimum transition levels.

Bibliography

Bakoglu, H. B. 1990. *Circuits, Interconnections and Packaging for VLSI.* Reading, MA: Addison-Wesley Publishing Co.

Booton, R. C. 1992. *Computational Methods for Electromagnetics and Microwaves.* New York: John Wiley & Sons.

Brench, C. 1994. "Heatsink Radiation as a Function of Geometry." *Proceedings of the IEEE International Symposium on Electromagnetic Compatibility.* New York: IEEE, pp. 105–109.

Brit, D. S., D. M. Hockanson, and F. Sha. 1997. "Effects of Gapped Groundplanes and Guard Traces on Radiated EMI." *Proceedings of the IEEE International Symposium on Electromagnetic Compatibility.* New York: IEEE, pp. 159–164.

Brown, R., R. Sharpe, W. Hughes, and R. Post. 1973. *Lines, Waves and Antennas.* New York: Ronald Press.

Collin, R. R. 1992. *Foundations for Microwave Engineering.* 2nd ed. Reading MA: Addison-Wesley Publishing Co.

Coombs, C. 1996. *Printed Circuits Handbook.* New York: McGraw-Hill.

Diaz-Olavarrieta, L. 1991. "Ground Bounce in ASIC's: Model and Test Results." *Proceedings of the IEEE International Symposium on Electromagnetic Compatibility.* New York: IEEE, pp. 387–392.

DiBene, J. T., and J. L. Knighten. 1997. "Effects of Device Variations on the EMI potential of High Speed Digital Integrated Circuits." *Proceedings of the IEEE International Symposium on Electromagnetic Compatibility.* New York: IEEE, pp. 208–212.

Dockey, R. W., and R. F. German. 1993. "New Techniques for Reducing Printed Circuit Board Common-Mode Radiation." *Proceedings of the IEEE International Symposium on Electromagnetic Compatibility.* New York: IEEE, pp. 334–339.

Drewniak, J. L., T. H. Hubing, T. P. Van Doren, and D. M. Hockanson. "Power Bus Decoupling on Multilayer Printed Circuit Boards." *IEEE Transactions on EMC* 37(2). New York: IEEE, pp. 155–166.

Erwin, V., and K. Fisher. 1985. "Radiated EMI of Multiple IC Sources." *Proceedings of the IEEE International Symposium on Electromagnetic Compatibility*. New York: IEEE, pp. 26–28.

Gerke, D., and W. Kimmel. 1994, January 20. "The Designers Guide to Electromagnetic Compatibility." EDN.

Gerke, D., and W. Kimmel. 1995. *Electromagnetic Compatibility in Medical Equipment*. Piscataway, NJ: IEEE Press and Interpharm Press.

Gerke, D. and W. Kimmel. 1987. "Interference Control in Digital Circuits." *Proceedings of EMC EXPO 87* (San Diego, CA), T13.

German, R. F. 1985. "Use of a Ground Grid to Reduce Printed Circuit Board Radiation." *Proceedings of the 6th International EMC Symposium* (Zurich, Switzerland), pp. 133–138.

German, R. F., H. Ott, and C. R. Paul. 1990. "Effect of an Image Plane on Printed Circuit Board Radiation." *Proceedings of the IEEE International Symposium on Electromagnetic Compatibility*. New York: IEEE, pp. 284–291.

Goel, A. 1994. *High-Speed VLSI Interconnections, Modeling, Analysis and Simulation*. New York: John Wiley & Sons.

Goulette, D., and R. Crawhall. 1996. "Quieter Integrated Circuits Ease EMI Compliance," Nortel Technology.

Goyal, R. 1994, March. "Managing Signal Integrity." *IEEE Spectrum*.

Hartal, O. 1994. *Electromagnetic Compatibility by Design*. West Conshohocken, PA: R&B Enterprises.

Herrell, D. H., and Benjamin Beker. 1997. "EMI and Power Delivery Design in PC Systems." IEEE Sixth Topical Meeting on Electrical Performance of Electronic Packaging, #97TH8318, Piscataway, NJ: IEEE Press, pp. 23–26.

Hsu, T. 1991. "The Validity of Using Image Plane Theory to Predict Printed Circuit Board Radiation." *Proceedings of the IEEE International Symposium on Electromagnetic Compatibility*. New York: IEEE, pp. 58–60.

Hubing, T., J. L. Drewniak, T. P. Van Doren, and D. M. Hockanson. 1995. "Power Bus Decoupling on Multilayer Printed Circuit Boards." *IEEE Transactions on EMC* 37(2): 155–166.

Hubing, T., T. P. Van Doren, and J. L. Drewniak. 1994. "Identifying and Quantifying Printed Circuit Board Inductance." *Proceedings of the IEEE International Symposium on Electromagnetic Compatibility*. New York: IEEE, pp. 205–208.

IPC-2141. 1996, April. *Controlled Impedance Circuit Boards and High Speed Logic Design*. Institute for Interconnecting and Packaging Electronic Circuits.

IPC-D-317A. 1995, January. *Design Guidelines for Electronic Packaging Utilizing High-Speed Techniques*. Institute for Interconnecting and Packaging Electronic Circuits.

IPC-TM-650. 1996, April. *Characteristic Impedance and Time Delay of Lines on Printed Boards by TDR*. Institute for Interconnecting and Packaging Electronic Circuits.

Johnson, H. W., and M. Graham. 1993. *High Speed Digital Design*. Englewood Cliffs, NJ: Prentice Hall.

Kaupp, H. R. 1967, April. "Characteristics of Microstrip Transmission Lines." IEEE Transactions," EC-16, No. 2.

Kraus, John. 1984. *Electromagnetics*. New York: McGraw-Hill.

Leferink, F. 1997. "Reduction of Printed Circuit Board Radiated Emissions." *Proceedings of the IEEE International Symposium on Electromagnetic Compatibility*. New York: IEEE, pp. 431–438.

Magnusson, P. C., G. C. Alexander, and Vijai Kumar Tripathi. 1992. *Transmission Lines and Wave Propagation*. 3rd ed. Boca Raton, FL: CRC Press.

Mardiguian, M. 1992. *Controlling Radiated Emissions by Design*. New York: Van Nostrand Reinhold.

Montrose, M. I. 1996. "Analysis on the Effectiveness of Clock Trace Termination Methods and Trace Lengths on a Printed Circuit Board." *Proceedings of the IEEE International Symposium on Electromagnetic Compatibility*. New York: IEEE, pp. 453–458.

Montrose, M. I. 1996. "Analysis on the Effectiveness of Image Planes Within a Printed Circuit Board." *Proceedings of the IEEE International Symposium on Electromagnetic Compatibility*. New York: IEEE, pp. 326–327.

Montrose, M. I. 1991. "Overview on Design Techniques for PCB Layout Used in High Technology Products." *Proceedings of the IEEE International Symposium on Electromagnetic Compatibility*. New York: IEEE, pp. 61–66.

Montrose, M. I. 1996. *Printed Circuit Board Design Techniques for EMC Compliance*. Piscataway, NJ: IEEE Press.

Montrose, M. I. 1998. "Time and Frequency Domain Analysis of Right Angle Corners on Printed Circuit Board Traces." *Proceedings of the IEEE International Symposium on Electromagnetic Compatibility*. New York: IEEE.

Motorola, Inc. 1996. *ECL Clock Distribution Techniques (#AN1405)*.

Motorola, Inc. 1989. *Low Skew Clock Drivers and Their System Design Considerations (#AN1091)*.

Motorola, Inc. 1988. *MECL System Design Handbook (#HB205)*. Chapters 3 and 7.

Motorola, Inc. 1989. *Transmission Line Effects in PCB Applications (#AN1051/D)*.

O'Sullivan, C. 1997. "Investigation of the Effectiveness of DC Power Bus Interplane Capacitance in Reducing Radiated EMI from Multi-Layer PCBs." *Proceedings of the IEEE International Symposium on Electromagnetic Compatibility*. New York: IEEE, pp. 293–297.

Ott, H., 1988. *Noise Reduction Techniques in Electronic Systems*. 2nd ed. New York: John Wiley & Sons.

Paul, C. R. 1984. *Analysis of Multiconductor Transmission Lines*. New York: John Wiley & Sons.

Paul, C. R. 1989. "A Comparison of the Contributions of Common-mode and Differential-mode Currents in Radiated Emissions." *IEEE Transactions on EMC* 31 (2): 189–193.

Paul, C. R. 1992, May. "Effectiveness of Multiple Decoupling Capacitors." *IEEE Transactions on Electromagnetic Compatibility*, EMC-34, pp. 130–133.

Paul, C. R. 1992. *Introduction to Electromagnetic Compatibility*, New York: John Wiley & Sons.

Paul, C. R. 1986. "Modeling and Prediction of Ground Shift on Printed Circuit Boards." *Proceedings of the Institute of Electrical Radio Engineers EMC Symposium* (York, England), pp. 37–45.

Paul, C. R., and K. B. Hardin, 1988. "Diagnosis and Reduction of Conducted Noise Emissions." *Proceedings of the IEEE 1988 International Electromagnetic Compatibility Symposium* (Seattle, WA), New York: IEEE, pp. 19–23.

Paul, C. R., K. White, and J. Fessler. 1992. "Effect of Image Plane Dimensions on Radiated Emissions." *Proceedings of the IEEE International Symposium on Electromagnetic Compatibility*, New York: IEEE, pp. 106–111.

Ramo, Simon, J. R. Whinnery, and Theodore Van Duzer. 1984. *Fields and Waves in Communication Electronics*. 2nd ed. New York: John Wiley & Sons.

Sadiku, M. 1992. *Numerical Techniques in Electromagnetics*. Boca Raton, FL: CRC Press, Inc.

Smith, Doug. 1993. *High Frequency Measurements and Noise in Electronic Circuits*. New York: Van Nostrand Reinhold.

Swainson, D. 1988. "Radiated Emission and Susceptibility Prediction on Ground Planes in Printed Circuit Boards." *Proceedings of the Institute of Electrical Radio Engineers EMC Symposium* (York, England), pp. 295–301.

Van Doren, T. 1995. *Circuit Board Layout to Reduce Electromagnetic Emission and Susceptibility*. Seminar Notes.

Van Doren, T. 1996. *Grounding and Shielding Electronic Systems*. Seminar Notes.

Van Doren, T., J. Drewniak, and T. Hubing. 1992, September 30. "Printed Circuit Board Response to the Addition of Decoupling Capacitors." Tech. Rep. #TR92-4-007, UMR EMC Lab.

Violette, J. L. N. and M. F. Violette. 1989. "An Introduction to the Design of Printed Circuit Boards (PCB's) with High Speed Digital and High Frequency System Performance Consideration." *Proceedings of EMC EXPO 89* (Washington, DC), A3.1–23.

Violette, J. L. N., and M. F. Violette. 1991. "EMI Control in the Design and Layout of Printed Circuit Boards." *EMC Technology* 5(2): 19–32.

Walker, C. 1990. *Capacitance, Inductance and Crosstalk Analysis*. Norwood, MA: Artech House, Inc.

Williams, Tim. 1996. *EMC for Product Designers*. 2nd ed. Oxford, England: Butterworth-Heinemann.

Witte, Robert. 1991. *Spectrum and Network Measurements*. Englewood Cliffs, NJ: Prentice-Hall.

Yuan, F. Y. 1997. "Analysis of Power/Ground Noises and Decoupling Capacitors in Printed Circuit Board Systems." *Proceedings of the IEEE International Symposium on Electromagnetic Compatibility*. New York: IEEE, pp. 425–430.

A

The Decibel

In the field of engineering, a common unit of measurement or reference is required. This often misunderstood unit, a logarithmic function, is the decibel (dB). This logarithmic function is required because of the scaling range of units involved. Most ratios are dimensionless, while some ratios are magnitudes expressed in dB (reference).

The basic unit of measurement is the logarithmic ratio of two products. Absolute power, voltage, or current levels are expressed in dB by giving their value *above* or *referenced* to some *base* quantity. The following describes power gain ($P_2 > P_1$) or loss ($P_2 < P_1$) in a system.

$$\text{Power Gain:} \quad \text{dB} = 10 \log \left(\frac{P_{\text{out}}}{P_{\text{in}}} \right)$$

In many situations, reference must be made for voltage, current, field strength, and the like instead of power. The following describes formulas for voltage and current gain ratios. The unit dB is dimensionless.

$$\text{Voltage Gain:} \quad \text{dB} = 10 \log \left(\frac{V_{\text{out}}^2/R}{V_{\text{in}}^2/R} \right) = 10 \log \left(\frac{V_{\text{out}}}{V_{\text{in}}} \right)^2 = 20 \log \left(\frac{V_{\text{out}}}{V_{\text{in}}} \right)$$

$$\text{Current Gain:} \quad \text{dB} = 10 \log \left(\frac{I_{\text{out}}^2 R}{I_{\text{in}}^2 R} \right) = 10 \log \left(\frac{I_{\text{out}}}{I_{\text{in}}} \right)^2 = 20 \log \left(\frac{I_{\text{out}}}{I_{\text{in}}} \right)$$

A pattern follows for voltage and current. An exception is the common reference of *dB above or below one milliwatt, denoted as dBm*. Radiated electromagnetic fields are described in terms of field intensity. These units are V/m (Volts per meter) for electric field strength or A/m (Amperes per meter) for magnetic field strength. The common units of measurement for the following voltage and current field strength intensities are

1μV/m = 0 dBμV/m
1mV/m = 0 dBmV/m
1μA/m = 0 dBμA/m
1mA/m = 0 dBmA/m
1mW = 0 dBm (Note the pattern difference.)

Most regulatory limits are described in μV/m. For example, 100 μV/m limit translates to 40 dBμV/m. The equations that describe this conversion are

$$\text{dB}\mu\text{V/m} = 20 \log_{10}\left(\frac{\text{V/m}}{1\mu\text{V/m}}\right)$$

$$\text{dB}\mu\text{A/m} = 20 \log_{10}\left(\frac{\text{A/m}}{1\mu\text{A/m}}\right)$$

Conversions between units are easy. For example:

1 μV = 0 dBμV = –107 dBm For a 50 Ω system
$V(\mu V) = 90 + 10\log_{10}(Z) + P(\text{dBm})$ For a given impedance Z in ohms

Five commonly used variations exist for the decibel. An example of these variations follows to present the concept of *dBs* using different units.

$$\text{dBm} = 10 \log\left(\frac{P}{0.001\text{W}}\right)$$

$$\text{dB}\mu\text{V} = 20 \log\left(\frac{V}{1\mu\text{Volt}}\right)$$

$$\text{dB}\mu\text{A} = 20 \log\left(\frac{A}{1\mu\text{Amp}}\right)$$

$$\text{dB}\mu\text{V/m} = 20 \log\left(\frac{V}{1\mu\text{Volt/meter}}\right)$$

$$\text{dB}\mu\text{A/m} = 20 \log\left(\frac{V}{1\mu\text{Amp/meter}}\right)$$

$$\text{dB}\mu\text{V/m/120 KHz} = 20 \log\left(\frac{A}{1\mu\text{Volt/meter}}\right) \text{ at a 120 kHz bandwidth}$$

Several pitfalls are related to use of the decibel, owing to the impedance of the system. Not all systems have the same impedance; hence, different values will be obtained under this situation.

- dBm = 10 log (P1/0.001 watts)
- 1 volt in a 50-ohm system is equal to:

$$\text{dBm} = 10 \log\left(\frac{1 \text{ volt}^2/50 \text{ ohms}}{0.001 \text{ watts}}\right) = 10 \log(20) = 13 \text{ dBm}$$

- 1 volt in a 600-ohm system is equal to:

$$\text{dBm} = 10 \log \left(\frac{1 \text{ volt}^2/600 \text{ ohms}}{0.001 \text{ watt}} \right) = 10 \log (1.67) = 2 \text{ dBm}$$

There is a common mistake most engineers make when performing decibel (logarithmic) math. This is known as the 6 dB problem. We must ask ourselves, "When does 6 dB not equal 6 dB"? Examples of this mistake follow. If the reference level is doubled, the logarithmic function increases by 6 dB. Three times the increase in the reference is a 9.5 dB increase.

- 1000 μVolts = 60 dBμV
- 1000 mVolts = 60 dBμV
  2000 μVolts = 66 dBμV
- 1000 μVolts = 60 dBμV
  3000 μVolts = 69.5 dBμV

B

Fourier Analysis

Every periodic signal is represented in both the time and frequency domain. Conversion between time and frequency domain is accomplished through use of Fourier analysis. Digital PCBs are always discussed in terms of operating frequency of the oscillator or processor. Although this frequency value is important for the speed of operation, the edge rate of the oscillator will determine the RF spectral distribution of energy created. The mathematics are complicated, but the results are simple. Most periodic continuous waveforms in the time domain will be observed in the frequency domain.

The same signal can be observed on both an oscilloscope (time domain) and a spectrum analyzer (frequency domain).

Periodic signals are represented by a series of sine and cosine functions.

$$f(t) = \frac{A_o}{2} + \sum_{n=1}^{\infty}(A_n \cos(n\omega_o t) + B_n \sin(n\omega_o t))$$

where $\omega_0 = \frac{2\pi}{T}$ = natural fundamental frequency

$$A_o = \frac{2}{T}\int_{t_o}^{t_o+T} f(t)dt$$

$$A_n = \frac{2}{T}\int_{t_o}^{t_o+T} f(t)\cos(n\omega_o t)dt$$

$$B_n = \frac{2}{T}\int_{t_o}^{t_o+T} f(t)\sin(n\omega_o t)dt$$

These equations show that a periodic signal is a summation of sinusoidal signals of multiple frequencies and amplitudes. Therefore, a periodic signal corresponds to a particular frequency range. A Fourier transform converts time-based signals to frequency domain energy. The Fourier transform is

$$F(\omega) = \int_{-\infty}^{\infty} f(t)e^{-j\omega t}dt$$

The Fourier envelope is used to quickly calculate the worst-case frequency spectrum envelope. For a given periodic square wave signal with a finite rise and fall time, the frequency spectrum envelope is calculated as

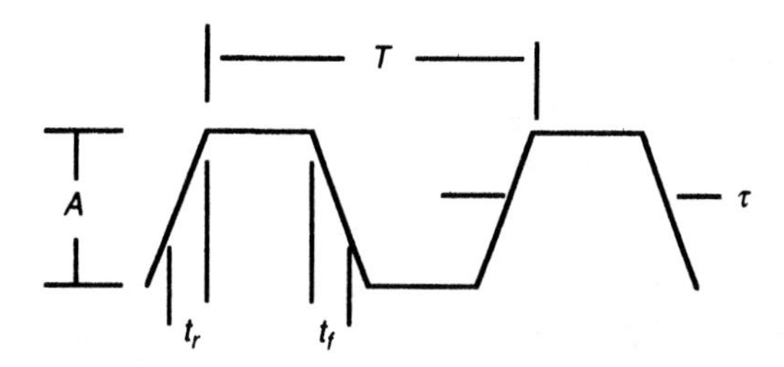

where A = peak amplitude (volts or amperes)
τ = pulse width (measured at half-maximum)
T = pulse period
t_r = rise time from 10–90% of the edge transition
t_f = fall time from 90–10% of the edge transition

If $t_r \neq tf$, the smallest of the two should be used.

The amplitude of the signal in the frequency domain, A_f, is calculated using

$$A_f = 2A\frac{\tau}{T}$$

where A is peak amplitude in the time domain. Corner frequencies, f_1 and f_2, are calculated using the following. Duty cycle δ is also shown.

$$f_1 = \frac{1}{\pi\tau} \quad f_2 = \frac{1}{\pi\tau_r} \quad \delta = \frac{\tau + \tau_r}{T}$$

Examining the corner frequencies, we note that the rising and falling edge of the periodic signal may not be the same and is often very different. With this situation, use of the faster of the two corner frequencies is required; this is generally the transition from high to low state.

The following figure illustrates a frequency spectrum representation of a periodic waveform (trapezoidal) with both corner frequencies identified. The amplitude of the signal (frequency domain) falls off at –20 dB per decade up to corner frequency f_2. Above f_2, the signal amplitude falls off at –40 dB per decade. The spectrum amplitude beyond f_2 is defined by

$$A_f = \frac{2A}{f^2 \pi^2 t_r T}$$

The current in the nth harmonic is

$$I_n = 2I_d \frac{\sin(n\pi d)}{n\pi d} \frac{\sin(n\pi t_r/T)}{n\pi t_r/T}$$

where I = peak-peak amplitude of the wave
d = duty cycle
t_r = rise time
T = period of the signal
n = harmonic number

The unit of I_n is the same as I_d; thus, the equation is dimensionless. For harmonic calculations, let's assume the rise and fall time edges are the same. If the two edges are different, the smaller of the two must be used for worst-case analysis. For a 50% duty cycle (d = 0.5), the first harmonic (fundamental) contains an amplitude of I_1 = 0.64I with only odd harmonics present. This is for the case where the rise time (t_r) is much less than period (T).

The following figure illustrates the envelope of harmonics for a symmetrical wave. The amplitude of the harmonics decrease with frequency at –20 dB per decade rate up to a frequency of $1/\pi t_r$. Beyond this point, the harmonics fall off at –40 dB per decade rate. As the rise time increases (becomes slower), energy in the higher order harmonics decreases.

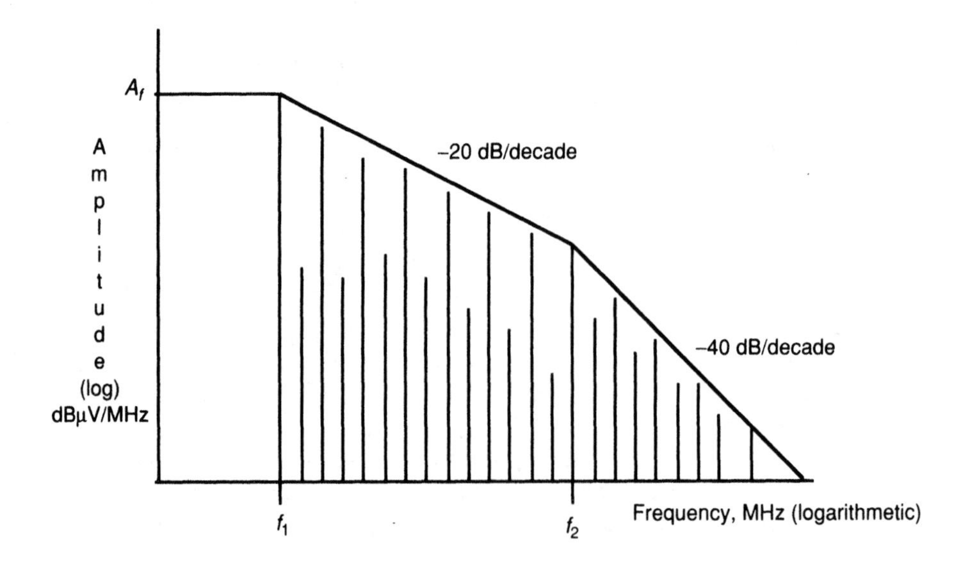

Note that changes in duty cycle and transition times will reduce the frequency spectrum envelope. As the frequency is doubled, the radiation increases by 6 dB if all parameters remain the same. If the frequency is doubled, we must cut the rise time in half to accommodate this faster signal. This decrease in edge rate will increase the amplitude of the signal by 12 dB.

In looking at the figure, for a 50% duty cycle, only odd harmonics are shown. For small duty cycles, that is, when the period becomes significantly long compared to the pulse duration, only a few harmonics within the envelope will reach the maximum level. There harmonics are observed as EMI because of their large amplitude.

An illustration of the spectral profile envelope of a signal does not show phase or polarity. At every multiple of $1/\tau$, there is a 180 degree reversal due to the fact that harmonics follow a sine or cosine function of frequency. Measurement equipment, such as spectrum analyzers, is insensitive to phase and will only display the absolute value.

When harmonics are spaced close to each other (many harmonics), they will not be added together. When measurements are performed using a spectrum analyzer, we tune the analyzer for a specific resolution bandwidth. Resolution bandwidth of a spectrum analyzer is the ability to display discrete frequency components within a specific frequency span (beginning and ending frequency range). The noise that is displayed from a periodic signal will appear as a narrowband signal. This means that the receiver will only see one harmonic at a time within the selected resolution bandwidth.

The designer must strive to limit and control the noise spectra of digital signals. Switching noise, typically in the MHz range, is a byproduct of digital electronics. This switch noise will find its way outside of the intended environment and cause EMC problems. In reviewing the spectra of these pulses, some parametric behaviors are observed. The spectrum is governed by amplitude (Af), which is in turn a function of the pulse amplitude, A. Limiting the pulse amplitude has a direct effect on the RF noise created. A slower edge transition time creates a smaller spectrum of RF energy. Wide pulses concentrate energy at lower frequencies than do narrow pulses.

Conversion Tables

Common Suffixes

Suffix	refers to
dBm	1 milliwatt
dBW	1 watt
dBμW	1 microwatt
dBV	1 volt
dBmV	1 millivolt
dBμV	1 microvolt
dBV/m	1 volt per meter
dBμV/m	1 microvolt per meter
dBA	1 amp
dBμA	1 microamp
dBμA/m	1 microamp per meter

Power and Voltage/Current Ratios

Ratio	*V* or *I* in dB	*P* in dB
10^6	120	60
10^5	100	50
10^4	80	40
10^3	60	30
10^2	40	20
10	20	10
9	19.08	9.54
8	18.06	9.03
7	16.9	8.45
6	15.56	7.78
5	13.98	6.99
4	12.04	6.02
3	9.54	4.77
2	6.020	3.01
1	0	0
10^{-1}	−20	−10
10^{-2}	−40	−20
10^{-3}	−60	−30

dB	Power Ratio	Voltage/Current Ratio
120	10^{12}	10^6
100	10^{10}	10^5
80	10^8	10^4
60	10^6	10^3
40	10^4	10^2
30	10^3	32
20	10^2	10
10	10.0	3.2
6	4.0	2.0
3	2.0	1.4
0	1.0	1.0
−3	0.50	0.71
−6	0.25	0.50
−10	0.10	0.32
−20	10^{-2}	0.10
−30	10^{-3}	0.03
−40	10^{-4}	10^{-2}
−60	10^{-6}	10^{-3}
−80	10^{-8}	10^{-4}
−100	10^{-10}	10^{-5}
−120	10^{-12}	10^{-6}

Conversion of dBV, dBmV, and dBμV

dBV	dBmV	dBμV
−120	−60	0
−100	−40	20
−80	−20	40
−60	0	60
−40	20	80
−20	40	100
0	60	120
20	80	140
40	100	160
60	120	180

Conversion of Volt/m to mW/cm^2 for Linear and dB Scales

V/m	dBμV/m	mW/cm^2	$dBmW/cm^2$
1.00×10^{-6}	0	2.67×10^{-16}	−155.8
1.00×10^{-5}	20	2.67×10^{-14}	−135.8
1.00×10^{-4}	40	2.67×10^{-12}	−115.8
1.00×10^{-3}	60	2.67×10^{-10}	−95.8
1.00×10^{-2}	80	2.67×10^{-8}	−75.8
1.00×10^{-1}	100	2.67×10^{-6}	−55.8
1.00	120	2.67×10^{-4}	−35.8
$1.00 \times 10^{+1}$	140	2.67×10^{-2}	−15.8
$1.00 \times 10^{+2}$	160	2.67	−4.2
$1.00 \times 10^{+3}$	180	267	−24.2
$1.00 \times 10^{+6}$	6	1.06×10^{-15}	−149.7
$2.00 \times 10^{+6}$	12	4.24×10^{-15}	−143.7
$6.00 \times 10^{+6}$	15	9.55×10^{-15}	−140.2
$8.00 \times 10^{+6}$	18	1.70×10^{-14}	−137.7

dBμV versus dBm for Z = 50Ω

dBμV	μV	dBm	Power Level
−20	0.1	−127	0.0002 pW
−10	0.316	−117	0.002 pW
0	1.0	−107	0.02 pW
5	1.778	−102	0.063 pW
7	2.239	−100	0.1 pW
10	3.162	−97	0.2 pW
15	5.623	−92	0.632 pW
20	10.0	−87	2.0 pW
30	0.03162	−77	0.02 pW
40	0.10	−67	0.2 pW
50	0.312	−57	2.0 pW
60	1.0	−47	20.0 pW
70	3.162	−37	0.2 μW
80	10.0	−27	2.0 μW
90	31.62	−17	20.0 μW
100	100.0	−7	2000.0 μW
120	1.0V	+13	20 mW

Frequency—Wavelength—Skin Depth

Frequency	λ	λ/2π	Skin Depth
10 Hz	30,000 km	4,800 km	820 mil
60 Hz	5,000 km	800 km	340 mil
100 Hz	3,000 km	480 km	260 mil
400 Hz	750 km	120 km	130 mil
1 kHz	300 km	48 km	82 mil
10 kHz	30 km	4.8 km	26 mil
100 kHz	3 km	480 m	8.2 mil
1 MHz	300 m	48 m	2.6 mil
10 MHz	30 m	4.8 m	0.8 mil
100 MHz	3 m	0.48 m	0.3 mil
1 GHz	30 cm	4.8 cm	0.08 mil
10 GHz	3 cm	4.8 mm	0.03 mil

λ = wavelength
λ/2π = near field to far field distance conversion
distance: metric (meters)

International EMC Requirements

BRIEF SUMMARY OF INTERNATIONAL EMC REQUIREMENTS

Basic Standards

General information about what is being measured and the test techniques. Prepared by the IEC (International Electrotechnical Commission) which oversees the work of CISPR (Comité International Spécial des Perturbations Radioélectriques, a.k.a. International Special Committee on Radio Interference). CISPR is responsible for establishing emissions limits, susceptibility levels, and test procedures.

Generic Standards

For use in specific environments (such as residential, commercial, light industrial, or heavy industrial) generic standards apply to all products or product families for which no dedicated or specific EMC standard exists. These standards are approved by CENELEC (Comité Européen de Normalisation Electrotechnique, or European Standardization Committee for Electrical Products), or ETSI (European Telecommunications Standards Institute) and submitted for publication in the *Official Journal of the European Union (OJ)* after adoption by the European Parliament. The standards submitted for publication are based on the work of the IEC and CISPR.

Product/Product Family Standards

For specific products or product families (such as Information Technology Equipment—ITE). Where a specific product family standard exists, it takes precedence over generic standards. Prepared by IEC, CENELEC, or CISPR.

Note: Test requirements are subject to change at the discretion of various regulatory agencies. The reader is urged to verify the applicable and current requirements that are in force at the time of product design and release.

Generic Standards (Sample List)

EN 50081-1	Electromagnetic compatibility—Generic emission standard Part 1: Residential, commercial, and light industry.
EN 50081-2	Electromagnetic compatibility—Generic emissions standard Part 2: Industrial environment.
EN 50082-1	Electromagnetic compatibility—Generic immunity standard Part 1: Residential, commercial, and light industry.
EN 50082-2	Electromagnetic compatibility—Generic immunity standard Part 2: Industrial environment.

Note: Due to constantly changing requirements, adoption, and publication of standards by the European Parliament, the issue date is not provided for all standard.

Product / Product Family Standards (Sample list)

EN 55011	Limits and methods of measurements of radio disturbance characteristics of industrial, scientific, and medical (ISM) radio-frequency equipment (CISPR 11).
EN 55013	Limits and methods of measurements of radio disturbance characteristics of broadcast receivers and associated equipment (CISPR 13).
EN 55014	Limits and methods of measurements of radio disturbance characteristics of household electrical appliances, portable tools, and similar electrical apparatus (CISPR 14).
EN 55020	Limits and methods of measurements of radio disturbance characteristics of broadcast receivers and associated equipment (CISPR 20).
EN 55022	Limits and methods of measurements of radio disturbance characteristics of Information Technology Equipment (CISPR 22).
EN61000-3-2	Mains harmonics.
EN61000-3-3	Mains flicker.

Note: The date of adoption or release is not listed. These standards have had amendments incorporated or have been reissued several times. Each action changes the effective date of issue. When use of these standards is required for compliance purposes, one should refer to the latest edition or release date that is appropriate for the product. Amendments may or may not be applicable to the product being certified at date of test.

DEFINITION OF CLASSIFICATION LEVELS—EMISSIONS

In North America (the United States and Canada) the Same Definition Exists

Class A: A digital device that is marketed for use in a commercial, industrial, or business environment, exclusive of a device which is marketed for use by the general public or is intended to be used in the home.

Products are self-verified for compliance.

Class B: A digital device that is marketed for use in a residential environment, notwithstanding its use in commercial, business, and industrial environments.

Products require certification from the Federal Communications Commission (FCC). Canada accepts FCC Certification.

International Definition, Defined Within EN 55022 and CISPR-22

Class A: Equipment is information technology equipment which satisfies the Class A interference limits but does not satisfy the Class B limits. In some countries, such equipment may be subjected to restrictions on its sale and/or use.

Note: The limits for Class A equipment are derived for typical commercial establishments for which a 30 m protection distance is used. The class A limits may be too liberal for domestic establishments and some residential areas.

Class B: Equipment is information technology equipment which satisfies the Class B interference limits. Such equipment should not be subjected to restrictions on its sale and is generally not subject to restrictions on its use.

Note: The limits for Class B equipment are derived for typical domestic establishments for which a 10 m protection distance is used.

FCC/ INDUSTRY CANADA EMISSION LIMITS

For FCC and Industry Canada, the frequency range to be measured is based on the highest fundamental internally generated clock frequency per the following list.

Less than 1.705 MHz	Test to 30 MHz
From 1.705 MHz to 108 MHz	Test to 1 GHz
108 MHz to 500 MHz	Test to 2 GHz
500 MHz to 1 GHz	Test to 5 GHz
Above 1 GHz	Test to 5th harmonic or to 40 GHz, whichever is lower

FCC/DOC Class A Radiated Emission Limits

Frequency (MHz)	Distance (meters)	Quasi-Peak Limit (dBμV/m)
30 to 88	10	39.0
88 to 216	10	43.5
216 to 960	10	46.5
Above 960	10	49.5

FCC/DOC Class A Conducted Emission Limits

Frequency	Quasi-Peak Limit
0.45 to 1.705 MHz	60.0 dBuV
1.705 to 30.0 MHz	69.5 dBuV

FCC/DOC Class B Radiated Emission Limits

Frequency (MHz)	Distance (meters)	Quasi-Peak Limit (dBμV/m)
30 to 88	3	40.0
88 to 216	3	43.5
216 to 960	3	46.0
Above 960	3	54.0

FCC/DOC Class B Conducted Emission Limits

Frequency	Quasi-Peak Limit
0.45 to 30.0 MHz	48.0 dBμV

Summary List—FCC and DOC

FCC/DOC Limits

Frequency	A Limit	B Limit
0.45–1.705	60 dBμV*	48 dBμV*
1.705–30 MHz	69.5 dBμV*	48 dBμV*
30–88 MHz	39 dBμV @ 10 m	40 dBμV @ 3 m
88–216 MHz	43.5 dBμV @ 10 m	43.5 dBμV @ 3 m
216–960 MHz	46.5 dBμV @ 10 m	46 dBμV @ 3 m
> 960 MHz	49.5 dBμV @ 10 m	54 dBμV @ 3 m

*This is the narrowband limit; the broadband limit is 13 dB higher.

INTERNATIONAL EMISSION LIMITS SUMMARY—SAMPLE LIST

Class B Limits for Light Industrial Equipment and Primarily Residential Areas

Frequency Range, MHz

0.15 -------------- 0.5 --------------- 5 ------------- 30 -------------- 230 -------------- 1000

SPECIFICATION	dBµV		dBµV		dBµV		dBµV/m	dBµV/m	Notes
	QP (1)	AVG (1)	QP	AVG	QP	AVG	QP (1)	QP	
EN 50081-1	66–56	56–46	56	46	60	50	30	37	@ 10 m, B limit
EN 55011*	66–56	56–46	56	46	60	50	30	37	@ 10 m, B limit
EN 55013 (2)	66–56	56–46	56	46	60	50	45–55 (3)	—	dBpW, Absorbing Clamp (3)
EN 55014	66–56	56–46	56	46	60	50	45–55 (3)	—	dBpW, Absorbing Clamp (3)
EN 55020	66–56	56–46	56	46	60	50	45–55 (3)	—	@ 10 m
EN 55022	66–56	56–46	56	46	60	50	30	37	@ 10 m
Class A Limits for Industrial Areas									
EN 50081-2	79	66	73	60	73	60	30	37	@ 30 m, A limit
EN 55011*	79	66	73	60	73	60	30	37	@ 30 m, A limit
EN 55022	79	66	79	66	73	60	30/40	37/47	@ 30 m/@ 10 m

Notes: (1) The dash between two numbers (e.g., 66–56) means the limit decreases with the logarithm of frequency.
(2) EN 55013 has other limits for emissions from receivers and televisions.
(3) Absorbing clamp measurement is for the frequency range of 30–300 MHz only.

*EN 55011 is for equipment covered under EN 50081-1 and EN 50081-2, and is the product standard for industrial, scientific, and medical equipment Group 1. This standard defines limits for radiated and conducted emission. Detailed specification limits for EN 55011 are shown in the next subsection for Group 2 products.

EN 55013 is for equipment covered under EN 50081-1, not intentionally generating RF from household electronics.

EN 55014 is for equipment covered under EN 50081-1, not intentionally generating RF such as brush motors and 50 Hz speed controls. This is also the emission product standard for Household Appliance Equipment (HHA).

EN 55020 is for equipment covered under EN 50082-1 for immunity from radio interference from household electronics.

EN 55022 is the emission requirement for products covered under both EN 50081-1 and EN 50081-2, which includes Information Technology Equipment (ITE).

Emissions—EN 55011 Industrial Scientific and Medical (ISM) Equipment

For all other EN 55 XXX specifications, refer to the *International Emissions Limits Summary.*

Special Note: Due to the unique specification limits for EN 55011, this section is provided for completeness only.

Classification of ISM Equipment

Group 1 ISM—Group 1 contains all ISM equipment in which there is intentionally generated and/or used conductively coupled radio frequency energy which is necessary for the internal functioning of the equipment itself.

Group 2 ISM—Group 2 contains all ISM equipment in which radio frequency energy is intentionally generated and/or used in the form of electromagnetic radiation for the treatment of material, and spark erosion equipment.

Line Conducted Emissions. Emissions levels less than Class A limits (Table IIA), or as agreed with the competent body. The need for mains terminal disturbance voltage limit for Class A equipment *in situ* is under consideration.

TABLE IIA Mains Terminal Disturbance Limits for Class A Equipment Measured on a Test Site

	Class A Equipment Limits dB(μV)			
	Group 1		Group 2*	
Frequency band (MHz)	Quasi-peak	Average	Quasi-peak	Average
0.15–0.50	79	66	100	90
0.50–5	73	60	86	76
5–30	73	60	90 Decreasing with logarithm of frequency to 70	80 Decreasing with logarithm of frequency to 60

*Mains terminal disturbance voltage limits for Group 2 Class A equipment requiring currents greater than 100A are under consideration.

TABLE IIB Mains Terminal Disturbance Limits for Class B Equipment Measured on a Test Site

	Class B Equipment Limits dB(μV)	
	Groups 1 and 2	
Frequency Band (MHz)	Quasi-peak	Average
0.15–0.50	66 Decreasing with logarithm of frequency to 56	56 Decreasing with logarithm of frequency to 46
0.50–5	56	46
5–30	60	50

TABLE IIIA Radiated Emissions for Group 1 Equipment

Frequency Band MHz	Measured on a test site		Measured *in situ*
	Group 1 Class A 30 m measurement distance dB (μV/m)	Group 1 Class B 10 m measurement distance dB (μV/m)	Group 1 Class A limits with measuring distance 30 m from exterior wall outside the building in which the equipment is situated dB(μV/m)
0.15–30	Under consideration	Under consideration	Under consideration
30–230	30	30	30
230–1000	37	37	37

TABLE IIIB Radiated Emissions Limit for Group 2 , Class A

	Limits with Measuring Distance 30 m	
Frequency Range (MHz)	From Exterior Wall Outside the Building in Which the Equipment is Situated dB(μV/m)	On a Test Site dB(μV/m)
0.15–0.49	75	85
0.49–1.705	65	75
1.705–2.194	70	80
2.194–3.95	65	75
3.95–20	50	60
20–30	40	50
30–47	48	58
47–68	30	40
68–80.872	43	53
80.872–81.848	58	68
81.848–87	43	53
87–134.786	40	50
134.786–136.414	50	60
136.414–156	40	50
156–174	54	64
174–188.7	30	40
188.7–190.979	40	50
190.979–230	30	40
230–400	40	50
400–470	43	53
470–1000	40	50

TABLE IIIC Radiated Emissions Limit for Group 2, Class B Equipment Measured on a Test Site

Frequency Band MHz	Measurement Distance 10 m dB(μV/m)
0.15–30	Under consideration
30–80.872	30
80.872–81.848	50
81.848–134.786	30
134.786–136.414	50
136.414–230	30
230–1000	37

SUMMARY OF CURRENT INTERNATIONAL IMMUNITY REQUIREMENTS

International standards for susceptibility (immunity) are provided by the IEC 1000-4-X series. This series describes the test and measurement methods for the Basic standards, which are specific to a particular type of EMI phenomenon. It is not limited to a specific type of product. Internal to this immunity series are the following.

- Terminology
- Descriptions of the EMI phenomenon
- Instrumentation
- Measurement and test methods
- Ranges of severity levels with regard to the immunity of the equipment.

The International Electrotechnical Commission (IEC) 1000-4-X series is based on the well-known IEC 801-X requirements. IEC requirements, when adopted by the European Parliament, are reissued with a new number, the EN 61000-X series. Currently, immunity tests are mandated in Europe, but only recommended in North America, and are optional worldwide. Not all IEC 1000-4-X standards have been converted to EN 61000-4-X standards. Minor differences exist between the IEC and EN series.

Note: Before starting any compliance test program, one should verify the need to apply a specific test and whether the standard being used has been updated by a newer version or has changed status from proposed to mandatory. Standards development in Europe has constantly been evolving.

BASIC IMMUNITY STANDARDS

- IEC 1000-1 General Considerations
- IEC 1000-2 Environment
- IEC 1000-3 Limits/Generic Standards
- IEC 1000-4 Test and Measurement Techniques
- IEC 1000-5 Installations and Mitigation Guideline
- IEC 1000-6 Miscellaneous
  - IEC 61000-6-1 Electromagnetic Compatibility (EMC) Generic Standard. Immunity for Residential, Commercial and Light Industrial Environments.

Under IEC 1000-4 are the EN 61000-4-X immunity specifications detailed in the following listing.

COMPREHENSIVE LIST OF IMMUNITY STANDARDS

Standard	Description
EN 61000-4-2	Electrostatic Discharge (ESD)
EN 61000-4-3	Radiated electromagnetic field
EN 61000-4-4	Electrical Fast Transient/Burst (EFT)
EN 61000-4-5	Surge
EN 61000-4-6	Conducted disturbance by RF
EN 61000-4-7	General guide on harmonics and interharmonics measurements and instrumentation (not a standard; procedure only)
EN 61000-4-8	50/60 Hz magnetic field
EN 61000-4-9	Pulsed magnetic field
EN 61000-4-10	Oscillatory magnetic field
EN 61000-4-11	Voltage dips and interruption
EN 61000-4-12	Oscillatory waves "ring wave"
EN 61000-4-13	Oscillatory waves 1 MHz
EN 61000-4-14	Harmonics, interharmonics, and main signaling
EN 61000-4-15	Voltage fluctuations
EN 61000-4-16	Flickermeter
EN 61000-4-17	Conducted disturbance in the range of DC to 150 kHz
EN 61000-4-18	Not assigned
EN 61000-4-19	Not assigned
EN 61000-4-20	TEM cells
EN 61000-4-21	Mode stirred chambers
EN 61000-4-22	Guide on measurement methods
EN 61000-4-23	Test methods for protective devices; HEMP radiated disturbance
EN 61000-4-24	Test methods for protective devices; HEMP conducted disturbance
EN 61000-4-25	Test methods for equipment and systems; HEMP
EN 61000-4-26	Calibration of probes and instrument for measuring electromagnetic fields
EN 61000-4-27	Unbalance in three-phase mains
EN 61000-4-28	Variation of power frequency

Note: Most of the EN 61000-4-x specifications have never been written or released. This includes standards identified as −12 and above. Titles have been issued and working groups have been assigned for many of these tests. When performing compliance testing, one should verify which standards are mandatory for a product along with required test levels and performance criteria.

PERFORMANCE CRITERIA FOR IMMUNITY TESTS

A functional description and a definition of performance criteria, during or as a consequence of EMC testing, shall be provided by the manufacturer and noted in a test report based on the following criteria:

Performance Criterion A. The apparatus *shall continue to operate as intended.* No degradation of performance or loss of function is allowed below a performance level specified by the manufacturer when the apparatus is used as intended. In some cases, the performance level may be replaced by a permissible loss of performance. If the minimum performance level or the permissible performance loss is not specified by the manufacturer, then either of these may be derived from the product description and documentation (including leaflets and advertising) and what the user may reasonably expect from the apparatus if used as intended.

Performance Criterion B. The apparatus *shall continue to operate as intended after the test.* No degradation of performance or loss of function is allowed below a performance level specified by the manufacturer when the apparatus is used as intended. In some cases, the performance level may be replaced by a permissible loss of performance. During the test, however, degradation of performance is allowed. No change of actual operating state or stored data is allowed. If the minimum performance level or the permissible performance loss is not specified by the manufacturer, then either of these may be derived from the product description and documentation (including leaflets and advertising) and what the user may reasonable expect from the apparatus if used as intended.

Performance Criterion C. Temporary loss of function is allowed, provided the loss of function is self-recoverable or can be restored by the operation of the controls.

INTERNATIONAL IMMUNITY REQUIREMENTS FOR MOST PRODUCTS

Specification	EN 61000-4-2 Electrostatic Discharge	EN 61000-4-3 Radiated RF Immunity	ENV 50204 Radiated RF Immunity	EN 61000-4-4 Electrical Fast Transients	EN 61000-4-5 Transients Signal Leads
EN 50082-1 Generic limit Light industrial equipment (Note 1)	8 kV (Air) 4 kV (Direct) Criterion B	80–1000 MHz 3 V/m 1 kHz, 80% AM Criterion A	900 ± 5 MHz Pulse modulated 50% duty cycle 200 Hz Criterion A	500 V, Signal 500 V, DC power 500 V, Process 1000 V, Power 5/50 ns, 5 kHz Criterion B	1.2/50μs AC Power: 1kV-CM 500V-DM DC Power: 500V-CM 500V-DM Process: 500V-CM Criterion B
EN 50082-2 Generic limit Heavy industrial equipment (Note 2)	8 kV (Air) 4 kV (Direct) Criterion B	10 V/m 1 kHz, 80% AM 80–1000 MHz except 3 V/m @ 87–108 MHz 174–230 MHz 470–790 MHz Criterion A	900 ± 5 MHz Pulse modulated 50% duty cycle 200 Hz Criterion A	1000 V, Signal 2000 V, Power 5/50 ns, 5 kHz Criterion B	1.2/50μs AC Power: 4 kV-CM 2 kV-DM DC Power: 500V-CM 500V-DM Process: 2 kV-CM 1 kV-DM Criterion B
EN 55014-2 Appliances and power tools	8 kV (Air) 4 kV (Direct) Criterion B	80–1000 MHz 3 V/m Criterion A	Not proposed	500 V Signal 1000 V, AC 5/50 ns, 5 kHz Criterion B	1000 V, DM 2000 V, CM on power only 1.2/50 μs Criterion B
EN 60601-2 Medical devices	8 kV (Air) 3 kV (Direct) Criterion B	26–1,000 MHz 3 v/M 80% AM–1 kHz Criterion A	Not Yet Proposed	500 V, Signal/IO 1000 V, AC 5/50 ns, 5 kHz Criterion B	1000 V, DM 2000 V, CM 1.2/50 μs power lines only Criterion B

Specification	EN 61000-4-6 Conducted RF Immunity	EN 61000-4-8 Radiated Magnetic	EN 61000-4-11 Voltage Dips, Interruption, Variation	EN 61000-3-2 Power Line Harmonics	EN 61000-3-3 Flicker
EN 50082-1 Generic limit Light industrial equipment (Note 1)	0.15–80 MHz 3 V 1 kHz 80% AM, 150 Ω Source Criterion A	3 A/m 50 Hz Criterion A	+10%, −15% (A) −30%, 10ms (B) −60% ,100ms (C) −95%, 5000ms (C) Criterion (x) above		
EN 50082-2 Generic limit Heavy industrial equipment (Note 2)	0.15–80 MHz 10 V 80% AM, 1 kHz 150 Ω Source Criterion A	30 A/m 50 Hz Criterion A	+10%, −15% (A) −30%, 10ms (B) −60% ,100ms (C) −95%, 5000ms (C) Criterion (x) above		
EN 55014-2 Appliances and power tools	0.15–230 MHz Category II 0.15–80 MHz Category IV 1 V, Signal 3 V, Power Criterion A	Not yet proposed	Not yet proposed	Not yet proposed	Not yet proposed
EN 60601-2 Medical devices	Not yet proposed	Not yet proposed	Not yet proposed	Not yet proposed	Not yet proposed

Note 1: Severity levels and frequency ranges are subject change. Consult test requirements for current values in effect at date of testing and certification.

Note 2: Additional test requirements exist and are not detailed above. Refer to EN 50082-2 for details.

Performance criterion:

Level A: The apparatus shall continue to operate as intended. No degradation of performance or loss of function is allowed.

Level B: The apparatus shall continue to operate as intended after the test.

Level C: Temporary loss of function is allowed, provided the loss of function is self-recoverable.

Index

ε_r. *See* dielectric constant
λ/20, 163
0V reference, 81, 113, 187
  ground plane, 56, 109
  ground traces, 113
  image plane, 56
  power return plane, 57
3-W rule, 100, 210
5/5 rule, 83
10-W rule, 177

A

absence of copper, 107
alternate return path, 83
alternate RF current return path, 94
Ampere's Law, 28–29, 36
amplitude, 16
analog and digital components, 122
analog
  circuits, 261
  ground, 111
  power, 111
antennas, 18
  intentional, 18–19
  unintentional, 18–19
anti-resonance, 128
aspect ratio, 95
asymmetrically placed components, 118

B

backplane grounding, 274
backward crosstalk, 205
balanced circuits, 269
basic concepts of radiation, 20
behavioral characteristics of components, 24, 77
bifurcated lines, 242
bond wires, 187
bridging, 108
  analog ground, 111
  analog power, 111
  ferrite bead-on-lead, 109
  grounding, 110
bulk capacitor
  definition, 126
  selection of, 152
buried capacitance, 141
bypass capacitors
  definition, 126
  selection, 148–152

C

capacitance
  buried, 141
  calculation of, 140, 143
  dielectric material, 143
  distributive, 141
  efficiency, 143

capacitance (*cont.*)
 lead length inductance, 143
 power and ground planes, 134, 138, 144
 relative permittivity, 140
capacitive
 coupling, 205, 261
 crosstalk, 57, 77
 decoupling, 108
 loading, 180
 reactance, 192
capacitors
 capacitance, 129
 dielectric material, 129, 140, 143
 energy storage, 131
 ideal, 130
 impedance, 129
 internal to a component, 155
 lead length inductance, 133
 parallel placement, 136
 peak transient current, 153
 physcial characteristics, 129
 placement, 147
 power and ground planes, 134, 138
 resonance, 132
 retrofit, 145
 self-resonance, 130, 133
causes of EMC, 19
characteristic impedance, 160, 216
clock
 distribution networks, 34
 drivers, 58
 skew, 58, 178
 skew buffers, 58
 speed, 185
closed loop
 boundary, 36
 environment, 28
CMRR. *See* common-mode-rejection-ratio
common-impedance
 coupling, 251, 256, 262, 266
 inductance, 266
 power and ground, 266
 traces, 262
common-impedance path
 avoiding, 264
common-mode
 chokes, 269
 currents, 41, 44, 53, 88
 decoupling capacitor, 73
 radiation, 46
 voltage, 45–46
common-mode-rejection-ratio, 270
component characteristics
 bond wires, 76
 capacitive crosstalk, 57
 capacitive overheads, 54
 drive current, 56
 edge rate, 54–55
 Fourier analysis, 56
 frequency domain, 56
 input power consumption, 54, 56
 inrush surge current, 54, 57
 interconnect pads, 77
 lead length inductance, 60
 logic crossover currents, 54
 output resistance, 56
 packaging, 60
 power peak inrush surge, 54
 radiated design concerns, 76
 rise and fall time, 56
 specific resonant frequency, 56
 time-domain, 56
component packaging, 60
 ground bounce, 60
 lead bond configurations, 14
 lead length inductance, 60
 small loop antenna, 60
 wire bond leads, 68
component placement
 radiated emissions, 77
conducted emissions (definition), 3
conductivity, 28, 39
conductive
 coupling, 13
 immunity, 3
connector pinout assignment, 112
containment (definition), 3
conversion between
 common-mode currents, 41, 44, 46
 differential-mode currents, 41–43, 45–46
copper wire, 40
coupling paths, 13–14
crosstalk, 203, 216
 backward, 205
 capacitive coupling, 205
 design techniques to prevent, 207
 far-end, 204
 forward, 205
 frequency domain, 205
 inductive coupling, 205
 mutual capacitance, 204
 mutual coupling, 204
 mutual inductance, 204
 near-end, 204

polarities, 206
time domain, 205
unit of measurement, 206
crosstalk in terminators, 237
how to remove, 238
lead length inductance, 237
multiple terminators, 237
signal bounce, 237
critical frequency, 49
current loop, 36
current ratios, 51

D

daisychaining, 242
data line filters, 108
common-mode, 108
DC resistance, 262
decoupling (definition), 126
power and ground planes, 134
decoupling capacitor
calculation of, 149
placement, 144
selection of, 148
device capacitive overhead, 54
die shrink, 83, 185
dielectric constant, 28, 48, 140, 166, 169, 172, 189
dielectric losses, 166, 169
differential microstrip/stripline
line-to-line impedance, 177
differential-mode
capacitor, 71, 73
currents, 22, 41–45, 88
radiation, 42, 53
differential traces, 177
digital and analog components, 122
digital-to-analog partitioning, 122
dimensions, 17–18
diode network, 236
distributed capacitance, 88, 161, 180
distributed capacitive load, 200
divergence theorem, 28
double-sided boards, 82, 114, 116
driver impedance, 162
dual stripline, 175–176
duty cycle skew, 59

E

edge rate, 53–55, 160, 186
effective relative permittivity, 166, 172
electric
charge, 28, 35
current, 35
dipole, 31
field coupling, 14
field strength (calculation), 62
fields, 28, 33, 46, 92
shock, 249, 253
sources, 30–31
electrically long trace, 163, 188, 195
electromagnetic compatibility (definition), 2
electromagnetic
field, 160
field coupling, 14
interference (definition), 2
waves, 166
electrostatic discharge (definition), 3, 5, 82, 94
embedded microstrip, 172–174
EMC. *See* electromagnetic compatibility
EMC environment, 6, 11, 13
EMI. *See* electromagnetic interference
emissions, 11, 15
end termination, 226
edge rate degradation, 226
equivalent series inductance. 26, 129, 143
equivalent series resistance, 26, 129, 143
ESD. *See* electrostatic discharge
ESL. *See* equivalent series inductance
ESR. *See* equivalent series resistance
external inductance, 263

F

Faraday cage, 28
Faraday's Law, 28–29
far-end crosstalk, 204
far-field effects, 31–32
ferrite beads, 26, 104
field transfer coupling, 273
flat straps, 266
flux cancellation, 35, 45, 50, 84, 89, 99, 114
FM radio band, 6
Fourier, Baron Jean Baptiste Joseph, 17
forward crosstalk, 205
FR-4 material, 167
dielectric properties, 167
frequency response, 168
laminate, 168
material, 168
resin system, 168

free space
impedance of, 32
plane field, 32
frequency, 16–17
frequency domain, 17, 36, 163
functional
partitioning, 97
subsections, 97
subsystems, 106
fundamental concepts of suppression
common-mode RF currents, 50
current transients, 50
pull-up/pull-down current ratios, 51
radiated emissions, 50
RF voltages, 50

G

Gauss's Law, 28
gridded ground system, 119
ground
bounce, 19, 60, 65, 66–68, 74, 93, 136
currents, 250
glitches, 65
grid structure, 112, 119
isolation, 269
loop, 260, 268
noise margin, 67
noise voltage, 51, 53, 63, 88, 110
pins, 112
reference system, 65
stitch, 259
straps, 264
voltage magnitude, 66
wire bond leads, 68, 77
ground planes, 45, 56
benefits, 134
capacitance, 136
discontinuity, 100
impedance, 202
placement, 144
structure, 87
grounded heatsinks, 70
common-mode decoupling capacitor, 71, 73
dielectric insulator, 73
differential-mode capacitor, 73
thermodynamic domain, 70
grounding
analog circuits, 261
backplane, 274
chassis, 254
definition, 247
digital circuits, 262
electric shock, 249
fundamental concepts, 249
ground currents, 250
ground loops, 268
ground noise voltage, 251
hybrid ground, 261
I/O connectors, 277
impedance, 252
metallic structure, 271
multipoint, 259
overview, 247
reference system, 165
resonance, 271
signal voltage referencing, 271
single-point, 17, 256
star, 264
stitch connection, 95
voltage referencing, 249
wires, 112

H

hidden schematic, 7, 24, 27
hybrid ground, 261
capacitive and inductive coupling, 261

I

I/O connectors, 277
image plane, 38, 56, 81, 87, 95
common-mode currents, 88, 99
definition, 9
differential-mode currents, 88, 99
distributed capacitive load, 88
ground plane structure, 88
interplanar capacitance, 88
mutual partial inductance, 88
violation, 99
voltage gradient, 88
immunity (definition), 3
impedance, 17–18
loop area, 144
free space, 32
inductance
copper planes, 260
definition, 84
mutual partial inductance, 86, 87, 89
partial inductance, 85, 87
self-partial inductance, 84, 86
trace lengths, 260
inductive coupling, 205, 261

inductors, 26
inrush surge current, 54, 57
intentional radiators, 17
interconnect pads, 77
Interconnecting and Packaging Electronics Circuits Organization, 167
interconnects, 112
  connector pinout assignment, 112
  ground pins, 112
  ground traces, 112
  ground wires, 112
  power distribution network, 112
  routing configuration, 112
internal inductance, 263
interplanar capacitance, 88
IPC. *See* Interconnecting and Packaging Electronics Circuits Organization
isolated
  area, 107
  plane, 105
isolation, 107, 269
  balanced circuit, 269
  common-mode choke, 269
  optical, 108, 269
  transformer, 108, 269

J

jitter, 74

L

layer jumping, 102
layout concerns
  asymmetrically placed components, 118
  double-sided boards, 114, 116
  flux cancellation, 114
  gridded ground system, 119
  radial routing, 118
  single-sided boards, 114, 115
  symmetrically placed components, 116
lead bond configurations, 64
lead length inductance, 39, 60, 63, 133, 143, 197
line filter, 254
line impedance, 162
loaded characteristic impedance, 180
loaded propagation delay, 180
localized ground plane, 120
logic crossover currents, 54
logic families
  CMOS, 54, 57, 59
  ECL, 54, 58, 59, 165
  GaAs, 58
  HCMOS, 57
  LVDS, 59
  TTL, 54, 57, 59, 165
loop
  antenna, 43, 60
  area, 91, 92, 192
  control, 94
  impedance, 144
  structure, 84, 92
Lorentz force relation, 28

M

magnetic
  coupling, 14
  field, 28, 33, 35
  field coupling, 14, 92
  lines of flux, 35
  sources, 30
Maxwell's equations, 28–29, 34–37
metal screws
  inductance, 271
microstrip topology, 171
  coated microstrip, 172
  dielectric constant, 48, 172
  embedded microstrip, 172–174
  impedance, 171
  intrinsic capacitance, 171, 173
  surface microstrip, 171–172
moat, 107
multiple terminators, 237, 239
  dual terminations, 239
  termination effects, 239
multipoint ground, 259, 271
  ground stitch location, 259
  resonance, 271
mutual
  capacitance, 204
  coupling, 204
  inductance, 204
  partial inductance, 84, 86, 87

N

nature of interference, 16
  amplitude, 16
  dimensions, 17
  emissions, 16
  frequency, 16
  immunity, 16

nature of interference (*cont.*)
  impedance, 17
  time, 17
negative reflections, 217
noise margin upset, 194
noise margin, 67
near-end crosstalk, 204
near-field effects, 31–32
noise coupling, 12–16,
  capacitive, 15
  conductive, 13
  coupling paths, 13
  electric field, 14, 33
  electromagnetic field, 14
  emissions, 12
  immunity, 12
  inductive, 15
  magnetic, 14, 33
  mechanism, 33
noise source, 8
Norton equivalence, 17

O

Omega layer, 228
optical isolators, 269
oscillators, 6, 75
output resistance, 56
output-to-output skew, 59
overdamped, 193
overshoot, 188, 216

P

parallel capacitors
  effectiveness, 136
parallel termination, 227
  analysis of effects, 230
  input shunt capacitance, 227, 230
  noise margin, 228
  power dissipation, 228
  resonance, 128
  when to use, 230
part-to-part skew, 60
partial inductance, 84–85, 87
partial split plane, 105
partitioning, 97, 106
  functional subsystems, 106
  quiet areas, 107
  quiet ground, 107
passive component behavior, 23
PCB traces, 25
  characteristic impedance, 160
peak power currents, 54
permeability, 28, 39
point discontinuities, 200
positive reflections, 217
power distribution
  network, 112
  system, 93
power disturbances, 5
power filtering for clock sources, 74
  ground bounce, 74
  transient current surges, 75
power peak inrush surge current, 54
power planes
  benefits, 134
  capacitance, 136
  placement, 144
Poynting vector, 32
propagation
  delay, 53, 55, 58, 166, 189, 195
  path, 9
  speed, 165
pulse
  skew, 59
  width, 59, 200

Q

quiet areas, 106–107
quiet ground, 107

R

radial routing, 118
radiated emissions, 77
  definition, 3
radiation resistance, 19
radio frequency (definition), 3
RC network, 234
  analysis of effects, 236
  propagation delay, 236
  time constant, 234
  trace impedance, 234
  when to use, 236
reflected
  pulse, 201
  voltage, 189
reflection equation, 190
reflections, 188, 216
  electrically long trace, 188
  reflected voltage, 189
  voltage margin, 188
regulations, 4
relative permittivity, 140, 166

resistivity of materials, 263
resistors, 25
  carbon composition, 25
  carbon film, 25
  film, 25
  leads, 26
  mica, 25
  wire-wound, 25
resonance, 132–134
  anti-resonance, 128
  parallel, 128, 140
  power and ground plane, 140
  review of, 126
  self-resonance, 133
  series, 127, 140
return
  current, 36
  plane, 56
RF (radio frequency)
  current, 36
  current return path, 45
  definition, 3
  energy, 17
  return current, 81
  voltages, 19, 57
RFI (radio frequency interference), 4
right hand rule, 35
ringback, 190
ringing, 188, 191, 216
  electrically long trace, 188
  reflected voltage, 189
  voltage margin, 188
round
  conductors, 266
  wire, 264
rounding, 192
routing concerns, 178
  clock skew, 178
  crosstalk, 178
  propagation time, 178
  trace impedance, 178
routing configuration, 112

S

safety ground, 249, 253
  electric shock, 253
  green wire, 253
  ground path, 254
  series choke, 254
secondary short-circuit fuse, 108
self-compatibility, 6
self-partial inductance, 84, 86
series resonance, 127
series termination, 221
  analysis of effects, 223
  edge rate degradation, 223
  output impedance, 222
  series resistor calculation, 221
  voltage level, 222
  when to use, 225
series-point grounding
  series and parallel connection, 256
shield ground, 108
shunt capacitance, 192
signal distortion, 191
signal integrity, 185–187
  edge rate, 186
  noise margin, 187
signal voltage referencing, 249, 254
single-point ground, 256
  hybrid, 264
  star, 264
simulation software, 193
skew
  duty cycle, 59
  output-to-output, 59
  part-to-part, 60
single stripline, 174–175
single-sided boards, 114, 115
skin depth, 202
skin effect, 29, 39, 89, 263
slots in planes, 101
slotted holes, 100
SMT. *See* surface-mount technology
sockets, 64, 181
source termination, 221
spectral plot, 6
split planes, 104
standoffs, 271
star ground, 264
static fields, 29
stray capacitance, 69
stray impedance, 253
stripline topology, 174–177
  advantages, 174
  dual stripline, 175–176
  impedance, 174, 177
  propagation delay, 175, 176
  single, 174–175
suppression, 8
  definition, 3
surface-mount technology, 63
susceptibility, 12, 15
  definition, 3

susceptor, 9
symmetricaly placed components, 116

T

T-stubs. *See* bifurcated lines
TDR. *See* time-domain reflectometer
TEM. *See* transverse electromagnetic field
termination impedance, 162
termination methodologies, 217
  diode network, 218, 236
  end termination, 226
  parallel, 218, 227
  RC network, 218, 227
  series, 218, 221
  source termination, 221
  Thevenin, 218, 230
terminator noise, 237
  lead length inductance, 237
  multiple terminators, 237
  signal bounce, 237
theory of electromagnetics, 28
  Ampere's Law, 28–29
  closed-loop environment, 28
  electric fields, 28
  Faraday's Law, 28–29
  Gauss's Law, 28
  Lorentz force relation, 28
  magnetic fields, 28
  Maxwell's equations, 28–29, 34–37
  static fields, 29
  time-varying currents, 29–30
thermodynamic domain, 70
Thevenin
  analysis of effects, 233
  equivalence, 17
  lumped capacitance, 233
  parallel equivalent resistance, 231
  termination, 230
  trace impedance, 231
  voltage transition point, 230
  when to use, 233
through-hole components, 100
time-domain, 17, 163
time-domain reflectometer, 48, 166, 190
time-varying currents, 29–30
trace discontinuities, 216
trace impedance, 178
trace routing, 242
  bifurcated lines, 242
  daisychaining, 242
  T-stubs, 242
trace violation, 108
transfer mechanisms, 13
transformers, 27
  common-mode isolation, 27
  parasitic capacitance, 27
transient current surges, 75
transmission line effects, 186
  overshoot, 188
  reflections, 188
  ringing, 188
transmission lines 17
  basics, 162
  crosstalk, 217
  dielectric constant, 166, 169, 172
  distributed capacitance, 161, 180
  distributed circuit, 164
  distributed line, 163
  effects, 163, 186, 216
  electrically long trace, 163
  electromagnetic field, 160, 169
  electromagnetic wave, 169
  lumped circuit, 164
  lumped elements, 163
  overshoot, 216
  PCB trace, 160
  point discontinuity, 200
  propagational speed, 165
  pulse width, 200
  reflected pulse, 201
  reflections, 216
  ringing, 216
  structure, 159, 200
  trace geometries, 169
  undershoot, 188, 216
transverse electromagnetic field, 36

U

underdamped circuit, 192
undershoot, 188, 216
unloaded propagation delay, 180

V

velocity of propagation, 47, 166, 168
vias, 181, 187, 197
voltage gradient, 88, 110
voltage reference, 19, 249

W

wire bond leads, 68
wires, 25

About the Author

Mark Montrose has numerous years of experience in the field of regulatory compliance, electromagnetic compatibility, and product safety. His experience includes extensive design, test, and certification of information technology equipment. He specializes in the international arena, including the European EMC, machinery, and low-voltage directives for light industrial, residential, and heavy industrial equipment.

Mr. Montrose graduated from California Polytechnic State University (1979), San Luis Obispo, CA, with a B.S. degree in electrical engineering and a B.S. degree in computer science. He holds an M.S. degree in engineering management from the University of Santa Clara (1983), Santa Clara, CA.

Mr. Montrose sits on the Board of Directors of the IEEE EMC Society; is a senior member of the Institute of Electrical and Electronic Engineers (IEEE); and is a life member of the American Radio Relay League with the Amateur Extra Class License, K6WJ. In addition, he is a distinguished lecturer for the IEEE EMC Society, a director of TC-8 (Product Safety Technical Committee of the IEEE EMC Society), and member of the dB Society. Mr. Montrose is an active participant in local, national, and international activities of the IEEE and the EMC Society.

He has authored and presented numerous papers in the fields of EMC and signal integrity for high-technology products, printed circuit board (PCB) design, and EMC theory at international EMC symposiums and colloquiums in North America, Europe, and Asia. Mr. Montrose is certified by California's Council for Private Postsecondary and Vocational Education in PCB technology and provides accredited PCB design and layout seminars to corporate clients worldwide.

In 1996, Mr. Montrose authored a best-selling reference textbook *Printed Circuit Board Design Techniques for EMC Compliance* published by IEEE Press and translated into Japanese by Ohmsha. This book is a companion to *EMC and the Printed Circuit Board: Design, Theory, and Layout Made Simple.*